DYNAMIC STRUCTURE OF DETONATION IN GASEOUS AND DISPERSED MEDIA

FLUID MECHANICS AND ITS APPLICATIONS
Volume 5

Series Editor: R. MOREAU
MADYLAM
Ecole Nationale Supérieure d'Hydraulique de Grenoble
Boîte Postale 95
38402 Saint Martin d'Hères Cedex, France

Aims and Scope of the Series

The purpose of this series is to focus on subjects in which fluid mechanics plays a fundamental role.

As well as the more traditional applications of aeronautics, hydraulics, heat and mass transfer etc., books will be published dealing with topics which are currently in a state of rapid development, such as turbulence, suspensions and multiphase fluids, super and hypersonic flows and numerical modelling techniques.

It is a widely held view that it is the interdisciplinary subjects that will receive intense scientific attention, bringing them to the forefront of technological advancement. Fluids have the ability to transport matter and its properties as well as transmit force, therefore fluid mechanics is a subject that is particulary open to cross fertilisation with other sciences and disciplines of engineering. The subject of fluid mechanics will be highly relevant in domains such as chemical, metallurgical, biological and ecological engineering. This series is particularly open to such new multidisciplinary domains.

The median level of presentation is the first year graduate student. Some texts are monographs defining the current state of a field; others are accessible to final year undergraduates; but essentially the emphasis is on readability and clarity.

For a list of related mechanics titles, see final pages.

Dynamic Structure of Detonation in Gaseous and Dispersed Media

Edited by

ANATOLY A. BORISSOV
Institute of Thermo Physics,
U.S.S.R. Academy of Sciences,
Novosibirsk, U.S.S.R.

KLUWER ACADEMIC PUBLISHERS
DORDRECHT / BOSTON / LONDON

Library of Congress Cataloging-in-Publication Data

Dynamic structure of detonation in gaseous and dispersed media / edited by A.A. Borissov.
p. cm. -- (Fluid mechanics and its applications ; v. 5)
Includes bibliographical references.
ISBN 0-7923-1340-2 (HB : acid free paper)
1. Combustion. 2. Detonation waves. 3. Explosions.
I. Borissov, A. A. II. Series.
QD516.D97 1991
541.3'61--dc20 91-22566
CIP

ISBN 0-7923-1340-2

Published by Kluwer Academic Publishers,
P.O. Box 17, 3300 AA Dordrecht, The Netherlands.

Kluwer Academic Publishers incorporates
the publishing programmes of
D. Reidel, Martinus Nijhoff, Dr W. Junk and MTP Press.

Sold and distributed in the U.S.A. and Canada
by Kluwer Academic Publishers,
101 Philip Drive, Norwell, MA 02061, U.S.A.

In all other countries, sold and distributed
by Kluwer Academic Publishers Group,
P.O. Box 322, 3300 AH Dordrecht, The Netherlands.

Printed on acid-free paper

Printed in the Netherlands

Contents

Preface

Of late the demands of industry in creating new composite and functional materials with present properties stimulated an increased interest to the investigation of processes which occur in the detonation technologies of complex chemical composition with an additive of disperse particles.

The collection includes a series of papers presented at the 3d International Conference "Lavrentyev Readings on Mathematics, Mechanics, and Physics" (Novosibirsk, 1990),was held by the Hydrodynamics Institute under the support of the Presidium of the Siberian Branch of the USSR Academy of Sciences to stimulate the international cooperation of the leading international centers. In the framework of this Conference the Round Table seminar was held by Prof. A. Borissov and Prof. V. Mitrofanov devoted to "Dynamic Structure of Detonation in Gaseous and Dispersed Media". The idea to hold such Round Table was supported by Chairman of Organizing Committee academician Prof. V.Titov from Hydrodynamics Institute, and academician Prof. V.Nakoryakov and also his Institute of Thermophysics.

The main ideas discussed at the Round Table were presented in the form of papers which reflected present situation of the problem of dynamic structure of the detonation waves in gaseous and dispersed media. The basic experimental facts concerning of complicated multi-dimensional non-stationary structure both of the detonation wave and its front surface, generation of the cell structure, the effect of transverse waves, obstacles, channel geometry etc. on the transition from dynamic regime to stationary structure are represented in the fist three papers.

One of the ideas in this part expressed in the work of J.H.Lee is a search for new approach for a description of dynamic structure. As a such approach the idea of the use of methods of the theory of non-linear waves in active media is expressed, which was already independently realized in the work of A.Borissov and O.Sharypov. New experimental facts confirming the idea of existence of continuous spectrum of the wave velocities in chemically reaction media from the deflagration velocity to the detonation one are presented in the experiments with porous media carried out by G.Lyamin, V.Mitrofanov, A.Pinaev and V.Subbotin. This information plays the important role in a practice for the estimation of the possibility of the deflagration transition into detonation both in small volumes and in large ones: at accidents at extraction and transport of oil and gas, chemical explosions at atomic power stations.

The next three papers contain new ideas of a description off the detonation wave structure from the point of view of non-linear wave dynamics, asymptotic methods and exact solutions using the methods of the differential equation group analysis. New non-stationary analytic solutions for the wave structure and its front surface have been derived in these works. The problems of further search of exact solutions in detonation also have been planned.

The next four papers reflect a new direction in the investigation of detonation of heterogeneous substances. The careful description of dynamic regimes and structure of the detonation waves in the mixtures of gas with metallic or organic particles has an enormous scientific value. It has the great interest connected with the problems of explosion hazard of modern technologies of new industrial material and food production.

On the whole the collection reflects the present situation of experiments on the detonation wave dynamics and transition from deflagration to detonation both in traditional gaseous and in dispersed media. After wonderful achievements in the field of numerical calculations of multi-dimensional dynamic structure of detonation the attempt to find out new theoretical approaches directed to the obtaining of analytical results describing the dependencies of detonation on basic thermochemical properties of reacting materials, for complicated spatio-temporal structure with non-linear heat release and non-linear interaction with gas-dynamic phenomena was made in this collection.

Taking into account the importance of the problems discussed at the Round Table, Kluwer Academic Publishers has kindly proposed to publish the Proceedings. This proposal was supported by participants of the Round Table.

The collection would be of the great interest for specialists in the field of the explosive substance application, explosion hazard at the accidents and modern technologies of new material production and also for scientists, engineers, post graduators and students studying the detonation processes.

The editor expresses his gratitude to all the authors of the presented papers as well as O. Marennikova, P. Kuibin, O.Sharypov and Prof. M. Torrisi for invaluable assistance in preparing the papers. He hopes that the publication of this book will serve as a good example of fruitful international scientific and technical cooperation.

Editor Professor A. A. Eorissov

Dynamic Structure of Gaseous Detonation

John H. Lee

McGill University, Montreal, Canada

Abstract. In contrast to the laminar one-dimensional ZDN model, the three-dimensional cellular structure of homogeneous gaseous detonation has been firmly established. The structure consists of an ensemble of interacting transverse shock waves sustained by the energy release from chemical reactions. The frequencies of the transverse fluctuations spread over a wide spectrum, however a dominant wavelength (cell size) can usually be identified from the triple point trajectories inscribed on a smoked foil. Regularity of the smoked foil pattern reflects the frequency (or wavelength) spectrum of the transverse wave oscillation. The recent results have demonstrated that the correlations between the dynamic detonation parameters and the dominant cell size alone are inadequate. The cell regularity, as characterized by a stability parameter, must necessarily also play an important role. Recent experiments, aswell as numerical simulations, have confirmed the essential role of transverse waves on the propagation mechanism. Damping of the transverse waves of an established C-J detonation by acoustic absorbing walls leads to decoupling of the reaction zone from the leading shock. Absorbing walls also suppress flame acceleration and transition to detonation even in the presence of obstacles. The role played by transverse shocks is credited to vorticity generation via initial interactions (Mach reflections and shear layers) and to the baroclinic mechanism of pressure and density gradient field interactions. The continuous spectrum of burning rates suggests no clear distinction can be made regarding the deflagration and the detonation

A. A. Borissov (ed.), Dynamic Structure of Detonation in Gaseous and Dispersed Media, 1–25.

mode of combustion. Shock waves due to pressure fluctuations are an integral part of the compressible turbulence in addition to velocity fluctuations from eddy motion. In terms of mechanism, there appears to be no sharp distinction between turbulent deflagration and detonation. However, the detonation is a unique self-sustained spatio temporal structure that is independent of boundary conditions. It is on this basis that one can define a detonation wave. Interpreting the detonation as an ordered structure in a highly non-equilibrium medium, it may be regarded as localized states of non-linear fields. The recent development in non-linear field theory may offer an interesting approach to further understanding of the fundamental physics involved. It appears that current experimental diagnostic techniques and numerical computation capabilities can in general give more detailed information on the detonation structure than can be utilized and interpreted adequately. It is suggested that future directions should aim at the choice of a novel global length scale (e.g. hydrodynamic thickness) other than the cell size to characterize the wave thickness. Together with an appropriate stability parameter that measures the spectrum of transverse fluctuations (or cell regularity), it is proposed that the data for the dynamic parameter be re-examined to achieve a more appropriate correlation.

1. INTRODUCTION

The objective of this paper is to review the progress made in the study of the propagation mechanism of detonation waves in gaseous mixtures and assess its current status. Propagation mechanism refers to the physical and chemical processes that permit the detonation to propagate at a velocity of a few thousand meters per second, three orders of magnitude greater than the laminar flame speed of the same mixture. Important milestones in detonation research can roughly be classified as follows: Discovery of the phenomenon by Mallard, le Chatelier, Vielle and Berthelot in the early 1880's, the theory of Chapman-Jouguet that successfully predicts the equilibrium detonation state (i.e. detonation velocity, pressure, specie concentration of products, etc.) about 20 years later, the model for the structure of the detonation proposed by Zeldovich, Döring and Von-Neumann in early 40's, the discovery of the details of the cellular multi-headed structure in the late 50's and 60's, the practical utilization of the detonation cell size as a characteristic length scale for the estimation of dynamic detonation parameters in the late 70's and early 80's, and the successful numerical simulation of two (and more recently, three-) dimensional cellular detonations also in the late 70's and 80's period. These advances have been well documented in a number of review papers and monographs [1-7]. The present paper emphasizes the results obtained in the past 5 years and summarized the current understanding of the detonation mechanism.

2. THE CHAPMAN-JOUGUET THEORY

Although this theory is well known, it is worthwhile to briefly restate the assumptions involved and what it can and cannot predict. The C-J theory does not require a knowledge of propagation mechanism (details of the structure are not involved). It is based solely on the three conservation laws (in the direction of propagation), the equation of state and C-J criterion that the downstream flow is sonic relative to the wave. The C-J criterion is a postulate, and its validity is based on the agreement of the C-J theory and experiments. The C-J theory also requires that the equilibrium conditions prevail upstream and downstream of the reaction zone. The C-J theory is based on energetics and the chemical rate processes are not considered. Hence, the C-J theory cannot predict the dynamic detonation parameters (i.e. initiation energy, critical tube diameter, limits, etc.).

3. THE ZDN STRUCTURE

The one-dimensional or laminar structure of the detonation wave as proposed by Zeldovich, Döring and von Neumann consists of a normal shock followed by an induction zone and a reaction zone. The termination of the reaction zone is the C-J or sonic plane where the equilibrium states are as predicted according to the C-J theory. The ZDN model provides a mechanism for the propagation of the detonation wave, i.e. autoignition by adiabatic shock compression. Knowledge of the elementary reactions and their respective rate constants permits the detailed time evolution of the various chemical species and the thermodynamic states in the reaction zone to be determined. The laminar ZDN structure is unstable for a self-propagating C-J detonation (i.e. not overdriven). Numerous theoretical stability analyses and numerical computations have demonstrated the instability of the ZDN structure to small perturbations. Experimentally, laminar ZDN detonations have not been observed for self-sustained C-J detonations. Only in overdriven and in transient decaying waves can the laminar ZDN structure be sometimes observed. The ZDN model permits a length or time scale (i.e. reaction zone thickness) to be determined from chemical kinetics. However, the use of the ZDN reaction length scale failed to predict the dynamic detonation parameters that are in accord with experiments. The ZDN reaction length is typically over an order of magnitude less than the effective reaction zone length for real detonations.

4. CELLULAR DETONATION (EXPERIMENTAL OBSERVATIONS)

The investigations of the 50's and early 60's have conclusively established the universal cellular or multi-headed structure of self-sustained C-J detonation fronts. The structure consists of an ensemble of transverse shock waves that propagate normal to the direction of motion of the detonation front itself. The upstream boundary of the detonation front is a cellular normal shock front where the cell boundaries are defined by the Mach intersections of the transverse waves with the normal leading shock. Intense chemical reactions occur at the Mach stems of the leading front and at the tips of the transverse shock near leading shock front. The entire structure is transient and tree-dimensional, although the entire complex propagates steadily at an averaged velocity very close to the theoretical C-J value. Locally, the velocity of the cellular front can fluctuate between 1.5 and 0.6 of the averaged C-J value. The Mach interaction between the transverse (or reflected shock) with the leading shock also produces a shear layer or slip line where the velocity is discontinuous. The trajectories of the triple points (or lines) traced out by the transverse shocks as they sweep back and forth across the leading shock front can be recorded on a glass or metallic foil coated with carbon soot deposit. The trajectories of these triple points or lines as recorded on the smoked foil as the detonation propagates by has a characteristic "diamond" or "fish scale" pattern. The averaged distance between any two transverse wave trajectories define the so-called dominant cell size of the detonation. In the literature, the "diamonds" on the smoked foil is usually called a cell, and a cell width and cell length can be defined based on the dimension of the "diamond" in the directions normal to and in the direction of propagation.

Strictly speaking, a detonation cell should be the characteris- tic pattern from a direct "head-on" view of the detonation front itself, but we shall use the diamond on a smoked foil as a cell in accordance with common definition. Chemical reactions are essentially completed within a length scale of the order of the length of a "diamond" on the smoked foil (which is approximately 1.6 times the width of the diamond or transverse waves spacing "s" or the cell size "λ"). However, the transverse shock waves take a few more cell lengths to dissipate its energy. Thus the effective length of the detonation wave or the hydrodynamic thickness, or the location of the C-J plane, is typically 3 to 1 cell lengths (as measured by Vasiliev et al. [8]) or between 2 to 4 cell lengths (as measured by Edwards et al. [9]). Near the front (within a cell length), the energy associated with the transverse waves can be quite significant [9] (between 10 to 60 % of the chemical energy density). However, as shown by Edwards et al. [9] the transverse fluctuations attenuate quite rapidly to negligible values at distances 2.5 to 4 cell lengths behind the front. As first pointed out by Wagner [10], the pressure amplitude of the transverse oscillation decreases with increasing number of spin modes. For well detonable mixtures, where the spin frequency is high (small cell size), the energy associated with the transverse waves is small and the transverse waves also attenuate very rapidly close to the front. The entire complex three-dimensional transient structure is thus well approximated globally by a one- dimensional thin reaction zone. However, the details of the internal structure of this zone still cannot be described by the one- dimensional ZDN model.

5. SUBSTRUCTURE OF CELLULAR DETONATIONS

The smoked foil pattern in general is quite irregular, indicating the presence of more than one dominant frequency of the transverse fluctuation as well as the excitation of higher harmonics which manifest themselves as fine diamond patterns superimposed on the dominant bands. It's an exception rather than a rule, that one can easily identify the dominant frequency or cell size from a smoked foil. Only in special mixtures such as $C_2H_2 - O_2$ or $H_2 - O_2$ mixtures highly diluted with Argon can regular cell patterns be obtained. Usually for high temperature systems such as fuel-oxygen mixtures and for overdriven detonations, the cell patterns are generally "good" (i.e. the dominant frequency can easily be identified if one is experienced). However, for the less sensitive fuel-air mixtures (and in particular CH_4 - air detonations), the cell patterns registered on the smoked foil can be extremely irregular and difficult to interpret (if at all possible). Normally, it is necessary to examine a long foil at a distance in order to pick out the dominant cell size. It is practically impossible to identify the cell size with confidence from just a small local region of a smoked foil. Shepherd et al. [11] have attempted to quantify the spectrum of frequencies of the transverse fluctuation by using digital-image-processing techniques. They analyzed the records from different mixtures having the same critical tube diameter so that the macroscopic characteristic length scale of these mixtures are the same. Quite different power spectral densities are obtained for these different mixtures, indicating that in general there exist a board spectrum of wavelengths rather than a single dominant one. Hence, by characterizing the cell pattern of a mixture by only a single cell size (or wavelength) from visual estimate that subjectively filters out other wavelengths on the frequency spectrum of the cellular structure, important information may be discarded. It is clear that in general, a spectrum instability modes are exited and this is manifested in the highly irregular nature of the smoked foil pattern. A single "dominant" cell size may quite often be determined, but it is obvious that the complex detonation structure cannot be described by this single parameter. The frequency spectrum itself may be of importance on the dynamic parameters.

Attempts have been made to characterize the "regularity" of the cell pattern by stability parameters. Ul'yanitskii [12] and Manzhalei [13] reported a qualitative correlation between cell regularity and the parameter E/RT (where E is global activation energy, R is the gas constant and T is the post shock temperature of a ZDN wave). Manzhalei noticed that for $E/RT < 6.4$, the cell patterns are fairly regular. Poor and irregular cell patterns are associated with higher values of E/RT . For high temperature detonations (fuel oxygen mixtures, overdriven detonations, high dilution with Argon), the value of E/RT is lower and thus a more "regular" cell pattern results in accordance with Manzhalei's criterion. The observations of Moen et al. [14] and Bull et al. [15] appear to support the E/RT correlation. The value of E/RT describes the steepness of the induction time-temperature dependence, and large values of E/RT give large fluctuations in induction time with small temperature fluctuations, thus rendering the system unstable. A more direct parameter that measures induction time fluctuations with Mach no. was also considered by Moen et al. [14] (i.e. $-\frac{\Delta\tau/\tau}{\Delta M/M}$). Again, they noted a qualitative correlation between regularity with the value of this parameter. In the later study of Shepherd et al. [11], they reported that the parameter $-\frac{\Delta\tau/\tau}{\Delta M/M}$ failed to correlate with the cell regularity results from their digital-image-processing technique of smoked foils. They proposed the use of critical overdriven parameter $f^*=(M^*/M_{CJ})^2$. M^* is the critical shock Mach no. of the overdriven detonation when the endothermic dissociation processes balance of exothermic recombination processes. The parameter f^* appears to correlate with their observations. It is doubtful that a single parameter can characterize the frequency spectrum of the instability mode of a cellular detonation, but qualitatively, any parameter that characterized the instability of the system will represent some measure of the regularity or the frequency spectrum of the transverse fluctuation.

6. NUMERICAL RESULTS

Numerical simulation using the method of characteristics of one-dimensional unstable detonations have been carried out by Fickett and Wood [16] as early as 1966. A more thorough study was later carried by Abouseif and Toong [17], and Moen et al. [18] with more advanced numerical methods. The results of these studies all showed that as the degree of overdrive f is decreased ($f = 1$ when $M \longrightarrow M_{CJ}$), the detonation becomes unstable with the shock pressure fluctuating periodically. A single dominant mode is observed when the degree of overdrive is high. As f decreases, a second unstable mode appears. Further decrease in the overdrive parameter leads to a distortion of the dominant mode, i.e. a steeper rise due to a reaction generated shock overtaking the main shock and subsequently a slower decay. As $f \longrightarrow 1$, higher frequencies and smaller amplitude modes appear during the decay phase of the first dominant mode. Similar results are also obtained by Moen et al. [18] who showed that as the degree of overdrive decreases, the regular oscillatory pattern of the first dominant mode becomes extremely complex as higher modes appear. Finally, as $f \longrightarrow 1$, the oscillations are so complex that identification of the dominant mode becomes very difficult. These numerical simulations essentially reproduce the main results of analytical stability analyses.

Two-dimensional simulations of detonation waves are successfully achieved by Taki and Fujiwara [19], and Oran et al. [20] in the late 70's. The starting point in the simulation is usually a one-dimensional ZDN detonation and the two-dimensional cellular structure is developed from perturbing the ZDN wave by spatial density inhomogeneities or by local explosions behind the shock. To initially obtain a stable ZDN wave [21] it is found that a careful choice and adjustment of numerical parameters (Courant number, artificial diffusion coefficients, grid size, order of accuracy, etc.) are necessary. The ZDN detonation obtained is then found to be only metastable. Imposing small one-dimensional periodic density perturbations will lead to a decoupling of the reaction zone from the leading shock. Similarly, in the two-dimensional case, small spatial density perturbations also lead to failure (i.e. decoupling of the shock from the reaction zone). However, if the perturbations are sufficiently strong, the a cellular detonation wave is formed which subsequently propagates at a steady averaged velocity equal to the *C-J* velocity of the mixture. A very regular cell pattern is usually obtained in these numerical simulations indicating that the choice of numerical parameters effectively suppress all the higher unstable modes but the dominant one. The cell size obtained from these simulations are found to correlate with the one- dimensional ZDN reaction length, and independent of the channel width. These detonations from the numerical experiments may well be failing detonations similar to the very regular two-dimensional failing waves observed by Strehlow [1] and Liboutin et al. [22], giving beautiful regular smoked foil patterns. There are strong indications that self-sustained detonation (like turbulence) must necessarily be three-dimensional.

When finer grid size is used in the numerical simulation [23], it is found that the regular structure becomes irregular with a finer substructure superimposed on the dominant cell pattern. The dominant cell pattern itself is also found to be unstable in that a pair of triple points may merge after a few cycles leading to a larger cell. The finer substructure can also amplify, become more prominent and form new cells. This kind of behavior begins to follow more closely to reality. A finer resolution essentially permits the higher frequency unstable modes to develop, thus leading to a more irregular and complex structure. Hence, without an "a priori" knowledge of the desired result, it is difficult to decide on the resolution needed to reproduce the phenomenon. The numerical simulation becomes more of a post-dicting than a predicting nature.

Little new insight into the complex detonation structure has been gained from these numerical simulations thus far. The cell dynamics (i.e. detailed structure of the reaction zones behind the transient incident shock, Mach stem and transverse reflected waves) are essentially the same as revealed by experiments. The important problem of the attenuation of the transverse shocks and the location of *C-J* surface has not been done. Even a check of the one-dimensional conservation laws for the averaged flow quantities across the cellular detonation has not been attempted. However, some interesting features of the reaction zone have been reported by Oran et al. [24]. In their numerical simulations, they found that large "pockets" of unburned mixture can escape processing by the Mach and transverse shock to autoignition and these "pockets" then get convected into the product regions. The eventual "explosions" of these pockets provide the perturbations leading to new cell formations and galloping waves. Of a more fundamental significance is that the existence of these pockets may then eventually burn via turbulent diffusion mechanism as in a deflagration, this breaks down the sharp distinction between deflagration and detonation regimes. Conventionally speaking, combustion in the deflagration regime is described to diffusion process (molecular and turbulent) and convective processes. The propagation mechanism in the detonation regime is due to autoignition by shock heating (either by a single planar shock as in the one-dimensional ZDN model or via incident, Mach stem and reflected shocks in the cellular model). The numerical results of Oran et al. [24] demonstrated that the complex wave dynamics at the front can lead to unburned gas pockets that eventually burn via turbulent mixing and diffusion mechanisms. Thus, detonations involves a hybrid mechanism of both shock ignition and turbulent diffusion and the sharp distinction between the two regimes disappears. Experimental proof of these unburned gas pockets are provided by the schlieren photographs of Oran et al. [24]. These experiments are carried out in very narrow channels and the self-sustaining nature of these waves is not clear. In well detonable mixtures, the three-dimensional nature of self-sustained waves makes it impossible to resolve the reaction zone from diagnostics that integrate along the line of sight. Future, larger sheet fluorescence (e.g. the *CH* radical) techniques may provide the details of the reaction process inside the three-dimensional zone.

The numerical simulation has recently been further extended to the three-dimensional case by Fujiwara and Reddy [25]. They computed the propagation of spinning detonation in a circular tube and demonstrated the presence of radial and azimuthal modes. The interpretation of the tremendous amount of information that is being generated is now a serious problem! It appears that current computing capability has exceeded our ability of cope with and interpret the results obtained. Future progress must be based on well posed specific questions where the numerical computations are used to provide specific answers. The feasibility of numerical computations of reacting flows to reproduce at least qualitatively cellular detonations, is now firmly established.

7. DYNAMIC DETONATION PARAMETERS AND STRUCTURE

Based on the cell size "λ" as a single global length scale that characterizes the structure of a cellular detonation, correlations with the dynamic parameters such as critical tube diameter, initiation energy and limits have been achieved in the early 80's [7]. The starting point is the $d_c = 13\ \lambda$ correlation between the critical tube diameter and cell size. It was recognized right at the beginning that this correlation is not exact. However, the error in the measurement of the cell size (which can sometimes be as great as a factor two) appears to cover up the discrepancies. However, the study of Argon diluted mixtures with very regular patterns (so that the cell size can be accurately determined) indicates that the $d_c = 13\ \lambda$ breaks down for these systems with very regular structure (e.g. for 75% Argon diluted $C_2H_2 - O_2$ mixtures, $d_c \simeq 24\ \lambda$) [14]. In a later study, Shepherd et al. [11] reported that $d_c \simeq 4$ to $10\ \lambda$ for 0% Argon dilution to $d_c \simeq 20 - 30$ for 8% Argon diluted $C_2H_2 - O_2$ mixtures. Digital image processing of the smoked foils for these mixtures shows a dramatic narrowing of the spectral content with increasing Argon dilution. These studies [11],[14] have firmly established that cells regularly play an important role in the transmission problem. Only for systems with poor or irregular cell patterns when the $d_c = 13\,\lambda$ holds. For increasing cell regularity, the factor of proportionality increases. This is reasonable since the critical tube diameter problem is essentially one of re-initiation of the detonation wave that is quenched by the lateral expansion waves as it emerges from the tube. For systems with irregular cells, more unstable modes are excited together with the dominant mode that defines the cell size of the system. For highly unstable systems, it is easier for the higher modes to be amplified forming new cells and re-initiate the detonation. It appears that the $d_c = A\,\lambda$ correlation must include a stability parameter, perhaps through the dependence of the factor of proportionality A. An improved form of the correlation has not been studied systematically as yet.

When reduced to the two-dimensional case of the critical slit width "W_c" for the transmission of a planar wave from a channel to a cylindrical wave, it is found that $W_c \simeq 3\,\lambda$ instead of $W_c \simeq 6\,\lambda$ [26] as dictated by geometrical considerations. This strongly suggests that the local re-initiation phenomenon is three-dimensional in accord with the view self-sustained detonations are three-dimensional waves instead of two.

Based on the critical tube diameter correlation, Lee [27] has developed analytical models for the critical energy for direct initiation of spherical detonations. A summary of these models is presented in the paper by Benedick et al. [28]. The surface energy model of Lee which is based directly on the $d_c = 13\,\lambda$ correlation appears to be quite good when compared to experimental data of critical energies for fuel-air mixtures. No direct measurement of critical energies for Argon diluted mixtures has been carried out, but if the $d_c = 13\,\lambda$ correlation is found to depend on cell regularity, the critical energy for direct initiation must also be a function of cell regularity (in addition to the cell size since the direct initiation phenomenon also requires the formation of cells behind the decaying blast. The highly unstable mixtures corresponding to irregular cells facilitates the generation and amplification of waves leading to cell formation and the onset of detonation. We would expect that for highly diluted mixtures with Argon (where the cells are very regular), the critical energy would be higher than (the same value of λ) for a mixture with irregular cells.

The correlation of limits (e.g. in circular tubes) with the cell size "λ" has been studied by Dupré et al. [29-31]. A simple universal d/λ criterion for the detonation limit in circular tubes cannot be found, indicating that stability parameters that characterize the regularity of the cell pattern also play a role in the limit criterion. In a more recent study [31], it is found that for a mixture with irregular cells, the value of d/λ is much lower than for a mixture with regular cells. Specifically for 14.3% H_2 with air (or with N_2 dilution), detonation is still possible for a value of $d/\lambda \simeq .147$, while for 13.6% H_2 + 6.8% O_2 with 79.6% Ar dilution, detonation failed when $d/\lambda \simeq 8.3$. In other words, for H_2- air mixtures with poor or irregular cells, detonation propagation is possible for a much smaller diameter tube for the same value of the dominant cell size λ. Thus, limits criterion must take into account cell regularity. In the study of Moen et al. [14], they compare velocity deficit in rigid tubes and find that mixtures with irregular cells are much less influenced by the boundary layer effects. In other words, mixtures with irregular cell patters demonstrate very small velocity deficit as compared to systems with regular cells. Setting a failure criterion (when the velocity deficit exceeds 10% of the C-J value) they observed that detonations in mixtures highly diluted with Argon (with regular cells) fail to propagate long before detonations in mixtures with irregular cells which failed at approximately the single headed spin limit. This result suggests that the propagation of detonation is greatly facilitated for irregular cell systems where a wide spectrum of unstable modes are excited. Moen et al. [14] also draw attention to the observation of Wolanski et al. [32] who managed to detonate methane-air mixtures in a 63.5 mm diameter tube with a velocity within 1.5% of the C-J value. The cell size for stoichiometric methane-air detonation is estimated to be about 300 mm and the small diameter of 63.5 mm tube used by Wolanski et al. is way past the single headed spin limit. The fact that steady detonation was observed is credited by Moen et al. [14] to the very irregular nature of the cell pattern for CH_4-air detonations.

For such a highly irregular cell mixture, the influence of tube boundaries is expected to be very small, thus permitting the detonation to propagate way beyond the limits.

The studies of the early 80's linked the dynamic detonation parameters to the structure via the dominant cell size. The results of the late 80's demonstrated that a single parameter (i.e. the cell size) is insufficient and that stability parameters that characterize the regularity of the cell pattern (or the complexity of the cellular structure) must also be used.

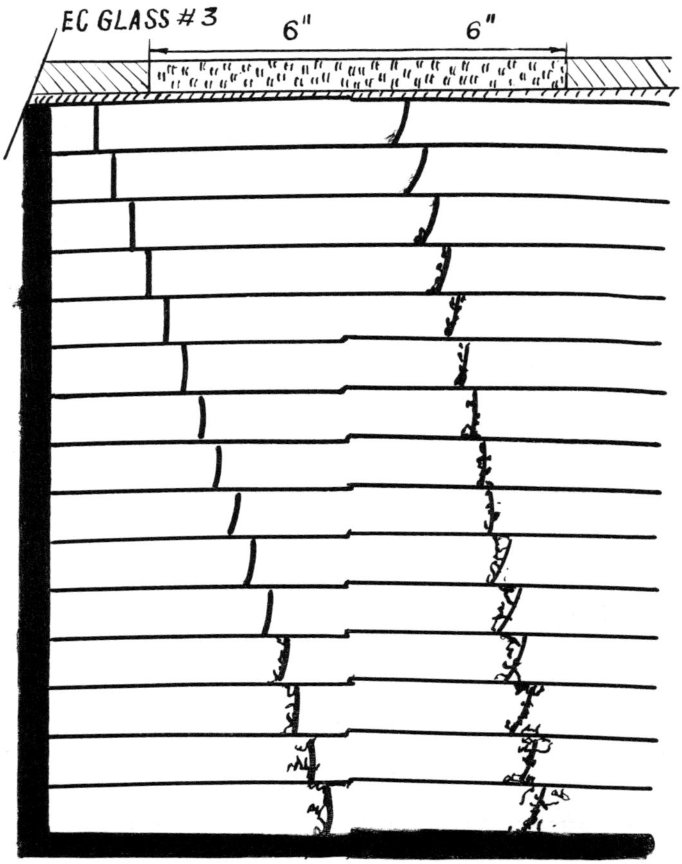

Fig. 1

8. ARE TRANSVERSE WAVES NECESSARY?

The one-dimensional ZDN structure is perfectly self-sufficient in that it provides the propagation mechanism of autoignition via shock heating. Although it is shown to be unstable theoretically and confirmed experimentally in that all self-sustained gaseous detonation have a three-dimensional cellular structure, the basic question regarding whether transverse waves are absolutely essential to the propagation of detonation is not resolved. Some recent experiments perhaps add more proof to the essential role of transverse waves. Dupré et al. [33] re-

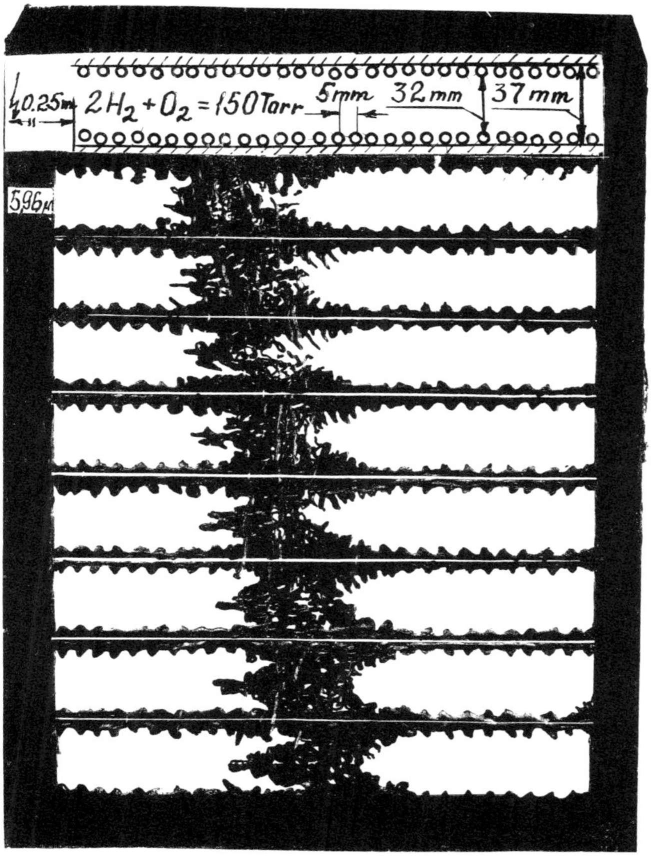

Fig. 2a

cently carried out an experiment whereby a fully established C-J detonation propagating in a rigid wall circular tube enters a section of a few tube diameters long in which the tube wall is replaced by an acoustic absorbing material (e.g. rubber, foam, porous medium). The transverse waves as they reflect off the wall in this acoustic damping section is greatly attenuated, resulting in larger and larger cells as the transverse waves are being eliminated. It is found that the detonation fails when the transverse waves are eliminated with the reaction zone decoupling from the leading shock front. The velocity of this decoupled shock-flame complex is about half the corresponding C-J detonation as it emerges from the acoustic damping section back into the normal rigid wall tube. Recent analysis by Chue et al. [34] shows that the resulting decoupled shock-flame complex after the transverse waves are damped corresponds to a C-J deflagration. Detonation failure by transverse wave damping is governed by the initial number of cells across the tube diameter and this number is a function of thee mixture. If the cells are too small compared to the tube diameter (i.t. too many cells across the diameter), then new cells can be regenerated as transverse waves at the wall are damped and the detonation will not fail. Cell regularity is again found to play a role here in that in highly irregular systems the critical number of cells across the diameter is much less than for highly regular systems for successful detonation propagation past the damping section. This is in accord with the observation of Moen et al. [14] who found that irregular cell systems are much less influenced by wall boundary layer. In fig. 1, a sequence of schlieren photographs of the failure of a detonation due to the damping of the transverse waves are illustrated. The mixture is $2H_2 + O_2$ at an initial pressure of *120 torr* and the acoustic damping material used is a fiberglass pad. Progressive thickening of the wave eventually leading to a decoupling of the reaction zone with the leading shock when the transverse waves are being eliminated upon reflection from the fiberglass layer can be clearly observed. a numerical simulation of this experiment was carried out by reddy et al. [35]. cell enlargement leading to eventual failure when the reaction zone decouples from the leading shock is reproduced numerically.

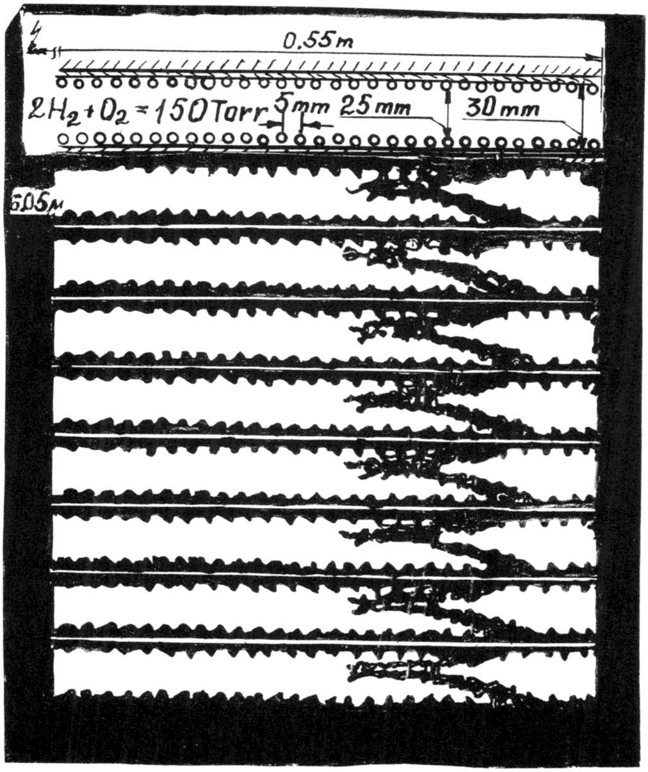

Fig. 2b

In a similar experiment, Teodorczyk [36] studied the transition of deflagration to detonation in a rough wall channel. in fig.2a, the structure of the accelerating turbulent flame brush just prior to transition is shown. One can see the intense pressure wave system generated by the obstacles. The mixture is $H_2 + O_2$ at an initial pressure of *150 torr*.

In contrast, Fig.2b shows the same experiment but acoustic absorbing material is now placed underneath the turbulent generating obstacles to damp out the transverse pressure waves. The intense pressure wave system associated with the flame zone is now gone. With the acoustic damping wall, the flame failed to accelerate to the high velocities (of the order of *800 m/s*) required for transition. Transition to detonation is not observed in the experiments with the damping wall. This indicates that turbulence alone, but without the transverse pressure waves fail to cause transition, proving indirectly the necessity to generate and amplify transverse waves if detonation is to be formed. Suppressing transition by lining the tube with acoustic damping material has been demonstrated long ago by Evan et al. [37] although the cellular structure of detonations was not widely established at that time. The use of obstacles to speed up to transition from deflagration to detonation has been used by Laffitte [38] and Shchelkin [39] many years ago. The mechanism is credited to the large scale turbulence that is generated by the obstacles, thus promoting flame acceleration. However, the recent results point out a more important role of wave generation by the obstacles. By eliminating the waves generated but not the eddies, it is shown that flame acceleration and transition to detonation is greatly suppressed.

The recent experimental results describing above contribute more proof to the essential role played by a system of interacting transverse waves to sustain the supersonic propagation of a detonation wave.

9. WHY TRANSVERSE WAVES ARE NECESSARY

It is interesting to note that for the same explosive mixture, the difference between the propagation velocity of the slowest (laminar flame) and the fastest mode (detonation) of combustion is about three orders of magnitude. Laminar flame propagates at about half a meter per second, while detonation propagates at about *2000 m/s* for fuel-air mixtures. In between these two limits there exist an almost continuous spectrum of turbulent flame speeds that depend on the boundary conditions [4]. The research in the past few decades contributed to the understanding of the mechanisms that enable the burning rate to change by three orders of magnitude. For a given system at prescribed initial and boundary conditions, it appears that nature tends to maximize the burning rate by rendering the combustion wave unstable forcing it to go to a faster rate compatible with the boundary conditions. The slowest mode of laminar flame propagation is governed by molecular diffusion mechanisms. There exist a number of instability mechanisms [4] (e.g. Landau-Markstein, Rayleigh-Taylor, selective diffusion, thermal diffusive instability, etc.) that can turbulize the laminar flame when the boundary conditions permit. The increase in subsequent burning rate is through the increase in flame area and local transport rates. The increase in specific volume across the flame results in a displacement flow of the unburned mixture ahead of the flame. Boundaries (i.e. wall and obstacles) produce velocity gradient fields which subsequently breaks down into turbulence when the Reynolds number exceeds a certain critical value. Thus, mean flow kinetic energy is randomized in the turbulent eddies which in turn couple to the flame front to augment the burning rate. Generally speaking, our notion of turbulence is derived from low speed or incompressible flows. Thus turbulence is associated with the velocity fluctuations of the eddy motion. However, when compressibility effects become important, pressure fluctuations set up. Pressure fluctuations result in the generation of acoustic waves which intensify to form shock waves. If one considers turbulence as a mechanism to increase viscous dissipation or energy degradation rate, then low speed turbulence involves the shear layers (i.e. boundary layer) and dissipation eddies (of the Kolmogorov scales), while high speed turbulence brings in shock waves as dissipative structures. The interaction of shock waves produce shear layers and thus a system of interacting shock waves can generate additional vorticity in the absence of wall. In a turbulent flame zone where a strong density gradient field exists, the baroclinic ($\vec{\Delta} p \times \vec{\Delta}\, {}^{1}/\rho$) term in the vorticity equation provides another source of vorticity. A model of compressible turbulence was advanced by Lighthill [42] is to consider in addition to the vortex motion, there exists also a statistical ensemble of *N*-waves interacting (i.e. union, regular and Mach intersections). Lighthill also stated that "the decision of the motion into turbulence on the one hand and sound or shock waves on the other, is almost without significance". In other words, for high speed flows shock wave interactions and vortex motion all contribute to the vorticity generation and dissipation of the system. With this view in mind, it is reasonable to consider that as a turbulent deflagration becomes more intense, strong pressure waves are generated which subsequently play important roles in the production of vorticity. With the transverse interacting waves playing a more prominent role, eventual resonant coupling of these transverse waves with the chemical energy release then result in the formation of a detonation. Thus a detonation in essence can be considered as a turbulent deflagration where compressibility effects (hence wave interaction process) are important. It is difficult to make a "clean cut" distinction between the two phenomenon, particularly through the manipulation of the boundary conditions, i.e. continuous spectrum of wave speeds can be achieved. A detonation however, is very unique structure in that it is boundary conditions independent (e.g. spherical detonation), quite unlike turbulent deflagrations. It is the resonant on-linear interaction between chemistry and gasdynamics that uniquely shapes the ultimate structure of the cellular detonation front. Nevertheless, within the structure the role played by the transverse waves (the vorticity generation mechanisms of compressible turbulence) is clear. Perhaps it is with this unique structure that one can make a distinction between detonation and deflagration. In terms of the mechanisms, there appears to be no sharp transition from vorticity production by shear and by waves as the speed increases.

10. CONCLUSION

We may conclude that the detonation structure is in essence an ensemble of interacting transverse shock waves sustained by chemical reactions. The transverse fluctuations like turbulent fluctuations, can have a wide spectrum of frequencies characteristic of a random process. Yet the detonation structure can have a high degree of spatio temporal order coexisting with the chaos. The degree of order in the chaos can perhaps be measured by a fractal dimension similar to recent attempts to characterize turbulent premixed flames [43]. The spatio temporal structure of a detonation can be regarded as a result of the coherent interaction of the numerous unstable modes. Cast in this light, the detonation structure is just another example of a wide class of phenomenon in chaotic dynamics (e.g. Rayleigh-Bernard convection, coherent vortices in shear flow, Faraday ripples, Langmuir waves in a plasma, etc.). Ordered structures may be considered as localized states of non-linear fields [44]. Recent advances in the study of solitons and dissipative localized structures in non-linear field theories may offer a novel approach to the study of detonation structure. At least developments in other fields may be brought in to elucidate the fundamental physics of the phenomenon.

Regarding future directions, recent advances in both experimental diagnostics and numerical computation have gone beyond the stage where we can cope with the information that is generated. In principle, we could obtain (with a great deal of effort perhaps) a detailed nap of the pressure velocity temperature density fields as well as the species concentration field of the spatio temporal detonation structure. The question is "how to use this data"? The ultimate objective of detonation research is perhaps to achieve the ability to predict, either by theories or semi-empirical correlations, the quantities of practical significance. We have already achieved the ability to predict equilibrium detonation states (i.e. detonation velocity, pressure, etc.) from the C-J theory. What remains to be done is essentially to refine the current empirical correlations (such as the $d_c = 13\ \lambda$ law) and analytical theories for predicting the dynamic detonation parameters (i.e. initiation energy, critical tube diameter limits, etc.). It is clear that the dominant cell size is not a good parameter to characterize the length scale of the structure. It is difficult to estimate from smoked foils and there appears to be no alternate simple ways to obtain the cell size. The critical tube diameter itself can be used as a length scale to characterize the detonation structure since it can be determined relatively easily experimentally without the subjectiveness of smoked foil interpretation for the cell size. Moen et al. [14] have already employed the critical tube diameter as a characteristic length scale in their studies. It is also clear from the research of the past decade that the cell size alone is insufficient and that a stability parameter that measures the cell regularity or the spectrum of the length scales must also be used to correlate the dynamic parameters to the detonation structure. If the critical tube diameter is used as a length scale, then both the regularity as well as an actual length scale of the detonation are both included. It would be better to seek a length scale independent of the stability of the system itself. In this regard, the so called hydrodynamic thickness (or the location of the C-J plane) seems attractive. However, previous techniques used by Vasiliev [8] and Edwards [9] are both insufficiently accurate and new methods must be devised. Numerical simulations may assist in this problem of determining the hydrodynamic thickness of a detonation wave. However, the resolution required as well as the correct modelling of the dissipation rate of transverse waves would render the task extremely difficult. With the successful determination of the length scale (e.g. hydrodynamic thickness) and also with an appropriate stability parameter chosen, the past data on the critical tube diameter, cell size, initiation energy, etc. (and perhaps additional data on these dynamic parameters) would have to be re-examined to achieve the necessary correlations. Analytical theories then follow these correlations quite readily.

We define a detonation as the unique self-sustained spatio temporal structure of interaction transverse waves that is independent of initial and boundary conditions. There are however the cases of quasi-detonations where boundary conditions play a dominant role in controlling the structure, hence the propagation mechanism itself. In quasi-detonations the generation and the dynamics of the transverse shock waves (with walls, obstacles and among one another) are completely controlled by the boundary condition. It is difficult in this case to develop generalized theories for their description. The same statement can be made for turbulent deflagrations where the velocity generation is again controlled by the boundary conditions. Although the mechanisms of propagation of turbulent deflagrations and quasi-detonations may be of a universal nature, it is doubtful that correlations of a general nature between the propagation velocity and the properties of the mixture and the boundary conditions or scaling laws can be made, except in very simple geometries.

Perhaps a statement often made by Manson that "a spherical detonation is a true detonation" is most appropriate to end our discussions. The very existence of a steady C-J spherical detonation demonstrates the necessity of a three-dimensional structure. The continuous change of curvature of the front would result in a non-steady ZDN structure. The propagation of a spherical detonation (unlike a planar wave) requires the continuous generation of cells as the front diverges so that the averaged cell dimension is preserved. The generation of cells rests on the excitation and amplification of the unstable modes of the system. Thus stability plays a obvious role in the case of a spherical wave. The complete absence of boundaries also requires that the complex spatio temporal structure of the spherical detonation to be maintained by it-self, completely making it truly a unique structure.

REFERENCES

1. Strehlow, R. "Fundamentals of Combustion" (1983) Robert Krieger Publishing Co. Inc., Chapter 9.
2. Fickett, W. and Davis "Detonation" University of California Press (1979).
3. Nettleton, M.A. "Gaseous Detonations" Chapman and Hall (1987).
4. Soloukhin, R.I. "Shock Waves and Detonations in Gases" Mono Book Corp. (1966).
5. Shchelkin, K.I. and Troshin Ya.K. "Gasdynamics of Combustion" Mono Book Corp. (1965).
6. Toong, T.Y. "Combustion Dynamics" McGraw Hill (1983).
7. Lee, J.H.S. "Dynamic Parameters of Gaseous Detonations" Ann.Rev. Fluid Mech. 16 (1984).
8. Vasiliev, A.A., Gavrilenko, T.P. and Topchian, M.E. (1972) Astronaut. Acta 17, pp. 499-500.
9. Edwards, D.H., Jones, A.T. and Phillips, D.E. J.Phys.D., Appl. Phys. vol.9 (1976), pp. 1331-1342.
10. Wagner, H.G. 9th Symp. (Int.'l.) on Combustion (1963), pp.454-60.
11. Shepherd, J., Moen, I., Murray, S. and Thibault, P.21st Symp. (Int'l.) on Combustion (1986), pp. 1649-1657.
12. Ul'yanitskii, V.Yu. Fizika Goreniya Vzryva 17, 127 (1981).
13. Manzhalei, V.I., Fizika Goreniya Vzryva 13 (3), pp. 47-4772 (1977).
14. Moen, I.O., Sulmistras, A., Thomas, G.O., Bjerketvedt, D., Thibault, P. Prog. in Astro. and Aero. Vol.106, pp. 220-243 (1986).
15. Bull, D.C., Elsworth, J. and Shuff, P. (1982) Comb. and Flame 45(1), pp.7-22.
16. Fickett, W. and Wood, W. Phys. Fluids 9: 903-916 (1966). See also Fickett, W., Jacobson, J.D. and Schott, G. AIAA Jourl.10: 514-516 (1972).
17. Abouseif, G. and Toong, T.Y. Comb. and Flame 45: 67-94 (1982).
18. Moen, I.O., Funk, J., Ward, S., Ruder, G. and Thibault, P.Prog. in Astro. and Aero. Vol. 94 (1985), pp. 55-77.

19. Taki, S. and Fujiwara, T. AIAA Jourl. vol.16, No.1, pp. 73-77 (1978).
20. Oran, E.S., Boris, J.P., Young, T., Flanigan, M., Burks, T.and Picone, M. 18th Symp. (Int'l.) on Comb., p. 1641 (1981).
21. Taki, S. and Fujiwara, T. AIAA Prog. in Astro. and Aero.Vol.94, pp. 186-200 (1984).
22. Liboutin, J.C., Dormal, M. and Van Tiggelen, P.J. Prog. in Astro. and Aero. 26, p. 358 (1981).
23. Sugimura,T., Fujiwara,T. and Lee,J.H. "Cellular Detonations-Instability and Substructure", paper presented at the 22nd Symp. (Int'l.) on Comb., Seattle, August 1988.
24. Oran, E.S., Young, T.R., Boris, J.P., Picone, J.M. and Edwards, D.H. 19th Symp. (Int'l.) on Comb., pp. 573-582 (1982).
25. Fujiwara,T. and Reddy,K.V. "Propagation Mechanism of Detonation - Three Dimensional Phenomenon" Memoirs of the Faculty of Engineering, Nagoya University, Vol.41, No.1(1989).
26. Benedick, W., Knystautas, R. and Lee, J.H. "Large Scale Experiments on the Transmission of Fuel-Air Detonations from Two-Dimensional Channels",Dynamics of Shock Waves,Explosions and Detonations, Prog. in Astro. and Aeron., Vol.94, pp.546-555 (1984).
27. Lee, J.H. in "Fuel-Air Explosions", p. 157. Univ. of Waterloo Press (1982).
28. Benedick, W., Guirao, C., Knystautas, R. and Lee, J.H. Prog. in Astro. and Aero., Vol. 106 (1985), pp. 181-202.
29. Dupre, G., Knystautas, R. and Lee, J. Pro. Astro. and Aero., Vol. 106,,p. 244 (1985).
30. Dupre, G., Peraldi, O., Joannon, J., Lee, J.H. and Knystautas,R. "On the Limit Criterion of Detonation in Circular Tubes",presented at the 12th International Colloquium on the Dynamics of Explosions and Reactive Systems, University of Michigan,23-28 July, 1989.
31. Dupre, G., Joannon, J., Knystautas, R. and Lee, J., "Unstable Detonations in the Near-Limit Regime in Tubes", Presented at the 23rd International Symposium on Combustion, Orleans, France, July 1990.
32. Wolanski, P., Kauffman, C.W., Sichel, M. and Nicholls, J. 18th Symp, (Int'l.),on Comb., pp. 1651-1660 (1981).
33. Dupre, G., Peraldi, O., Lee, J.H. and Knystautas, R. Prog. in Astro. and Aero., Vol. 114, p. 248 (1988).
34. Chue, R., Clarke, J. and Lee, J. "On C-J Deflagrations", to be published.
35. Reddy, K.V., Fujiwara, T. and Lee, J.H. Memoirs of Faculty of Engineering, Nagoya University, Vol.40, No.1 (1988).
36. Teodorczyk, A. Private Communications.
37. Evan, M.W., Gwen, F.I., Richeson, W.E. J. App. Phys. 26, pp. 1111-1113 (1955).
38. Laffitte, P. Compt. Rend. 186, 95 (1928).
39. Shchelkin, K.I. Soviet Phys. JETP, 10, p. 823 (1940).
40. Lee, J.H. in "Advances in Chemical Reaction Dynamics", Ed. Rentzepis, P. and Capellos, C.D. Reidel Pub. (1985).
41. Sivashinsky, G.I. Ann. Rev. Fluid Mech., Vol. 15, pp. 179-199 (1983).
42. Lighthill, M.J. "Effect of Compressibility on Turbulence" in Proc.

of Cosmical Gas Dynamics (1949).
43. Gouldin, F.C., Hilton, S.M. and Lamb, T. 22nd Symp. (Int'l.) on Comb., pp. 541-550 (1989).
44. Ezerskii, A.B., Rabinovich, M., Rentov, V.P. and Starobinets, L.M. Soviet Phys. JETP, 64(6) (1986).

The Limits of Stationary Propagation of Gaseous Detonation

A. A. Vasiliev
Lavrentyev Institute of Hydrodynamics
Siberian Division of the USSR Academy of Sciences
Novosibirsk 630090, USSR

Abstract. The intent attention of the researchers from many countries to the problems of detonation is sure to play an important role in study and modern understanding of many aspects of this field of knowledge. The structure of detonation wave (DW) in gaseous and heterogeneous systems, detonation initiation in chemically active media, DW behavior in near-critical regimes of propagation and many other problems were studied during the last forty years. Some results of these investigations for gaseous mixtures are presented in the reviews by J.H.Lee [1,2] and A.A.Vasiliev, V.V.Mitrofanov and M.E.Topchian [3].

Increasing use of hydrocarbon raw material in industry and in life, its storage and transportation makes more possible the appearance of accident situation when the large quantity of fuel-air mixture is created. In this case it may lead at occasional ignition to great losses of supplies and equipment, ecological catastrophes and even to human victims. Besides only scientific interest such cases must cause the particular attention of the researchers in studying many unsolved problems concerning the detonation of gaseous and heterogeneous systems (fuel at liquid or solid phase). These unsolved problems are suggested to be discussed during "Round Table" being held in the framework of the III International Lavrentyev Reading in mechanics, mathematics and physics (Novosibirsk, 1990).

1.

The problem on detonation limits, including two main aspects, is

A. A. Borissov (ed.), Dynamic Structure of Detonation in Gaseous and Dispersed Media, 27–49.

one of the most important problems of hydrodynamics of chemically active media. The first aspect of the problem concerns the determination of concentrating fuel-oxidant relations when the stationary propagation regime of detonation wave is still observed in the system. The second aspect is caused by the boundary conditions influence on DW. The author should like to attract attention namely to the geometric limits, i.e. to the definition of boundary sizes of experimental apparatus where the detonation regime is observed.

The main geometric characteristic of multiwave gaseous detonation with a pulsating front is the cell size - typical rhomboid imprint, which is recorded by DW on soot surface. For fixed mixture composition and the diameter of detonation tube the cell increases in size with decrease of initial pressure P_0, and the transition from multifrontal regime to spinning one is observed [5,6]. Spinning detonation is characterized of a single transverse wave (TW) on the DW front, axially rotating around the periphery of detonation tube. The spinning detonation in circular tubes is the limit regime of stationary propagation of detonation.

In the case of further decrease of P_0 spinning detonation falls off and under the certain conditions the quasistationary regime of wave propagation was observed. This regime was characterized by high longitudinal (along the axis) pulsation and it was called a galloping one [7-11]. During one pulsation the strongly over-driven wave with multiheaded structure attenuates while propagating. The quantity of TW on the front reduced to the single TW with the transition of spinning regime. Then spinning regime is broken down itself and the wave attenuation is observed. In the latter case the shock wave (SW) is separating and removing from the flame front as long as the period of induction τ of shock-compressed explosive mixture is finished. When τ is finished, the explosive initiation of mixture in the induction zone occurs with formation of a new over-driven detonation wave. During the next pulsation the process is repeated. For a galloping regime the velocity pulsation is comparative with average velocity of DW.

When P_0 decreases further, the quasistationary regimes of explosive mixture combustion with high supersonic velocities (~1000 m/s) can be observed. For this case there exists a low level of pulsations of flame front and the absence of cellular structure [12].

The multitude of the processes observed makes the concept of limiting regime of wave propagation to the nonuniform. What criterion must be at the basis of limits? In the framework of classical one-dimensional model the detonation wave represents the gas-dynamic complex comprising plane shock wave, induction zone, zone of chemical reaction and the region of adiabatically expanded products of detonation. The propagation velocity of such wave is stationary and the detonation limit is determined by some maximum level of energy and pulse losses from the zone of chemical reaction, after which the propagation of stationary wave becomes impossible [4].

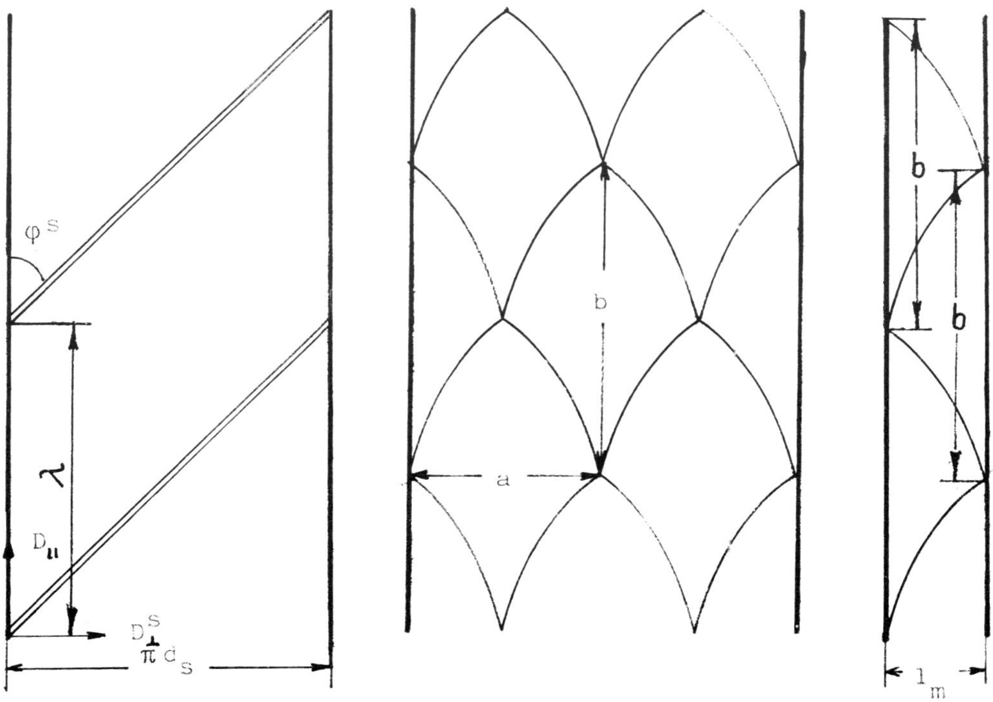

Fig.1. Smoken-foil imprints of spinning (a), multiheaded (b) and marginal detonations (c).

Multifrontal detonation is not, strictly saying, the stationary one, as far as the elements of its front execute the periodically repeating pulsating motions, which amplitude changes rather widely in a time of one pulsation. The velocity of self-sustaining wave, defining as

$$\bar{D}=\frac{1}{T}\int_0^T D(t)dt \quad (1)$$

averaged over pulsation period is, in principle, the analog of stationary value. As a rule $\bar{D}\approx D_o$ for the multifrontal regime.

The spinning detonation is the unique one not only due to its structure, but also because it represents the single case of stationary DW propagation. For different mixtures $D_s\cong(0.8-1.0)D_o$.

In terms of velocity the galloping regime coincides completely with the multifront one, DW quasisteadiness is determined by averaging over the period of longitudinal pulsation, $D_g\cong(0.7-0.85)D_o$.

As for high-velocity regimes of combustion ($D_f\cong0.5\ D_o$), they may be classed with quasistationary regimes similar to the multifront and galloping regimes of detonation and may pretend to be referred to the detonation regimes due to their supersonic nature. Thus, the velocity of the process in unlikely to be the basis for the definition of limiting regime. In the author's opinion the relation between the characteristic scale of pulsation and the detonation channel size is more preferential for this purpose. The multifront regime far from the limiting one is characterized by the value $b/d\ll1$; here the longitudinal size of cell b is naturally to be chosen as a pulsation scale. The spin step $\lambda\simeq d$ may be chosen as a characteristic scale for spinning regime. Although the velocities of spinning and galloping regimes differs, insufficiently, this difference is rather sufficient in pulsation scale - the duration of gallop pulsation is some tens of calibers. Thus, if the limiting regime is determined as a regime, which scale of pulsations is commensurate with the channel size, the spinning regime will be the proper one..

The attempts to explain theoretically the spinning detonation are connected with acoustic theory. According to it TW velocity on the detonation front is connected with acoustic velocity of reaction products in detonation wave [13-15]. For spinning detonation with single TW, its axial velocity $D_{\perp}^{S}=1.84\cdot c$, where c is the sound velocity in detonation products, and 1.84 is the value of major root of the Bessel function of 1 order, which describes the radial component of velocity potential of a gas for two-dimensional acoustic equation. The time of one revolution of TW is $t^o=\pi\cdot d/D_\perp^s$, d is the tube diameter. The trajectory of spinning TW represents a spiral line with the step $\lambda=D_\parallel t^o\simeq\pi\cdot d$ and with the slope to the tube axis $tg\ \varphi_s=D_\perp^s/D_\parallel=1.84\ c/D_\parallel\simeq1$, i.e. $\varphi_s\simeq45^o$, since $c/D_\parallel \simeq 0.55$, $D_\parallel$ is the longitudinal (along the tube axis) velocity of detonation. At multifrontal detonation the transverse a and longitudinal b are the sizes of elementary cell and are connected by the relation $a = b\cdot tg\ \varphi$. Carrying out the analogy between the imprints of spinning detonation (Fig. 1a) and multifrontal one (Fig. 1b), the longitudinal size of a cell is determined logically as the analog of the spiral step of spinning detonation $b=\lambda$ (the trajectories of TW are shifted along the tube axis at a distance b). Then the condition of spinning regime realization may be written as

$$a =\pi\cdot d_s \qquad (2)$$

Given the dependence $a(P_o)$ for explosive gas mixture, the diameter of circular tube d_s is determined according to (2) for the limiting regime - spinning - of stationary detonation propagation.

In experimental investigation the circular detonation tubes and the detonation channels of rectangular cross-section are widely used. The transition from multifrontal detonation to the regimes with lesser quantity of TW is also observed here when P_o decreases. The regime with single TW is the analog of spinning detonation in the circular tube and was named a marginal one for rectangular channel. In this case the transverse wave moves periodically between the lateral walls of rectangular channel. The spiral track of TW on the lateral surface of circular tube, characteristic for spinning detonation is transforned into rhomboid imprint, typical for the cellular structure (Fig. 1c). The channel width l is connected with transverse size of cell by the relatio

$$l = l_m = a/2 \qquad (3)$$

According to the dependence $a(P_o)$ and using (3) the width of rectangular channel l_m, where the marginal regime is achieved, may be determined. It should be noted that through the marginal regime in rectangular channel is the analog to the spinning regime in circular tube, the principle difference exists between them. It is connected with the character of transverse wave motion; at spinning detonation TW rotates around the tube periphery with constant velocity and at marginal detonation TW velocity during the impacts with lateral walls of the channel is variable. The latter is typical for the multifrontal detonation. Actually, the regimes observed in the rectangular channels are always of pulsating character, and their quasisteadiness is determined by the velocity averaged over the pulsation period as in (1). Can the marginal regime be considered to be the limiting one for the rectangular channels? Only from the point of view of comparability of pulsation scale with the size of channel.

However, relation (3) is not sufficient criterion for determination the limit of quasistationary propagation of gas detonation in rectangular channel, as far as the depth of channel δ is negligible in (3).

Assuming the detonation limits to be determined by equal level of losses in the circular and rectangular channel, the interconnection of geometric parameters of the channel with physical-chemical parameters of explosive mixture may be estimated for the case when the limit of stationary (from the viewpoint of (1)) detonation propagation is observed. For this estimation the hydraulic analogy is used. According to it the hydraulic diameter d^o plays the role of the diameter of circular tube for the pipes with cross-section different from a circular one. For the rectangular channel with the size $l \cdot \delta$

$$d^o = 2 \cdot l \cdot \delta / (l + \delta) \qquad (4)$$

The detonation limits will be described by relation: $d_s = d^o$

$$a/\pi = 2 \cdot l \cdot \delta / (l + \delta) \qquad (5)$$

The main principle consequences from the limit criterion of (5) are the following:

1) for the square channel ($l = \delta$) the limit is described by the relation $l = a/\pi$ typical for spinning detonation. Moreover, the circular tube with the diameter d and the channel of square cross-section with the side equals d, proved to be equal with the hydraulic diameter $d^o = d$. It may be expected that the quasispinning regime in the channel with square

section will correspond to limiting regime in the circular tube (spinning detonation). The relation between the channel sizes and elementary cell for limiting regime in the channel of square section is not integer: $l/a=1/\pi\neq n$, where n is the integer (whole) number.

2) marginal detonation means the single TW in the channel, determined by the relation $l_m=a/2$, will be achieved according to (5) only under the condition:

$$l/\delta \leq (\pi-1) \neq n \qquad (6)$$

i.e. when the relation between the sides of rectangular channel is not integer.

3) in the channel with the width l coinciding with the size of elementary cell a, the detonation limit according to (5) is observed when

$$l/\delta \leq (2\cdot\pi-1) \neq n$$

4) when $l/\delta \gg 1$, $l/a \gg 1$ follows from (5), i.e. unlike circular tubes, where the spinning detonation with single TW is observed at the limit of stationary propagation in the rectangular channel the limits depend on the relation of the channel sizes. Particular emphasis should be placed upon this fact which results in the conclusion that when $l/\delta \gg (\pi-1)$, the multifrontal regime will be the limiting one.

Some ways are available for experimental testing (5):

i) $l/\delta=const=k$, then from (4)

$$\delta = a\cdot(k+1)/2\cdot k\cdot\pi \qquad (7)$$

i.e. the interconnection between δ and a is linear. At large k $\delta\to a/2\cdot\pi$, i.e. the limiting depth of the channel is $2\cdot\pi$ times less than the cell size. In its turn $a\to 2\cdot\pi\cdot\delta$, i.e. when P_o is varied the size of cell a in rectangular channel cannot exceed the value $2\cdot\pi\cdot\delta$. Of course, it concerns the cell in quasistationary wave. So,

$$\delta_{min} = a/2\cdot\pi$$

$$(8)$$

$$a_{max} = 2\cdot\pi\cdot\delta$$

ii) $l=const$. In this case from (5):

$$\delta = l\cdot a/(2\cdot l\cdot\pi-a) \qquad (9)$$

when $l \gg a$ $\delta\to a/2\cdot\pi$ as (8).

iii) δ=const. From (5):

$$l = a\cdot\delta/(2\cdot\pi\cdot\delta-a) \qquad (10)$$

that also gives

$$\delta_{min} = a/2\cdot\pi$$

The experimental investigations of the limits of stationary propagation of gas detonation are carried out on the rectangular channels with different l and δ. Explosive chamber 3 (Fig.2) was the system of co-axial channels of rectangular cross-section. The optical glass 1, through which the photographing of the process fluorescence in each

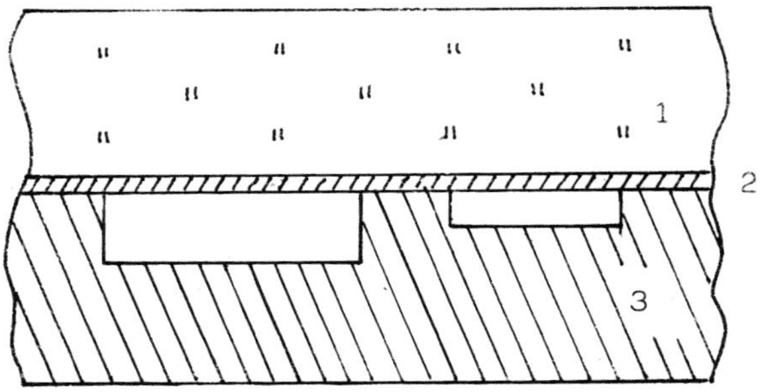

Fig.2. Scheme of experimental apparatus for investigation of detonation wave in co-axial rectangular channels.

channel on moving film is made, is the general wall for all channels. When studying the track imprints along the glass surface the smoked foil 2 was set up along the whole length of explosive chamber. The experiments have been carried out on the mixtures $2H_2+O_2$, $C_2H_2+2.5O_2$ and $C_2H_2+2.5O_2+10.5Ar$. The mixture composition was controlled by partial pressures of individual gases. The initiation was carried by high-voltage discharge ($CU^2/2 \sim 40$ J). The explosive chamber was evacuated prior to each experiment. The initial pressure P_o was the main varied parameter.

The limiting conditions in rectangular channels with k=const==6, typical for plane channels for stationary propagation were studied in the first series of experiments. The channel length was L=300 mm, δ=5,4,3,2,1 and 0.5 mm. The channel wigth l=3 mm was constant for second series of experiments and δ=1.0, 0.5, 0.4, 0.2 and 0.1 mm. The results of these studies for mixtures $2H_2+O_2$ and $C_2H_2+2.5O_2$ are shown in Fig.3-4 respectively: light points mean the breakdown of stationary regime, the dark ones mean stationary detonation. The averaged dependences of cell size on initial pressure $a(P_o)$ are indicated by lines 1, lines 2 and 3

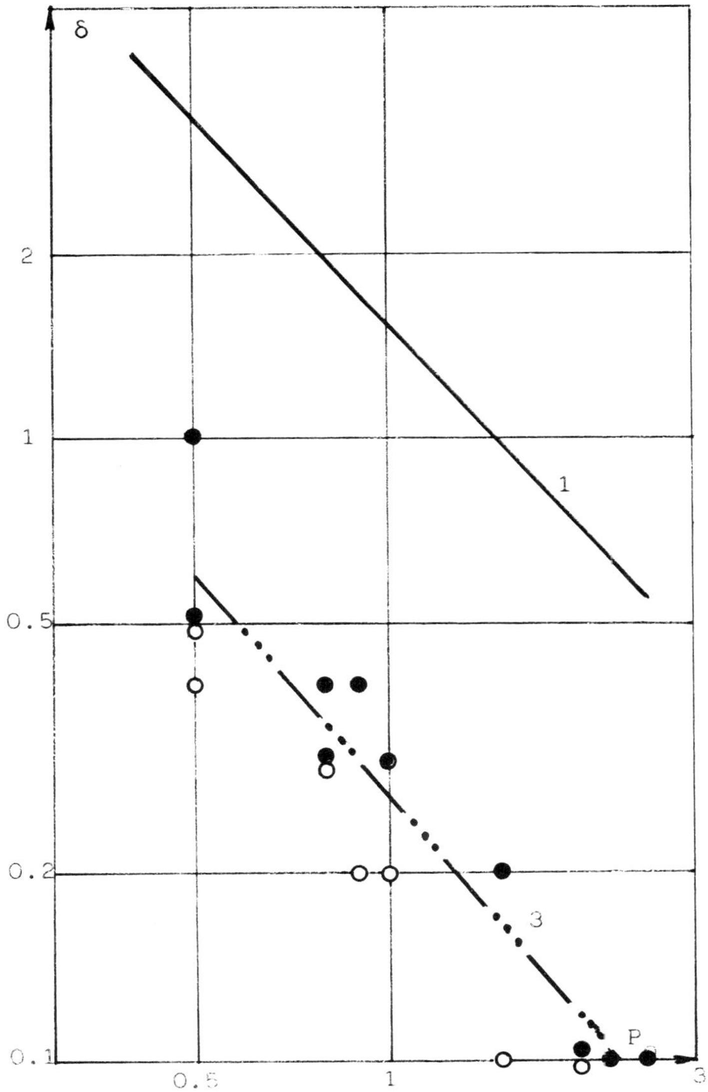

Fig. 3. Dependence $\delta(P_o)$ for $2H_2+O_2$ and $C_2H_2+2.5O_2$: δ-mm, P-atm.

indicate the calculated lines of the limits according to (7) and (9). Both in qualitative relation and quantative one the calculated lines are in good agreement with experiment. In rectangular channels with large

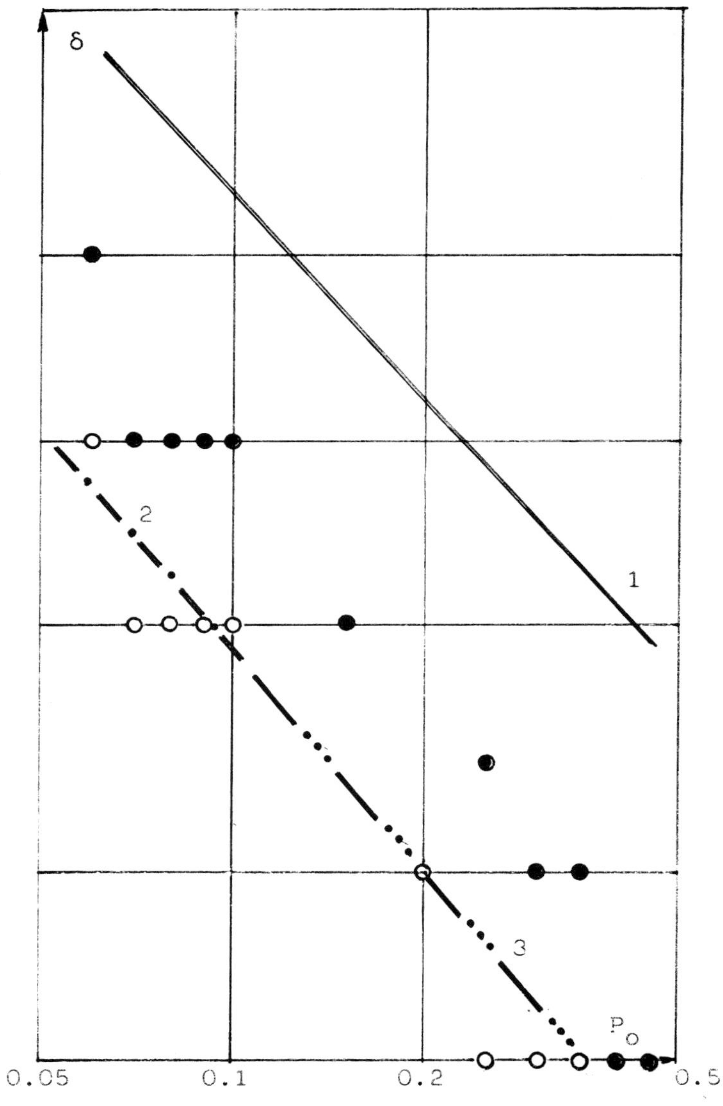

Fig. 4. Dependence $\delta(P_o)$ for $2H_2+O_2$ and $C_2H_2+2.5O_2$: δ-mm, P-atm.

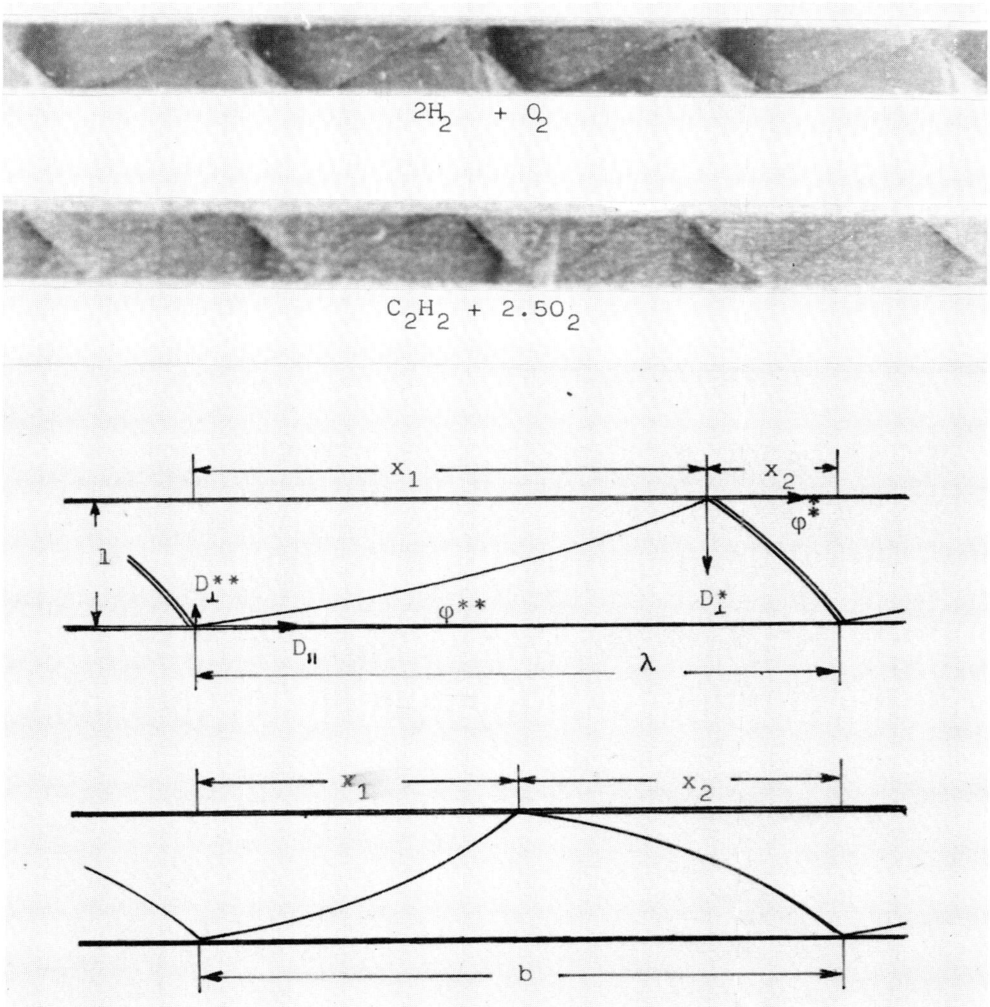

Fig. 5. Smoken-foil imprints of quasispinning regime in rectangular channel.

relation l/δ, the limit is determined only by the channel depth δ. In this case the limiting depth is approximately $2 \cdot \pi$ times less than the size of elementary cell of detonation wave.

2.

In the square channels, the channel depth δ at the limit is π times less than a (according to (4)). The experiments have been carried out in the explosive chamber with the square channels δ=16, 9, 6, 4, 3 and 2 mm, the channel length L ~600 mm. The smoke-foil imprint of limiting detonation regime in the square channel 4×4 mm and the enlarged scheme of interpretation of detonation imprints in square and rectangular channels are shown in Fig. 5.

The main peculiarity of the imprints is the disarrangement of the

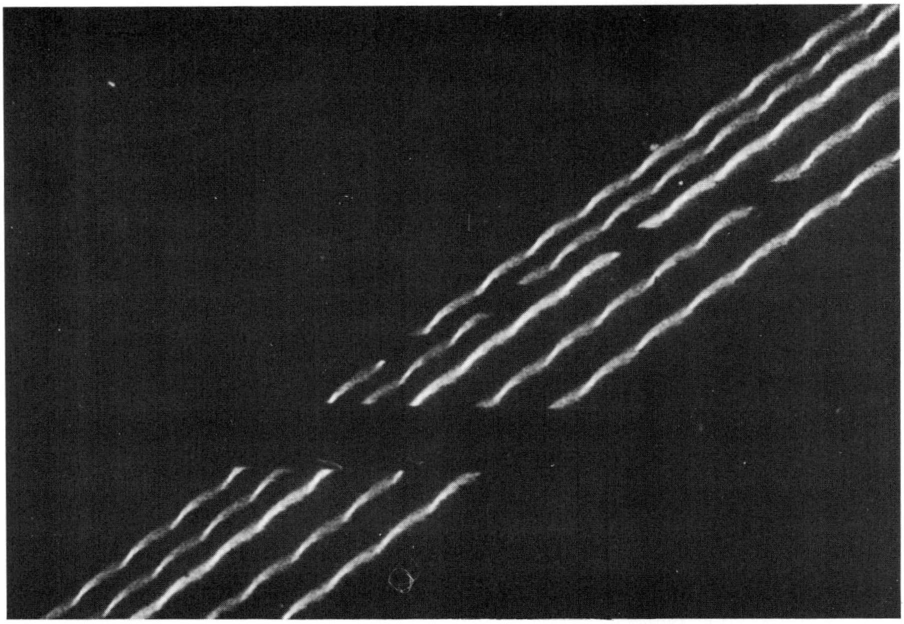

Fig. 6. Selfluminocity streak-photographs of near-spinning regime in rectangular channels.

trajectory symmetry at direct and return motion of the wave between the opposite sides of square channel. At marginal detonation at rectangular channel ($l/\delta\approx 2$) the similar trajectories are symmetrical (Fig. 5c). The explanations of this phenomenon is connected with the fact that "quasi-spinning" regime of detonation propagation is limiting at rectangular channels. In this case TW executes a rotatory-translatory motion at DW moves along the channel. It should be mentioned that in the scope of acoustic model the rotating mode of the solution for rectangular channel is absent [15], but it is valid only for plane acoustic waves with the surface parallel to the walls of a rectangular channel. For the inclined wave such mode is quite natural. The streak-recorder photograph of detonation wave propagation in co-axial channels 4×4 mm (in details this photo will be described in 3) is shown in Fig.6, where the quasispinning rotating character of wave propagation along each channel is fixed sharply. Such regime is likely to be mentioned for the first time in [16]. In this case a wide line on the smoke-foil imprint with large inclination to the channel axis is the trajectory projection of TW itself at its motion along a side of square channel. A narrow line corresponds to the shock wave, which occurs at TW turn about 90° for the motion along the following wall of the channel and its partial reflection from this wall. Asymmetry of these lines is connected with different velocity of wave propagation: TW velocity is connected with fundamental mode of axial oscillations of detonation products ($D^S_\perp = 1.84 \cdot c$ for round tube), the velocity of reflected wave is not connected with this mode.

For estimation the equivalency of the channels of round and rectangular cross-sections may be used (the circle diameter and square side are equal d) from the point of view of hydraulic diameter and losses. TW in a round tube covers $\pi \cdot d$ at one revolution and in a square channel - 4d. If the rotatory mode of fundamental oscillation is the same in the channels of round and square cross-section, so the linear velocity of TW motion along the square side must be higher as much as the square perimeter is more than a circle perimeter, so one can obtain the following: $D^*_\perp = D^S_\perp \cdot 4/\pi$. Therefore, TW trajectory moving along a square wide will have the greater inclination than the inclination of a spinning wave in round tube ($\varphi^S \approx 45^O$), as far as tg $\varphi^* = D^*_\perp / D_\parallel = 4/\pi \cdot$ tg φ^S, i.e. $\varphi^* \approx 52^O$. The step of "quasispinning" detonation λ in the square channel l×l, as the experiment shows, is in perfect correspondence with the relation $\lambda = \pi \cdot l$, which results from the condition of limits (5). The velocity of reflected wave is reasy determined with scheme of Fig.5: $x_2 = l/\text{tg}$

$\varphi^{**}=l\cdot\pi/4$, $x_1=3\cdot l\cdot\pi/4$, then $tg\ \varphi^{**}=1/x_1=4/3\cdot\pi=tg\ \varphi^{*}/3$, i.e. $\varphi^{**}\approx 23^{o}$ and $D_{\perp}^{**}=1/3D_{\perp}^{*}$. The difference between spinning regime in a round tube and quasispinning one in a square channel resides not only in the fact that the reflected waves appear in square channel together with the major TW. Their main difference is the steadiness of spinning regime at any moment of time for a round tube and the steadiness "in average" for a square channel. Due to the reflections at the angles the rotary velocity of TW is nonuniform, the velocity of reflected wave turned to be also nonuniform. The complete description of the processes which are observed at the limit of detonation propagation in a square channel, should be expected to follow from the theoretical calculation of three-dimensional problem of TW propagation. This problem is developed in [17,18].

3.

In spite of the evident interaction of the energy and pulse losses from the chemical reaction zone with existence of DW limits the physical mechanisms responsible for these losses are required to be investigated. The losses connected with friction against the channel walls, boundary layer formation, mixture turbulization, etc. are the traditional ones. The sufficient non-uniformity and unsteadiness of the gaseous multifront wave with pulsating front is the consequence of transverse wave appearance and their propagation along DW front. The importance of their role in the processes of combustion intensification, detonation initiation, etc. for gaseous mixtures is without a doubt. What is the role of transverse waves in the near-limiting regimes? How sensitive is the detonation wave to the boundary conditions in near-critical regimes (NCR)? NCR is the intermediate regimes which are laid between the limiting regime from one side and the multifront one from another side, for which the various geometrical effects on DW do not influence on its self-sustaining regime of propagation. As far as unambiguously are the limiting conditions of DW propagation determined? The investigations of specific effect on the transverse waves and the following DW reaction on

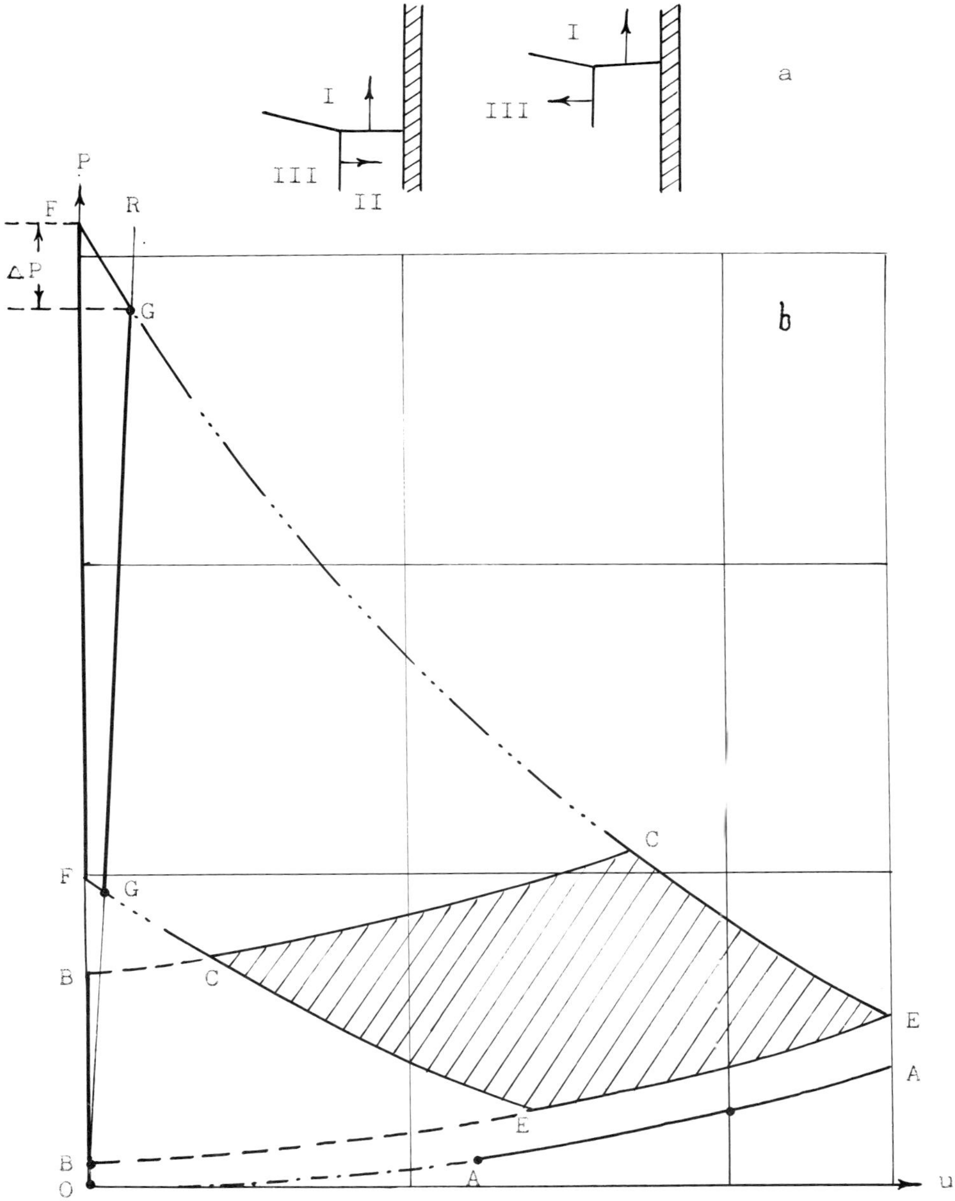

Fig.7. (P,u)-diagram of TW reflection from acoustic rigid wall.

this effect may give the answer on these questions. The attenuation of transverse waves in near-critical regimes and even their total elimination (as, for example at DW diffraction on the convex angle) leads to the breakdown of self-sustaining detonation regime and to the transition to the damping non-stationary one which in dependence of the boundary conditions may either decay further or reinitiate again in the form of detonation. Near the limits such mechanisms may play no less important role than the traditional mechanisms of losses. The possibility to influence on DW through TW attenuation on the channel walls is shown by G.Dupre et. al. [19], where the metal fine grid is used as a reflecting wall. For the regimes close to the spinning one, the transverse waves were attenuated so much after DW passing by the grid that the decay of detonation regime and the transition to the regime of high-velocity combustion with the velocity $\sim 0.5D_0$ were observed. The grid played the role of porous medium which absorbs the transverse waves. The grid model as a system of capillary tubes perpendicular to the tube axis and its influence on DW propagation are analyzed by K.V.Reddy, T.Fujiwara, J.H.Lee [20]. The effect of transverse waves attenuation at their collision with the wall is shown also by A.A.Borisov et. al.[21],where the tube wall was covered by thin film of liquid.

In the channels with constant cross-section the transverse waves attenuation may occur at their interaction with the walls at the expense of non-elastic collision. Let us consider the idealized scheme of TW interaction with a wall shown in Fig.7a: DW front transfers the mixture from state 1 into state 11, the state on TW before the reflection on a wall is denoted as 111. TW reflection on a wall is one of the classical problems of gas-dynamics on the disintegration of arbitrary discontinuity [22]. Although the pressure amplitude behind DW and TW in gaseous mixtures may change from tens to hundreds atmospheres, such wave for the reflecting wall may be considered with high accuracy as the acoustic one. In this case the amplitude of reflected wave depends on the relation of the acoustic impedances $\mathfrak{K}=\rho\cdot c$ of contacting media. The diagram (P,u) of the transverse wave reflection from a wall is shown in Fig.7b. Here 0 is the initial state of the mixture before DW front, AA is the region of possible states of gas behind DW caused by nonstationary character of some front elements in time duration of separate pulsation; BB are the initial states before TW (the absence of mass flow in a wall direction); ECCE is the range of possible parameters behind the transverse wave before its collision with the wall; FF are the parameters behind the reflected TW (under the condition of flow stop) after TW interaction with ideally rigid wall ($\mathfrak{K}=\infty$); GG are the parameters at TW reflection from the wall with finite $\mathfrak{K}$. Nevertheless, even small decrease of the parameters of reflected TW may influence sufficiently on DW propagation, especially near the limits by virtue of the exponential dependence of chemical reaction velocity on 1/T.

How important is the role of the mechanism of TW attenuation among the others? How sensitive are TW to small perturbations? These problems may be solved in terms of experimental investigations of DW propagation when the acoustic impedance of the channel walls $\mathfrak{K}$ was the variable parameter. The explosive chamber comprised 5 co-axial channels with the same cross-section 4×4 mm (as in Fig.2) which are milling (with the precision ∇6) in preliminary collected package of 20×20 mm plates of the materials with different $\rho \cdot c$. The streak-recorder photographs in $C_2H_2+2.5O_2$ mixture, when the regimes close the quasispinning one are observed in all five channels, is shown in Fig.6 (mentioned above in 2.). As shown in Fig.8, the decrease of P_0 does not lead to simultaneous breakdown of detonation in all channels. In turn, it means that the detonation limit is determined not only by the channel size but by the wall material, and the transverse waves in near-critical regimes are rather sensitive even to such insufficient effects as $\rho \cdot c$ change. The problem on the role of walls is the principally important one in experimental determination of critical diameter of the gaseous charge [23].

4.

The quasistationary regime of high-velocity combustion is observed below the quasispinning and galloping regimes. The dependence of average velocity of such process in the mixture $C_2H_2+2.5O_2$ on initial pressure is shown in Fig.9: the line 1 corresponds the calculated Chapman-Jouguet detonation velocity; the line 11 is the experimental dependence for the conditions far from the limit (*a* « is a character channel size), the line 111 is the velocity in the channel corresponding to the limiting conditions; the lines 1V are the high-velocity regimes of combustion (below the limit). The abrupt decrease of wave velocity in transition from the limiting detonation regime to the regime of high-velocity combustion is the most interesting peculiarity of this plot [12]. It should be mentioned that the high-velocity regime has been fixed in the channels with constant cross-section as shown in [12] and later in [19]. The

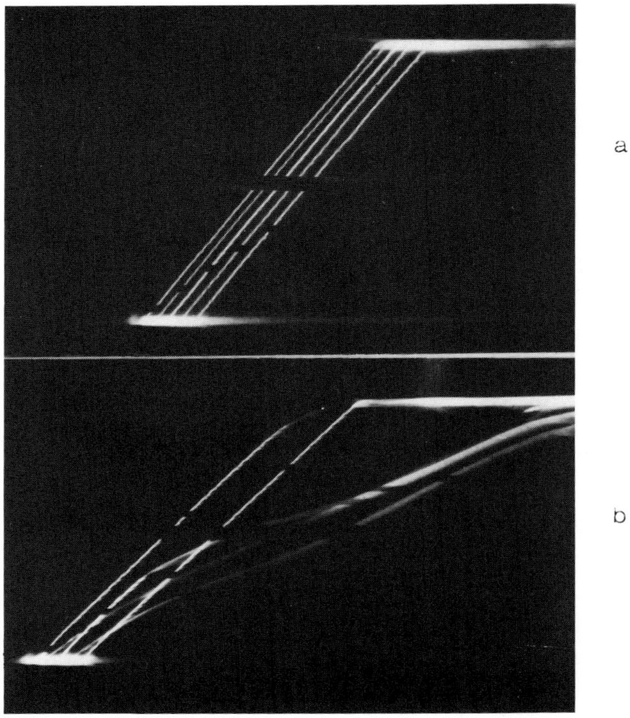

Fig.8. Selfluminocity streak-photographs of detonation wave in difficult ρ *c*-channels: a) multiheaded detonation; b) near-limiting regime.

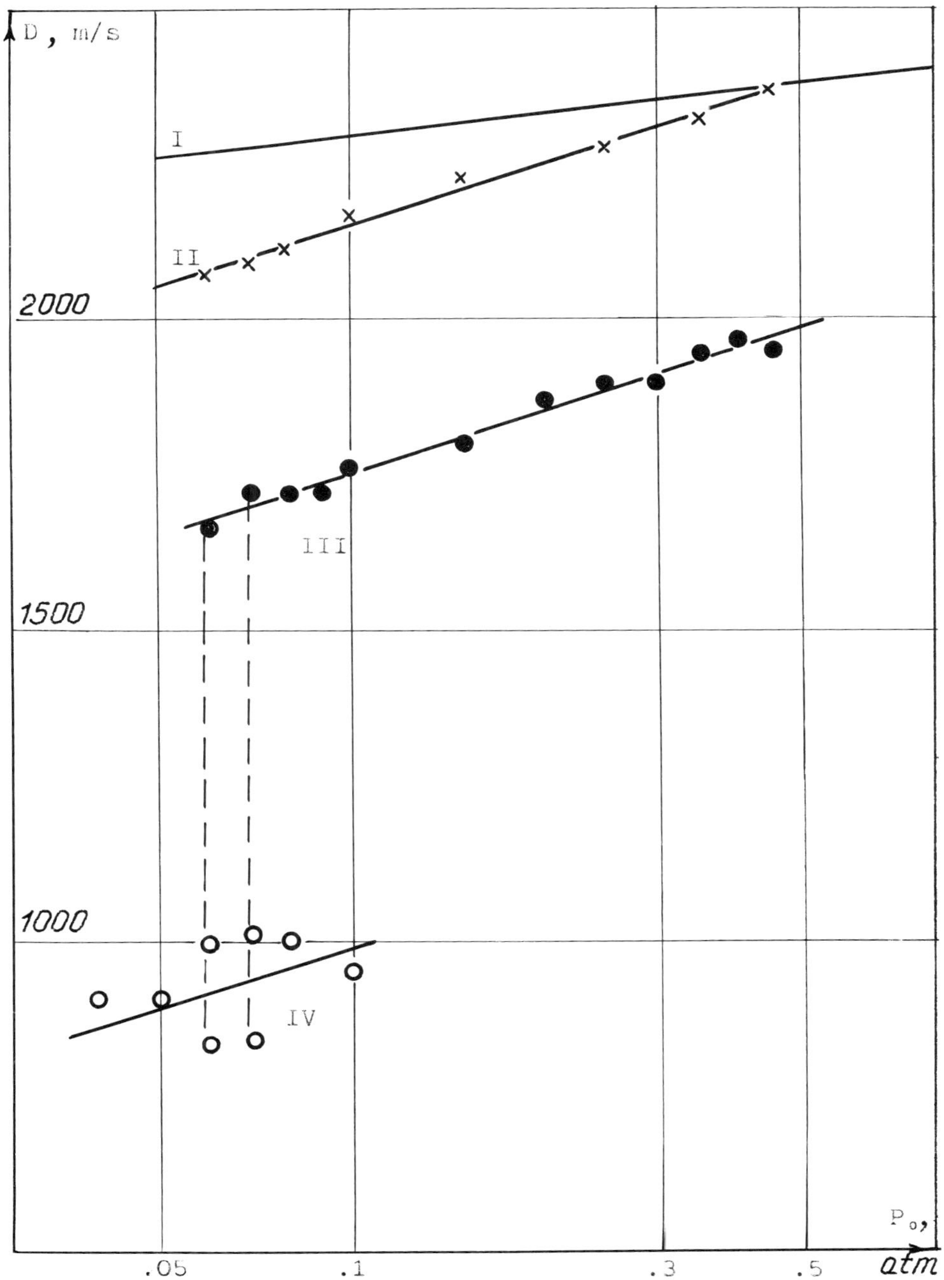

Fig. 9. Dependence $D(P_o)$ for critical detonation regime and high-velocity combustion process.

the leadership of J.H.Lee, concerning the study of detonation in the channels with obstacles (for example, see [19]). Such regimes are of great interest and the high-velocity regimes of combustion can be expected to attract intent attention of many researchers in recent years.

Some additional results concerning the regime observed are given in this section. As far as the high-velocity regime in [12] and [19] was observed at attenuation of strong detonation wave, the problem is natural whether the same quasistationary regime of velocity combustion would be observed at the weak initiation and further acceleration of combustion front from low-velocity laminar flame to supersonic turbulent one. The photographs in Fig.10 are the positive illustrations for this problem: "a" corresponds to the regime of flame acceleration up to the quasistationary velocity; "b" is the formation of adiabate explosion before the flame front, similar to that observed at the deflagration to detonation transition or reinitiation of galloping regime.

Uncertainity of the criterion for rough and smooth walls of detonation channel and their role in the process observed causes the second problem. The rectangular channel with the walls of polished metal plates has been formed for this purpose. The experimental observation in the channels produced by using qualitative milling or polished plates are proved to have no principle difference.

Conclusion.

Thus, in the given paper the criterion of limit of stationary detonation propagation in the channels of arbitrary cross-section is suggested. It relates the interaction of main sizes of the channel with the most important characteristic of detonation wave - the size of elementary cell. In the rectangular channel with large l/δ the limit is determined only by the channel depth $\delta_{min}=a/2\cdot\pi$. $\delta_{min}(P_o)$ is defined according to $a(P_o)$ dependence. In the square channel the limit is determined by the relation $\delta_{min}=l_{min}=a/\pi$, quasispinning regime with the single transverse wave, executing the complicate rotary-translation motion along the square sides, is the limiting one in the square channel. Such

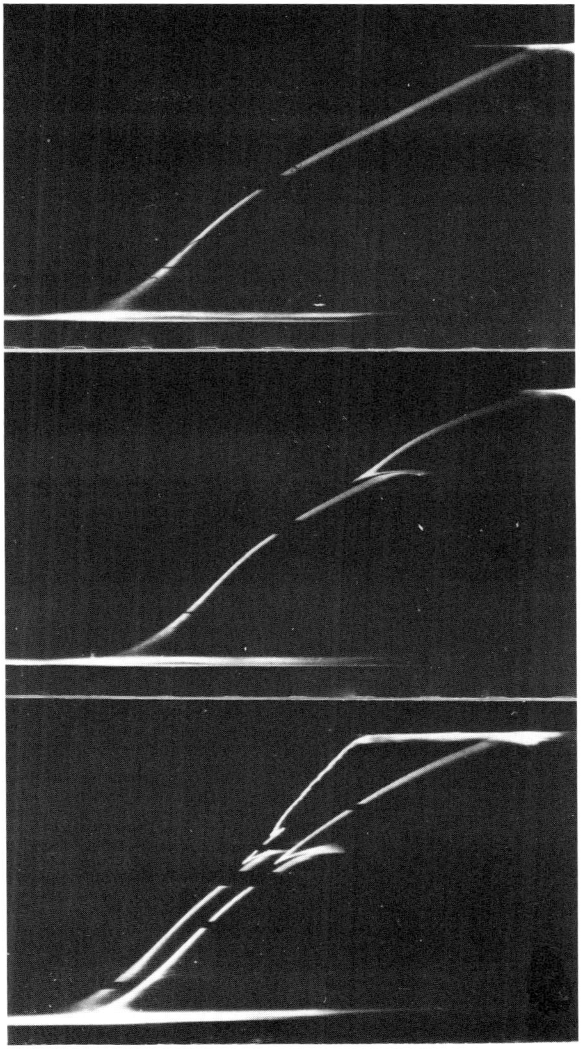

Fig. 10. Self-luminocity streak-photographs of high-velocity combustion regime: a) acceleration of laminar flame to quasistationary high-velocity combustion one; b) quasigalloping regime of combustion; c) reinitiation of detonation and combustion waves.

same dependences were obtained by A.V.Pinyaev and G.V.Lyamin [24] while studying the detonation in porous media and in the works, carried out by regime corresponds to the classical spinning regime as the limit of detonation in a round tube. The quasispinning regime may be thought to be present in the channels with the cross-section in the form of regular polygon.

The possible mechanism of detonation wave attenuation at the expense of partial decrease of the transverse waves amplitude at their reflection from the walls, having the final value of acoustic resistance, is analyzed.

The high-velocity quasistationary combustion regimes are established as representing a new class of supersonic regimes of propagation. There are practically independent of the method of excitation of these regimes (from the detonation or the laminar flame) and small scaled roughness of the walls.

Fig. 11. Acceleration of combustion wave in quasisteady high-velocity regime of free explosive gaseous jet.

The experimental investigations have confirmed the principal conclusions.

References

1. J.H.Lee. Initiation of gaseous detonation. Ann.Rev.Phys.Chem., 1977, 28: 75-104.

2. J.H.Lee. Dynamic parameters of gaseous detonation. Ann. Rev. Fluid Mech., 1984, 16, 311-336.
3. A.A.Vasiliev, V.V.Mitrofanov, M.E.Topchian. Detonation waves in gases. FGV, 1987, 5,109-131.
4. Ya.B.Zel'dovich, A.S.Kompaneets. Theory of detonation. M., Gostekh-izdat, 1955.
5. K.I.Schelkin, Ya.K.Troshin. Gasdynamics of combustion. M., Izdatel'-stvo AN SSSR, 1963.
6. B.V.Voitsekhovsky, V.V.Mitrofanov, M.E.Topchian. Structure of detona-tion front. Novosibirsk, izdatel'stvo SO AN SSSR,1963.
7. A.J.Mooradian, W.E.Gordon. Gaseous detonation - initiation of detona-tion. J.Chem.Phys., 1951, 19, 1166.
8. J.P.Saint-Cloud, G.Guerrand, G.Brochet e.a. Quelques particularities des detonation tres instables dans les melange gazeux. Acta Astr., 1972, 17, 4/5, 487.
9. N.Manson, G.Brochet, J.Brossard e.a. Vibration phenomena and instabi-lity of self-sustained detonation in gases. 1X-th Symp. (Internat.) on comb., 1963, 461.
10. V.Yu.Ulianitsky. Investigations of galloping detonation for gaseous mixture. FGV, 1981, 17, 1, 118.
11. A.A.Vasiliev, V.Yu.Uljanitsky. Near-limit regimes of gaseous detona-tions. 1X-th ICOGERS, USSR, 1981.
12. A.A.Vasiliev. Geometrical limits of gas detonation propagation. FGV, 1982, 18, 2, 132.
13. N.Manson. Propagation des detonations et des deflagrations dans les melanges gazeux.Comp.Rend., 222, 46 (1946).
14. J.A.Fay. A mechanical theory of spinning detonation. J. Chem. Phys., 20, 6, 942-950.
15. B.T.Chu. Vibration of the gaseous column behind a strong detonation wave. Proc.Symp.Aerothermochem., 1955, Evanson (1956).
16. J.H.Lee, R.I.Soloukhin, A.K.Oppenheim. Current views on gaseous deto-nation. Astr.Acta, 1969, 14, 565-584.
17. S.U.Schoffel. A computational approach explaining universal empirical correlations between the fundamental gaseous detonation parameters. X1-th Internat. Symp. on Comb.Processes, Poland, 1989.
18. T.Fujiwara, K.V.Reddy. Propagation mechanism of detonation - three-dimensional phenomenon. Mem.Fac.Nagoya Univ., 1989, 41, no.1, 93-111.
19. G.Dupre, O.Peraldi, J.H.Lee, R.Knystautas. Propagation of detonation waves in an acoustic absorbing walled tube.Progress in Astronautics and Aeronautics. v.114, Washington, 1988, 248-263.
20. K.V.Reddy, T.Fujiwara, J.H.Lee. Role of transverse waves in a detona-tion waves - a study based on propagation in a porous wall chamber. Mem.Fac. Nagoya Univ., 1988, 40, 1, 149-162.
21. A.A.Borisov, A.E.Malikov, V.V.Kosenkov, V.S.Aksenov. Propagation of gaseous detonations over liquid layers.12-th ICDERS, USA, 1989.
22. L.V.Ovsiannikov. Lectures on the basis for gaseous dynamics. M., Nau-ka, 1981.
23. A.A.Vasiliev. Critical diameter of gas-mixture detonation. FGV, 1982, 3, 98-104.
24. A.V.Pinaev, G.A.Lyamin.The principal regularities of subsonic and de-tonation combustion of gases in inert porous media.FGV,1989,4,75-85.

Propagation of Gas Explosion in Channels with Uneven Walls and in Porous Media

G.A.Lyamin, V.V.Mitrofanov, A.V.Pinaev, V.A.Subbotin
Lavrentyev Institute of Hydrodynamics
Novosibirsk 630090 USSR

Abstract. Various regimes of propagation of chemical reaction waves in combustible gas mixtures along the channels with complex configuration were investigated. Tubes of 35 and 20 mm i.d. packed with river sand, gravel or metal balls and two-dimensional models of a porous medium in the form of planar channels with regular 4-fold widening and narrowing of the width were used in the experiments. The sizes of grains of the porous medium δ ranged from 10^{-4} to 10^{-2}m. The mixtures under study were C_2H_2, H_2, C_3H_8 with either oxygen or air, at different concentrations. In somecases, other fuels and non-reactive diluents were employed. Wave propagation velocities, pressure and radiation profiles were registered. Channels with variable width had one or two transparent plane walls allowing the flow structure to be observed by photography. The dependences of steady wave propagation velocities D on the initial pressure of mixtures p, which were varied in the range of 10 - 5 MPa, were studied in more detail.

For active fuel-oxygen mixtures with the composition close to a stoichiometrical one, only detonation regimes of combustion were found to be steady both in porous media and in channels with variable width with $D \geq D_{min} \simeq (1.5 \pm 0.3)c_0$. In this case, the flame is always accelerated to detonation combustion or is damped down, the dependence $D(p_0)$ for all mixtures in the porous medium being continuous within the interval from D_{min} to the ideal Chapman-Jouguet detonation velocity for agiven mixture

A. A. Borissov (ed.), Dynamic Structure of Detonation in Gaseous and Dispersed Media, 51–75.

D_{C-J}. In contrast, in the channels with variable width $D(p_0)$ has a break in the vicinity of 0.5 D_{C-J} and, therefore, high-speed and low-speed detonation regimes of combustion can be distinguished, which differ in the mechanism of flame transfer.

For the pressure p_0 higher than the p_{cr} value, critical for non-quenchin transmission of detonation from neckings to widenings (pores), a shock-wave mechanism of ignition takes place and the flame front has an ordinary cellular structure with transverse waves. If $p_0 < p_{cr}$, alternating disruptions and restorations of the shock wave ignitionare observed. With decreasing average relative time of action of the latter, the velocity D decreases. The disruption manifests itself in separating the shock front from the reaction zone with simultaneous disappearance of the cellular structure, when the wave comes out into a widening. The restoration takes place either in the same "pore" as a result of the secondary local detonation wave initiation in the layer of compressed mixture or in one of the following pores, where the flame penetrates by a convective mechanism: with the jet of hot products leading to the accelerated burning in the zone of intense mixing of initial mixtures and products.

At low-speed detonation regime of combustion, only a convective mechanism of the flame transfer works with the same processes of accelerated mixing and combustion as at high-speed one, but without generating internal local detonation waves. Low-speed detonation regime of combustion in the channel with various width and porous medium corresponds to a "choking" regime in tubes with circular obstacles investigated by R. Knystautas et. al. (1985).

For compositions far from stoichiometrical ones or in the case of dilution of the mixture, the detonation regime of combustion remains in a wide range of concentrations passing to the regions of higher pressures. At the same time, at lower pressures the steady-state regime of turbulent combustion with a velocity ranging from 10^{-1} to 10^{2} m/s appears. The curve $D(p_0)$ for most such mixtures has either one break in the velocity from about (1-70) m/s to 500 m/s, or two breaks with typical values of (2-10) m/s → 100 m/s and 250 → 500 m/s. For some mixtures a continuous variation of D over the range 100 to 400 m/s vs p_0 was obtained. It means that a continuous transition of the velocity of the steady-state wave of the explosive transformation through the speed of sound in gas c_0 is achieved, implying the division into the waves of detonation and the waves of combustion to be relative. In maximally diluted mixtures, the flame velocity in porous medium did not exceed 1 m/s in the whole pressure range under investigation.

At $D < c_0$, the pressure profile of the wave has a smooth rise of the pressure ("forerunner") which shrinks and turns into shock at $D \rightarrow c_0$. At the same time, the height of the pressure peak increases: at $D < 20$ m/s, $\Delta p/p_0 < 0.2$; at $D \simeq c_0$, $\Delta p/p_0 = 4\text{-}5$; at $D > c_0$, Δp increases up to the values exceeding the pressure of the explosion within a constant volume. A Peklet criterion calculated in terms of the pore size, normal flame velocity and initial parameters of the mixture at the lower limit of the existence of a steady state combustion wave, is found to be in the range of 20 to 110 for all mixtures and porous media used irrespective of the combustion regime and the limit velocity value. In all regimes of gas explosion propagation through the porous medium, non-complete burn-out was revealed.

INTRODUCTION

For explosive gaseous mixtures, there exist two fundamental velocities of chemical reaction wave propagation. They are normal velocity of laminar flame U_n and Chapman-Jouguet detonation velocity D_{C-J} slightly dependent on initial pressure p_0. The characteristic values of U_n are 10^{-1}-10^{1} m/s, and D_{C-J} exceeds sonic velocity in undisturbed gas, c_0 by factor of 4 to 7,5. When there are obstacles in the path of the wave, the detonation velocity may decrease due to different actions to the reaction zone, and the combustion velocity may increase due to flow turbulization. Although these factors are generally known, quantative characteristics of the phenomenon are often unexpextable, the details of the processes are very complicated and continue to attract attention of many researchers. Among the problems of particular interest are as follows: (i) down to what limiting values the turbulent flow velocity may increase given the obstacles in the channel, (ii) how the transition between these regimes occurs and whether it is always possible to discriminate them, and (iii)how much steady-state regimes may exist under given conditions.

A number of available works is related to the study of combustion and detonation in the tubes with periodical obstacles. As early as 1926, W.R.Chapman and R.N.Wheeler revealed in such channels the combustion waves propagating with the velocity higher than 100 m/s. A plate with orifices was used as an obstacle. A substantial and early contribution into the study of this problem was made by Shchelkin (1940, 1949, 1963) who placed wire spirals into a tube along its inner wall. He established that the detonation velocity in such "rough" channels may decrease to, approximately, 1 km/s. Velocities of about 100 and 200 m/s, those ranging over 500 to 900 m/s and higher than 1100 m/s were achieved in the channels with periodical annular obstacles (J.Lee et. al., 1984; R.Knystautas et. al., 1985).

The flame acceleration up to dozens and several hundreds meters per second was also observed by H.G. Wagner (1985), I.O. Moen (1985) and by other researchers. Along with the velocity, the pressure increase up to several atmospheres and even higher was registered in the wave. Almost all the recent investigations in the tubes with obstacles were performed under atmospheric pressure; the compositions of mixtures, mainly fuel-air ones, and the scales of experimental arrangements were varied. In the works by Ya.B. Zel'dovich et al (1955, 1984, 1985), a theoretical model of "quasi-detonation combustion" favorable for the velocities $D >$ 1 km/s was determined. Within the framework of this model, the mixture ignition occurs in the zone of the leading shock wave reflection from obstacles at the tube walls, and then the wave propagates towards the axis according to the turbulence mechanism. Such a structure of combustion zone was observed in the experiments by V.S. Babkin and L.S. Kozachenko (1960). The model of high-speed turbulent combustion in a tube with obstacles was developed in early works by K.I. Shchelkin and later by S.S. Rybanin (1969). One of the most recent achievements in this field is the model suggested by B.H. Hjirtager (1982), based on the k - ε model of turbulence and on the assumption that a chemical reaction occurs only in the regions where a characteristic time of its development is less than that of turbulence energy dissipation ($\sim k/\varepsilon$). The model used in two-dimensional numerical calculations of the combustion wave propagation from the closed end of the tube through 5 annular obstacles shows a satisfactory agreement with the experiment.

Granular or cemented, if though permeable, porous medium consisting of contiguous solid particles ("grains") or solidified foams is regarded as a space ultimately blocked up with obstacles. The porous medium may be used to retard the flame in apparatus of some chemical productions. The ultimate conditions for overtravelling the flame through the porous medium were studied by I.I. Strizhevsky and V.F. Zakaznov (1973), who revealed that the critical Peclet number coincides with that in narrow tubes. The gaseous detonation in the porous medium of expanded polyurethane was investigated by O.E. Popov et al (1974), in a medium packed with large spheres (δ = 38 mm and 19 mm) by C W. Kauffman et al (1982). In both this works, active fuel-oxygen mixtures were considered, and all the detonation waves propagated with a velocity higher then 1000 m/s.

There are a number of investigations of filtration combustion of fuel-air mixtures in porous media under the conditions when the mixture flow velocity with respect to the medium before the flame front was artificially varied.The results were summed up by V.S. Babkin and Yu.M. Laevsky (1987).In this case the combustion front velocities were ranged over -0.1 to 10 m/s.

A systematical investigation of fast combustion waves in porous media was performed at the Lavrentyev Institute of Hydrodynamics, Novosibirsk.

G.M. Mamontov and his co-authors (1980) were first who succeeded in reducing the detonation velocity to, approximately,500 m/s in the channels filled with river sand or stainless steel fillings. The same authors and then V.A. Subbotin and A.Ya. Kuznetsova (1984) revealed some details of the flow structure on detonation in planar channels with variable width modelling a porous medium. The combustion regimes in porous media, both supersonic and subsonic, were investigated by G.A.Lyamin and A.V.Pinaev (1984-1989) over a wide range of compositions, pressures and pore sizes.

This paper presents the generalized results, obtained by the authors, which are supplemented by the experimental data enabling a deeper insight into the phenomena.

EXPERIMENTAL DETAILS

The experiments were conducted in vertical tubes of several sizes, filled with pre-washed and dried fractions of river sand or gravel, or steel balls. The sizes were (Fig. 1a): L = 3 m for d = 35 mm, L = 0.8 - 1 m for d = 22 mm and L = 0.7 m for d = 20 mm. Steel balls with an exact diameter δ = 2, 3, 4 mm; steel shot with $\delta = 3 \pm 0.5$ mm were used. The sand or gravel particle sizes ranged in different experiments over 0.06 to 12 mm. Within one sifted fraction, the sizes δ differed by a factor of 2 and not more. The tubes were vacuumized, with removing water from the surface of particles, and then they were filled with gaseous mixture. Initiation was realized by a wire, heated or exploded by a pulse of electric current, in the small space free of particles at the upper end of the tube. The time of passage of the wave and the profiles of corresponding wave parameters were recorded at different points along the tube by the photomultiplier (3) through the glass-fibrous light guiders 1 (Fig. 1a), by several piezoelectric pressure gauges 2 and ionization probes (the latters are not shown). In some experiments, the tube was equipped with a special organic glass section to record continuously the velocity with the use of a streak camera. The gaseous mixtures, such as $C_2H_2 + 2.5O_2$, $C_2H_2 + 2.5O_2 + 3.5N_2$, $x\%\ H_2 + O_2 (x = 4 - 50\%)$, $C_3H_8 + 5O_2$, $4\%\ C_3H_8$ + Air, $x\%\ C_2H_2$ + Air (x = 4 - 10 %) were used. The initial pressures varied from 10^{-2} to 5 MPa.

The flow structure was studied in the channels with variable width with the distance between plane-parallel walls being h = 1.0 and 1.5 mm, one or both of the walls being transparent. The profiles of all the

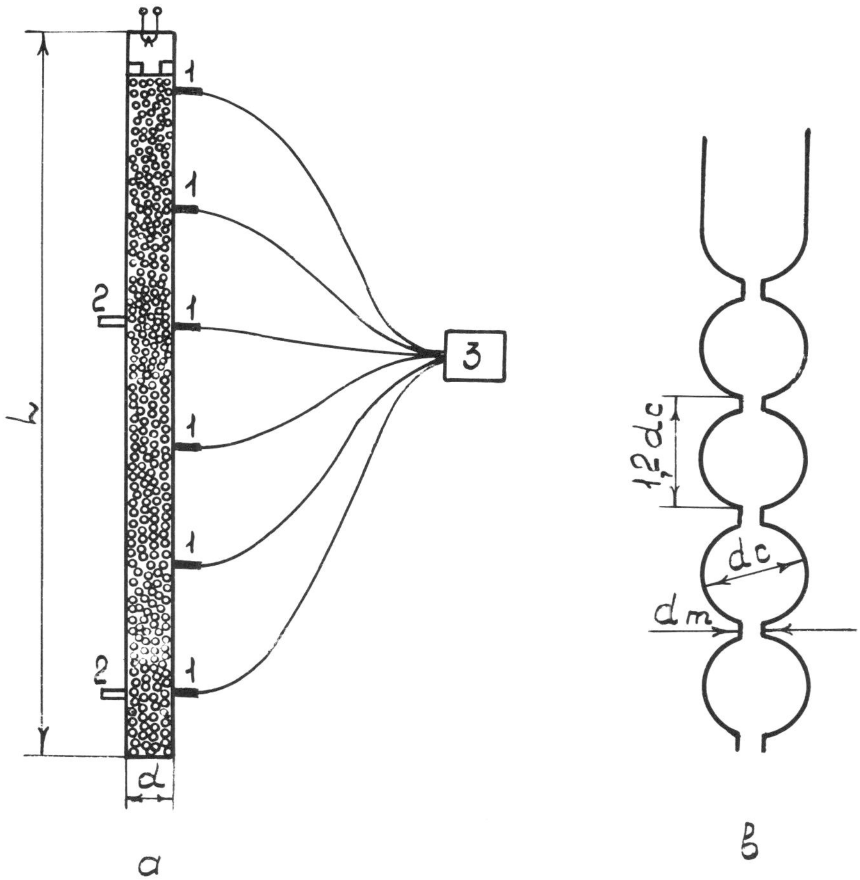

Fig. 1. Sketch of the experimental channels:
a - tubes filled with solid particles,
b - profile of the flat channel of variable width.

channels with variable width were geometrically similar (Fig. 1b). The "pore" diameters d for different channels were 4, 8 and 16 mm. A relative necking of the channel, $d_1/d = 0.25$, was invariable and corresponded to the ratio between cross-sectional areas of the neckings and widenings packed with the balls of one and the same size. There was a section of the channel 0.1 - 0.2 m long of constant width d. The length of the channel section with variable width was 0.5 m. The channel was filled with C_2H_2, H_2, C_3H_8 or the mixture of acetone vapour and O_2; sometimes Ar or He was added. The pressure was ranged over 2×10^3 to 3×10^5Pa. Initiation was realized by an electric spark at the beginning of the constant-width section of the channel; in some experiments this was done in an opposite direction.

The processes occurring in the channel with variable width were recorded by photographing in darkness, the film being immovable and the shutter of the objective being opened, by selfluminescence or schlieren pattern sweep on the moving film, or by a streak-camera photography through a schlieren system. The ionization probes equipped with electronic chronometers were used to measure velocity.

RESULTS AND DISCUSSION

Processes in porous medium. In the experiments with active mixtures close to stoichiometric ones, only detonation combustion may be in a stationary regime in porous media. The velocity of such a process, D, continuously increases with increasing p_0 and δ from $(1.5 \pm 0.3)c_0$ up to the values close to ideal Chapman-Jouguet detonation velocity, D_{C-J}, in the mixture in a free space (Fig. 2, Curves 1 and 2). The estimates show that the detonation combustion velocity approaches D_{C-J} when the gaseous detonation cell size a is less by an order of magnitude and even more than the characteristic size of the "pore", 0.3δ. For the abovementioned mixtures, the flame does not propagate in porous media below the detonation existence limit. The characteristic pressure and luminescence profile in the detonation combustion wave, obtained in the same cross-section of the tube, are illustrated in Fig. 3a. The pressure profile at the front is of shock-wave form almost coinciding with the luminescence origin. A further quick pressure decrease is due to the cooling of combustion products in porous media and correlates in the order of magnitude with the characteristic time of thermal relaxation of gas in "pores" taking into account a relative motion. A relative pressure drop

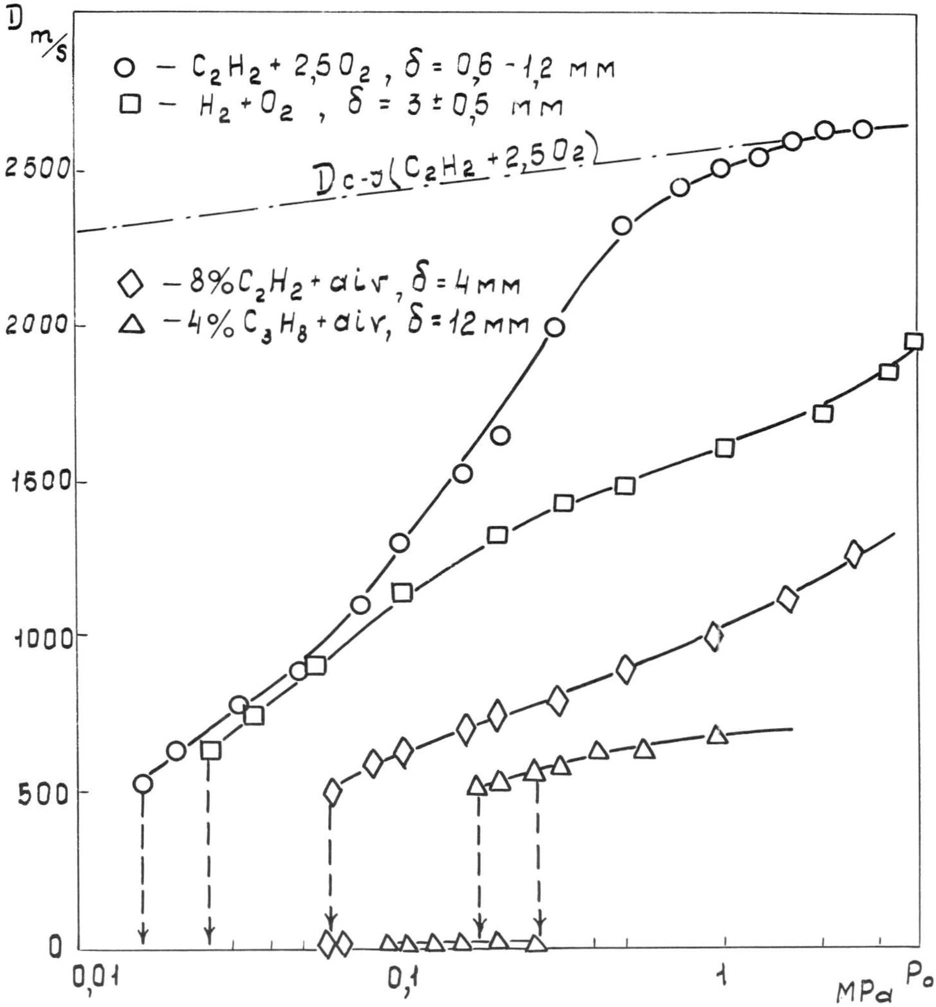

Fig. 2. The dependences of flame velocity D in porous medium on the initial pressure p_0 for detonation regimes.

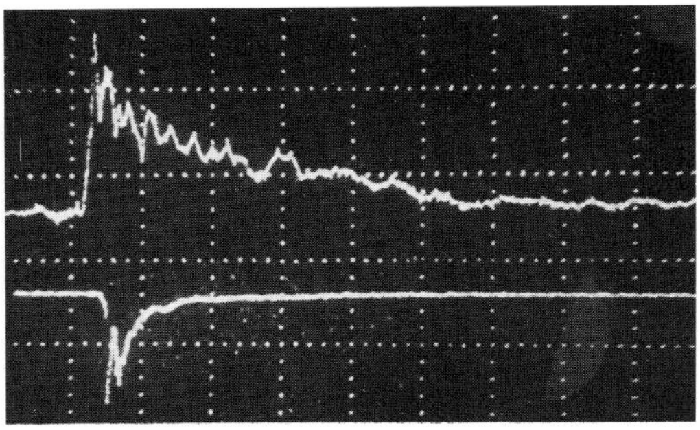

a) $2H_2 + O_2$
δ = 12 mm
p_0 = 0.06 MPa
D = 1820 m/s
X - 50 μs/div
Y - 0.42 MPa/div

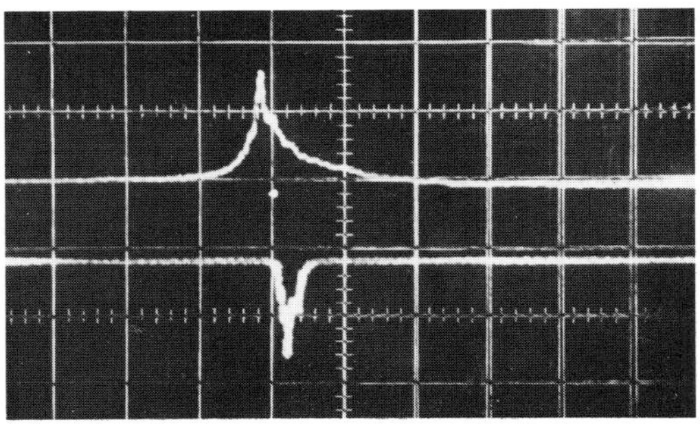

b) 18% $H_2 + O_2$
$\delta = 3 \pm 0.5$ mm
p_0 = 0.15 MPa
D = 160 m/s
X - 500 μs/div
Y - 0.2 MPa/div

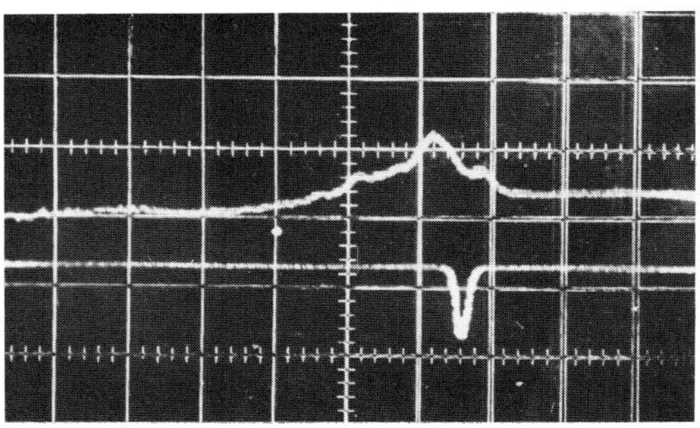

c) 7% C_2H_2 + air
δ = 0.7 - 1,2 mm
p_0 = 2.1 MPa,
D = 50 n/s
X - 500 μs/div
Y - 1 MPa/div

Fig. 3. Typical signals from pressure gauge (upper trace) and photomultiplier (lower trace) for different combustion regimes in porous medium.

at the front, $\Delta p/p_0$, increases with increasing p_0 and δ from 4-8 near the limit up to the value significantly exceeding (by 50 % and even mo-

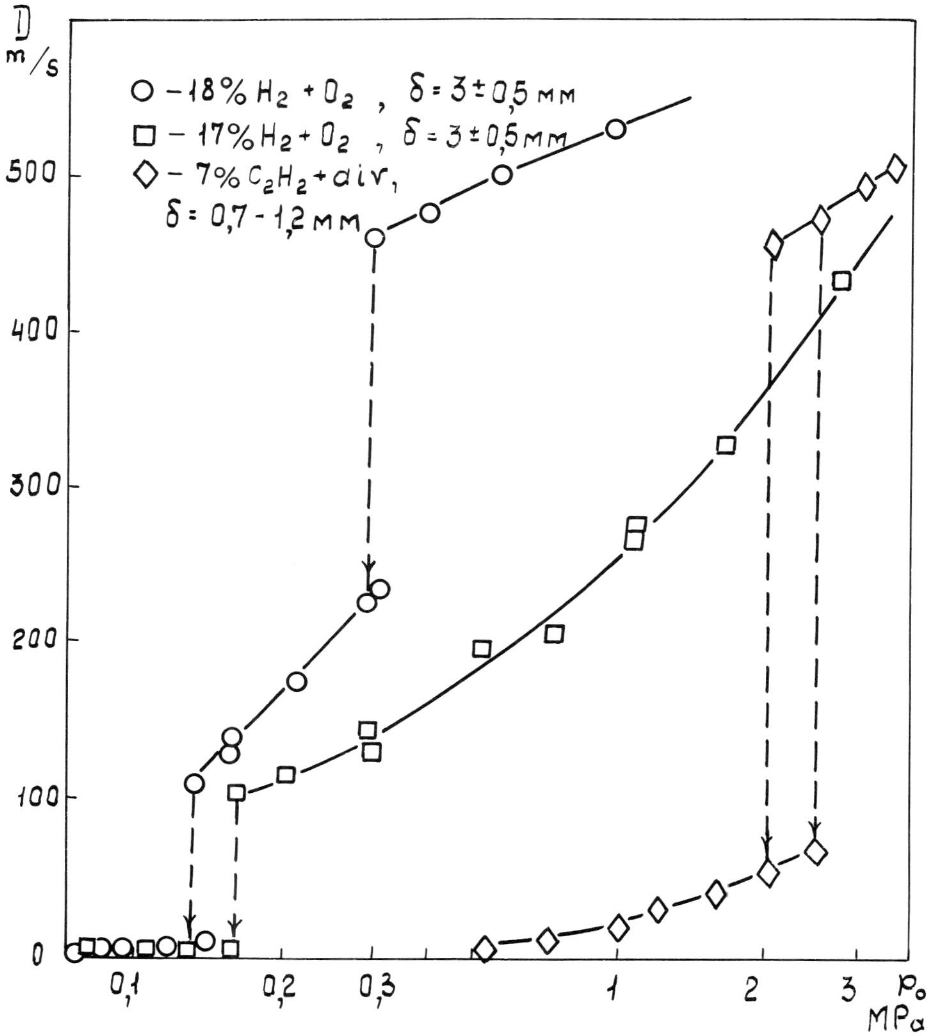

Fig. 4. The dependences of flame velocity D in porous medium on the initial pressure p_0 in poor gas mixtures.

re) the combustion pressure within the closed volume.

The detonation regime of combustion in the mixture diluted by inert gas takes place at higher pressures and pore sizes. Below the detonation limit, there develops the region in which the flame propagates with a significantly lower subsonic velocity (Fig. 2, Curves 3 and 4). Such quasi-stationary regimes of rather slow combustion with $D < 10$ m/s were previously studied by V.S. Babkin et al (1983). In our work, however, along with analogous regimes, that one with much higher velocities, $D \geq 20$ m/s, was observed (Fig. 4), which was called the fast combustion regime. It is steady-state, since gasdynamical disturbances penetrate the bounded region before the flame front due to high interfacial friction in porous medium. The distinguishing feature of fast combustion is an excessive pressure $\Delta p > 0$ in the wave (Fig. 3b,c). When $\delta = 1 - 3$ mm, the wave width is 0.1 - 0.5 m. When the pressure reaches maximum, a chemical reaction starts, then the pressure gradually decreases. A maximum relative value of excessive pressure in the wave, $\Delta p/p_0$, varies from 0.1 - 0.2 when $D \simeq 20$ m/s to 4 -5, when $D \simeq c_0$. When D increases, the forward front pressure steepness increases too, and near c_0 it almost degenerates into a shock jump.

The transition from one combustion regime to another by varying p_0 may occur both gradually or by jump (Fig. 4). The case with the 17% H_2+ O_2 mixture (Curve 2), when a continuous transition from subsonic combustion to supersonic one (with respect to c_0) may be realized by varying p_0 , is of particular interest. Such a behavior of the curve points to the fact that the gas burning in porous media is divided into detonation and deflagration only conditionally. A gradual transition from slow combustion to fast one in the 7% C_2H_2+ Air mixture (Fig 4, curve 3) is indicative of the absence of the differences in principle also between these two regimes. A qualitative difference consists only in the absence of a full stationarity of slow combustion wave due to pressure variation in the whole volume of the tube as the mixture is burning. This difference is, however, associated with the dimensions of experimental facilities: if an effective fast combustion wave length is not in excess of 0.5 m in the case under consideration and the stationarity of the process may be provided by facility several meters long, the slow combustion stationarity may be provided by the facility as long as tens and even

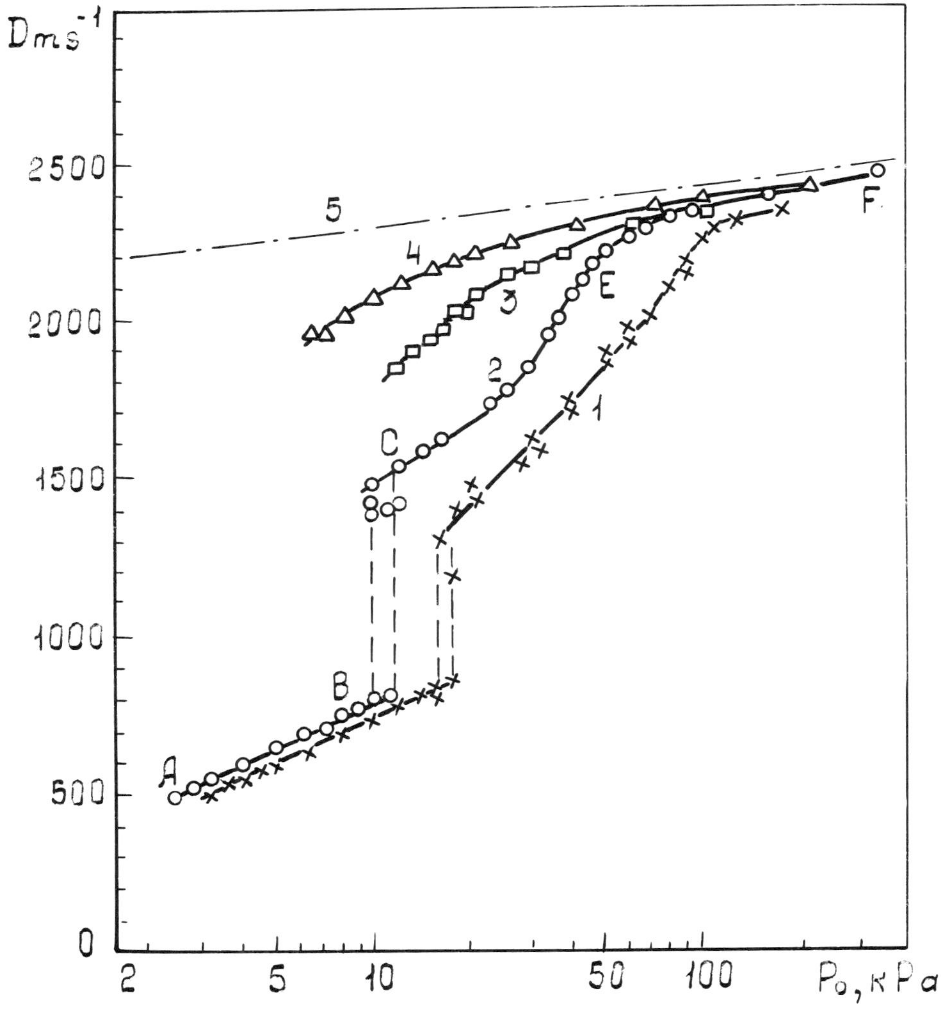

Fig. 5. The dependences of detonation velocity on the initial pressure in a channel of variable width; mixture $C_2H_2 + 2,5\ O_2$, h = 1,5 mm;
1 - d_c = 4 mm, **2** - d_c = 8 mm, **3** - channel 2x1.5 mm^2,
4 - channel 8 x 1,5 mm^2, **5** - Chapman-Jouguet velocity.

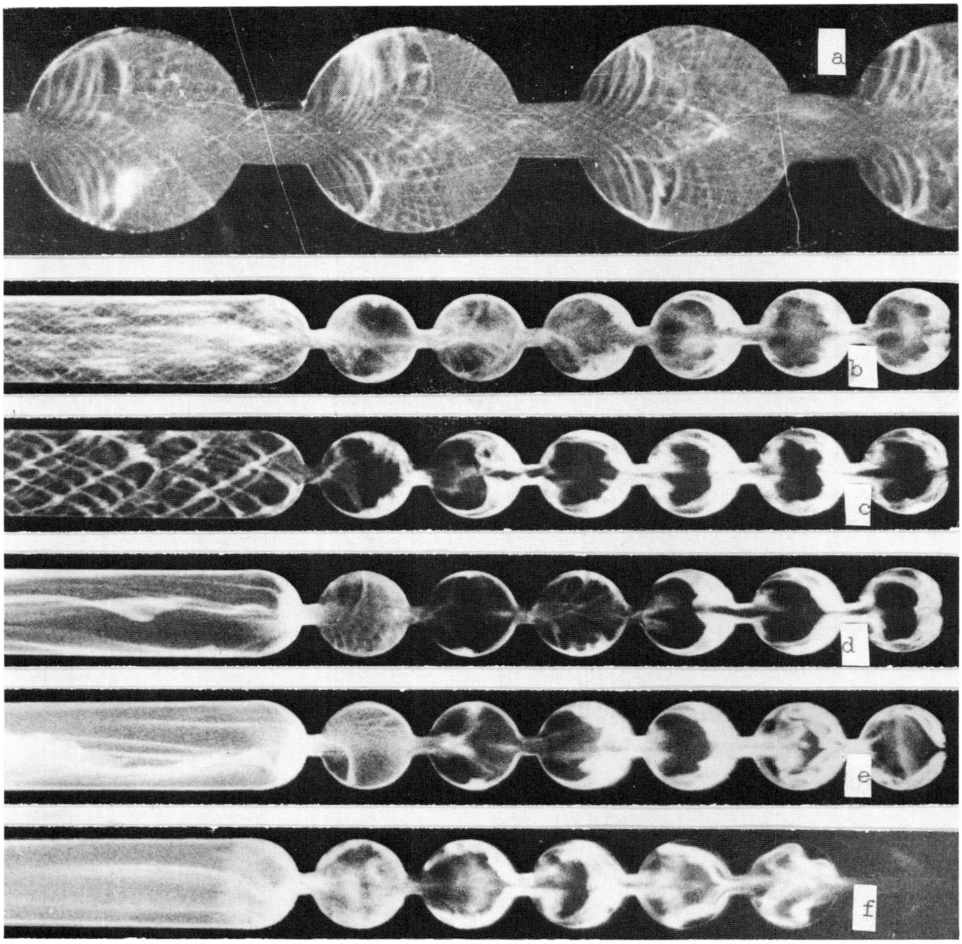

Fig. 6. Open-shutter photographs of detonation in a channel of variable width, d_c = 16 mm;

a) h = 1,0 mm, C_2H_2 + 1,5 O_2, p_0 = 20 kPa;

b) h = 1,5 mm, C_2H_2 + 2,5 O_2, p_0 = 9,2 kPa;

c) h = 1,5 mm, C_2H_2 + 2,5 O_2, p_0 = 5,2 kPa;

d) h = 1,5 mm, C_2H_2 + 2,5 O_2, p_0 = 4,4 kPa;

e) h = 1,5 mm, C_2H_2 + 2,5 O_2, p_0 = 3,6 kPa;

f) h = 1,5 mm, C_2H_2 + 2,5 O_2, p_0 = 2,8 kPa.

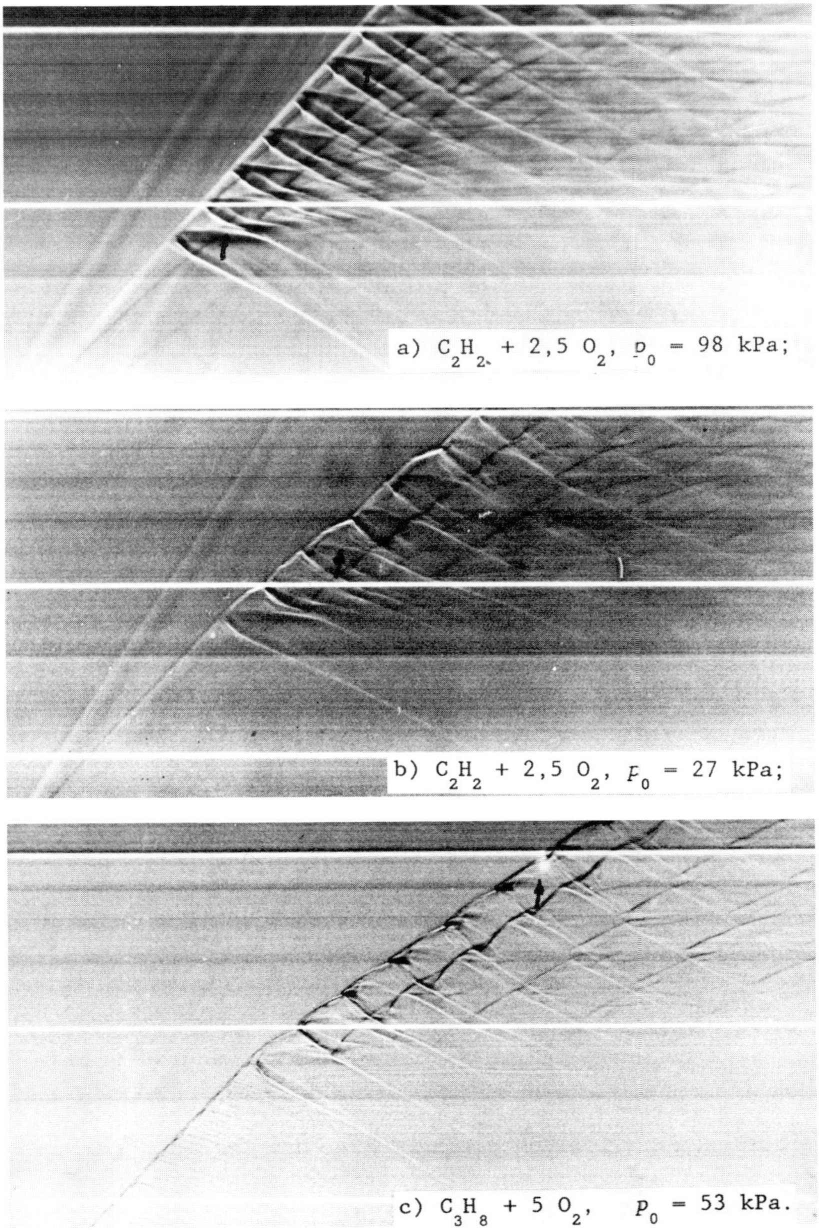

Fig. 7. Shlieren-streak records of high speed detonation regimes in a channel of variable width, d_c = 8,0 mm, h = 1,0 mm; a) C_2H_2 + 2,5 O_2, p_0 = 98 kPa; b) C_2H_2 + 2,5 O_2, p_0 = 27 kPa; c) C_3H_8 + 5 O_2, p_0 = 53 kPa.

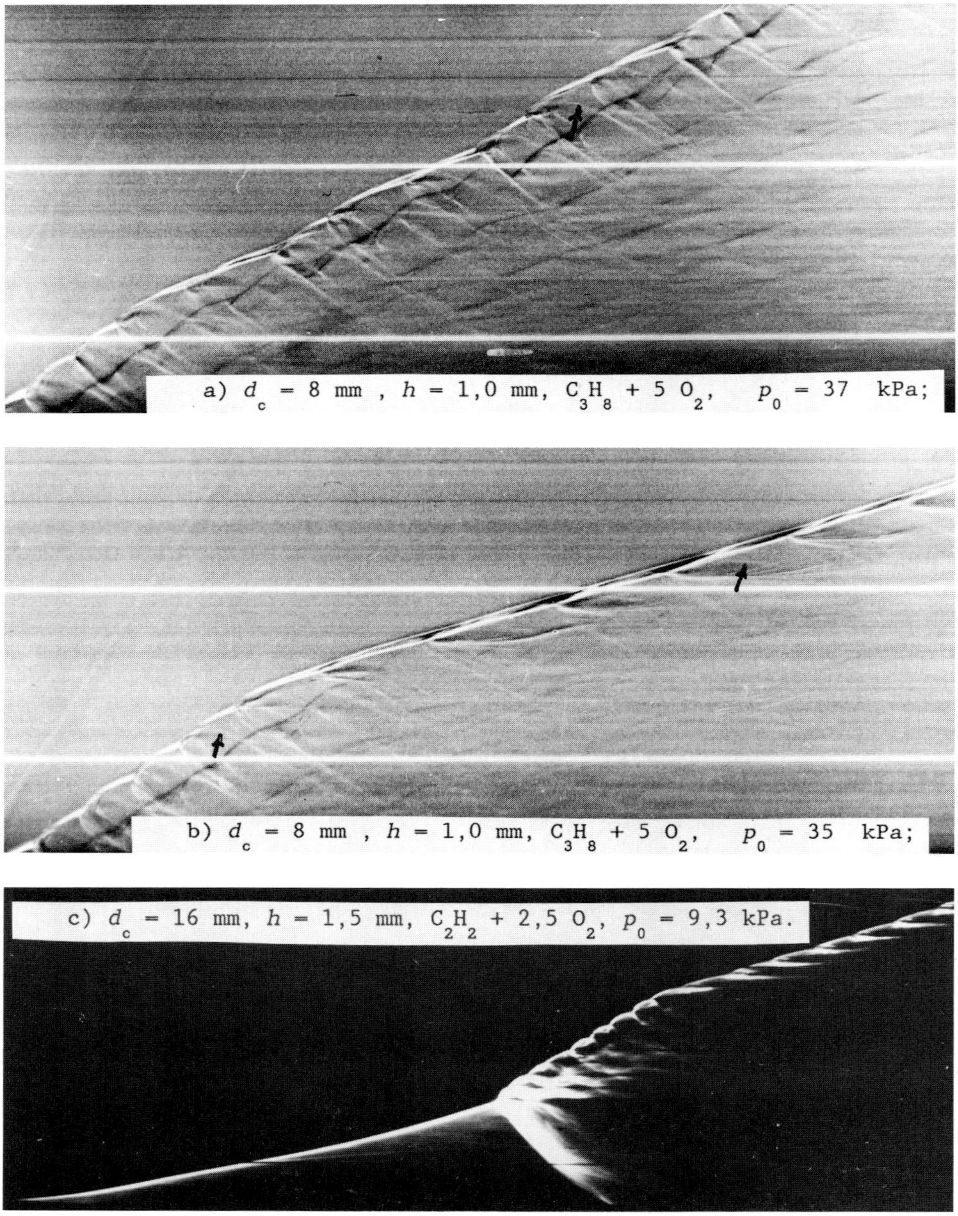

Fig. 8. Shlieren-streak records (a,b) and streak record (c) in a channel of variable width;
a) d_c = 8 mm , h = 1,0 mm, $C_3H_8 + 5\ O_2$, p_0 = 37 kPa;
b) d_c = 8 mm , h = 1,0 mm, $C_3H_8 + 5\ O_2$, p_0 = 35 kPa;
c) d_c = 16 mm, h = 1,5 mm, $C_2H_2 + 2{,}5\ O_2$, p_0 = 9,3 kPa.

hundreds meters. The increase in the pressure in the wave by 10-20 % over p_0 is assumed to be the conditional interface between slow combustion and fast one. As far as the interface between fast combustion and detonation combustion at the point $D=c_0$ is concerned, it should be noted that sonic waves in gas within a porous space are affected by strong dispersion and dissipation: high-frequency oscillations with the velocity $\sim c_0$ quickly decay, and weakly decaying low-frequency waves propagate with the velocity $c \to 0$ when frequency $\omega \to 0$ (V.N.Nikolaevsky et al, 1970).

A low limit of flame propagation in porous medium is satisfactorily described by Peclet criterion $Pe=U_n d_m/\kappa$,based on normal velocity of the flame U_n, maximum size of pores, $d_m \sim 0,3\ \delta$ and thermal diffusivity κ of an initial mixture. For all the mixtures under consideration with limit parameters $Pe=65\pm45$, whichever of the regimes and velocities is reached near the limit. As the measurements of final pressure of products show, in all the cases an essential amount of non-reactive gases is observed. Its value increases near the limit and reaches 50% and more.

Processes in planar channels of variable width. Only detonation (supersonic with respect to c_0) regimes of combustion were realized in planar channels with variable width due to the use of high active fuel-oxygen mixtures with a composition close to a stoichiometric one. Fig. 5 displays typical dependencies of the velocity of the process on initial pressure. The dependencies are the same for all the mixtures under consideration.

There are two characteristic sections on the curves. They are the section **AB** corresponding to low-velocity detonation combustion and **CE** corresponding to high-velocity detonation combustion. The latter in its turn may be divided into two parts: **EF** where the velocity slightly depends on pressure, and **CE** where the velocity more strongly depends on p_0. Between the low-velocity detonation region and high-velocity one there is the interfacial zone **BC** characterized by the regime instability. Here the transitions from low velocities to high ones and vice versa are observed.

Presented in Fig.6 are the photos of self-luminescence obtained on an immovable film, in Fig.7 is the schlieren-streak record of high-velocity detonation, Fig.8 illustrates the same supplemented with the transition into slow-velocity detonation. The region illustrated in Fig. 6a corresponds to the right-hand vicinity of the point **E** (Fig.5), where p_0 is close to the critical pressure of the non-decaying detonation passing from the necking to the widening of the channel. To the right of the point **E**, there exists an ordinary multi-headed detonation, as in a flat channel. The schlieren picture of such detonation is shown in Fig. 7a. Fig.7b corresponds to the regime to the left of the point **E**. Here periodical (after each second widening of the channel) destructions of the multi-headed structure of the detonation front occur with a subsequent dividing of it into a shock jump and a combustion front. Then re-initiation of multi-headed detonation in the same "pore" occurs. At lower pressures the destructions and re-initiation take place in each "pore" of the channel (Fig.7c). Fig.8a illustrates a process corresponding to the boundary region of **BC**. In this process, the complex consisting of a shock wave and the separated combustion front passes through two widening sections of the channel without re-initiation of local detonation. Then the process is accelerated for a short period of time and the combustion wave and the shock front interflow. Shown in Fig.8b is a low-velocity detonation regime which is reached after high initiation an initial section of the channel. An analogous process is illustrated in Fig.8c (self-luminescence). In this case, high-velocity detonation combustion is excited by the combustion-to-detonation transition at the joint of a flat channel and a variable-width one.

The processes shown in Fig.8c and Fig.6e,d are similar. The detonation passing in the channel with variable width is shown in Fig.6b,c. It may be established from the trajectories of transverse disturbances visible in the first three sections (Fig. 6b, d), that high-velocity detonation occured there, and that low-velocity detonation took place in the next sections. Fig. 6f shows the decay of low-velocity detonation.

The main difference between low-velocity detonation and high-velocity one visible in the schlieren-streak records is that for low-velocity process the shock front and the combustion front interflow at

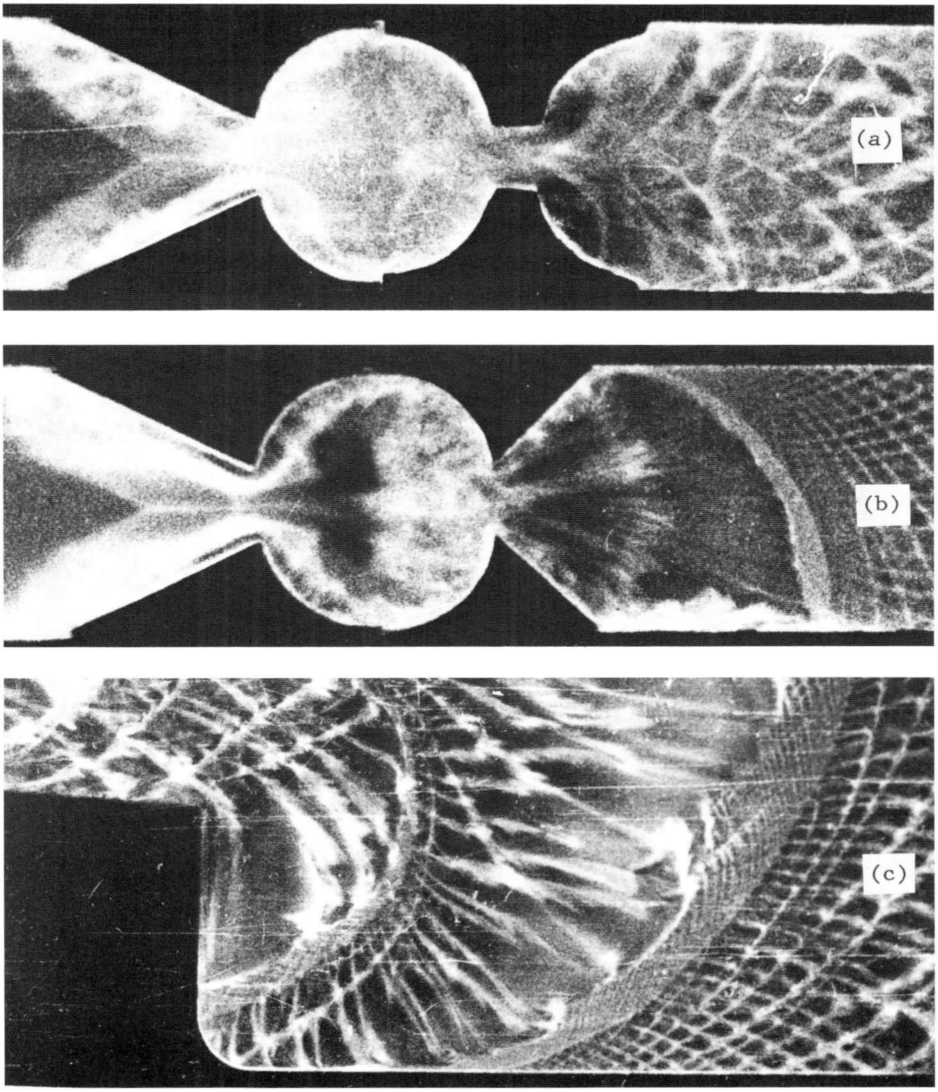

Fig. 9 Origin of detonation in the widening of a channel:
a - in expendingg stream,
b, c - by shock wave reflection at a ccrner (b),
and a flat wall (c).

none of the points of the trajectory, since everywhere between them there is a layer of unreacted gas. The trajectories originating near the front and initially almost parallel to the time axis are seen in the schlieren-streak records on all the original films and are shown by arrows in some photos (Fig. 7 and 8). They correspond to secondary shock waves appearing after gas has begun to outflow from the narrowing section. Then the jump moves upstream and passes through the necking. After that gas flow in an opposite direction sometimes may be observed. The presence of the retardation shock wave even at the limiting parameters of low-velocity regime states to a supersonic character of compressed unburned gas outflowing from the necking. On the basis of this fact, one may assume that a supersonic regime of flow of flame-compressed unburned gas through the same sections of the channel is the important condition for the detonation process in the channels with variable cross-sectional areas to exist.

Propagation mechanisms. When a maximum cross-sectional area of neckings in a porous medium is higher by an order of magnitude and more than the transverse size of a detonation cell in a free space ($d_m \gg a$), the velocity deficit is low $(D_{C-J}-D)/D_{C-J} \leq 0.1$ and the front remains to be of cellular structure. Thereby a classical shock-wave mechanism of detonation is supplemented by transverse waves, leading shock pulsations, "islands" of slowly burning gas and other accompanying details (Voitsekhovsky B.V. et al, 1963; Vasiliev A.A. et al, 1987). Non-ignited islands behind obstacles, i.e. the regions free of cells with weakened shock heating, appear additionally in porous medium. Due to this fact and a large cold surface of pores as well, a non-complete burning of gas is observed, i.e. the islands are frozen. The non-complete burning due to cooling action of the surface of a solid porous skeleton takes place, as was mentioned above, under all other reaction wave propagation regimes.

An essential wave retardation is observed at the decay of the cellular structure within the section, where the wave passes from the necking to the widening of the "pore" channel. The decay as above was investigated by many authors (see, for example, the last references). An inverse re-initiation of a wave having a cellular structure, called the local detonation here, may take place, first of all, as a result of Mach or regular reflection of the leading shock wave of sufficient intensity from the inner wall of the widening. A high power transverse detonation wave is thereby initiated in the layer of compressed mixture and itself initiates detonation of adjacent incompressed mixture passing into the next necking. Fig. 9, obtained analogously to Fig. 6, but in larger single "pore", illustrates this process. The above-mentioned mechanism of local detonation re-initiation is not, however, a unique one. In particular, for the regime shown in Fig. 8a, where the local detonation is observed only in one of three subsequent "pores", it is impossible.

The other re-initiation mechanism is associated with a rapid acceleration of combustion in the zone of mixing of initial gas composition and the jet of hot products ejected through the necking from the previous widening of a porous medium. This mechanism was first discussed and studied by Knystautas R. et al (1978). This and some other works consider mixing in small-scale vortices generated by large-scale vortices. The authors of the present paper revealed new important details of this phenomenon.

It is very essential that the flow of non-burning mixture pushed by the flame at the exit from the necking is supersonic and the head of hot jet moves through the necking with acceleration, then quickly expanding also with acceleration in different directions. On the contact surface, a Rayleigh-Taylor instability is developed, causing intense mixing. The subsequent interaction of this zone with detached drag shock, which can be observed in Fig. 7 and 8, accelerates mixing and the following combustion by the Markstein instability, in addition to the previous effect. Shock waves reflected from the walls make a contribution too. Combustion in the pore, intensified by these processes may lead to the local detonation of the mixture and to the restoration of the shock-wave mechanism of the flame transfer to the next pore. Details of the transition to the detonation process are not seen. However, undoubtedly, there can take place an acceleration of the reaction wave to detonation due to a gradient of ignition delays, theoretically considered for the first time by Ya.B. Zel'dovich et al (1970), and recently by G.E. Gel'fand et al (1989). An analogous SWACER-mechanism was analyzed by R.Knystautas et al (1978). It should be noted that in our experiments, large-scale vortices in the widenings of the channel were observed as early as after the occurrence of accelerated combustion in the mixing zone and they did not affect the re-initiation process.

In the case of low-velocity detonation in the channels with variable width, re-initiation of local detonation characterized by a cellular structure and local velocity near D_{D-J} are not observed. However, the above-described convective mechanism consisting in the formation of hot jets with supersonic zones before them, accelerated mixing and combustion provides the combustion wave transfer with a supersonic velocity from one "pore" to another. In porous media, the velocity jump observed sometimes near $D \simeq c_0$ may be also explained by the disappearance of supersonic zones in the flame-pushed initial gaseous composition and probably by the termination of self-ignition in the zones of mixing and compression of gas. A low-velocity detonation in the channels with variable width and in porous media is likely to be corresponding to the "chocking" regime in the tubes with annular obstacles used by R. Knystautas et al (1985).

The absence of the velocity discontinuity in porous medium between high-velocity detonation and low-velocity one as well as in some mixtures over the range $D \leq c_0$ may be explained by a simultaneous action of different ignition transfer mechanisms in different pores in non-ordered medium.

The processes under consideration are certain to be studied in more detail.

CONCLUSION

1. Chemical reaction waves in explosive gaseous mixtures in rigid porous media have been investigated over the pressure range 10^{-2} to 5 MPa for the sizes of particles of porous media from 10^{-4} to 10^{-2} m and different mixtures. It has been shown that any gaseous explosion propa-

gation velocity, from normal velocity of laminar flame to Chapman-Jouguet detonation velocity, may be achieved by varying the above parameters. The wave surplus pressure may be varied from 0 to the values 1.5 times as high as the explosive pressure in a constant volume.

2. The flow structure is investigated in the case when the explosive waves propagate in planar channels with transparent walls, a porous medium being modelled by a periodical width variation It has been established that as the detonation velocity decreases, the shock wave reaction initiation mechanism is transformed into a convective one, such that the flame is transferred to the next pore by the jets of hot products. The Rayleigh-Taylor instability on the surface of the head of hot jet leading to its fast mixing with an initial mixture and subsequent acceleration of gas burning in a pore is thereby of particular importance.

3. The detonation velocity in porous media gradually decreases to ~500 m/s with varying the initial pressure, then for most mixtures it jumpy decreases to subsonic velocities or the process completely decays (the latter is observed mainly in active fuel-oxygen mixtures). For some compositions, however, the velocity continuously decreases down to ~ 100 m/s. When transition through a sonic velocity threshold, the pressure profile in the wave becomes smooth, without sharp breaks. Due to this fact, dividing the chemical reaction waves into detonation and combustion may be assumed to be conditional.

4. In very poor mixtures with almost limiting concentrations, the combustion velocity in porous media slightly differs from a normal velocity of a laminar flame and is as low as 1 m/s for the pressure range under investigation. The mixture enrichment leads to the increase in combustion velocity with pressure, for all the mixtures the velocity jump from 2-80 m/s to 100-150 m/s or directly to supersonic velocities of 500-600 m/s is observed. For some mixtures, the second velocity jump, from, approximately, 250 to 500 m/s, is observed.

5. At a lower limit of self-sustained reaction wave propagation over porous media, the Peclet number is within the range 20 to 100 m/s, whatever the marginal regime and value of velocity.

REFERENCES

Babkin V.S., Kozachenko L.S. Mechanism of pre-detonation flame propagation in rough tubes, Dokl. Akad. Nauk SSSR, v. 134, 3, pp. 591-592 (1960).

Babkin V.S., Laevsky Y.M. Filtration combustion of gases. Fiz. Goreniya Vzryva, 23, No.5, p. 27 (1987).

Chapman W.R., Wheeler E.N.V. J. Chem. Soc., p.2139 (1926) and J. Chem. Soc., p. 38 (1927).

Gel'fand B.E., Frolov S.M., Tsyganov S.A. Criterion of shock and detonation wave excitation in reacting media. Khimicheskaya Fizika, No.5, pp. 655-666 (1989).

Hjertager B.H. Numerical simulation of turbulent flame and pressure development in gas explosions/ Fuel-Air Explosions.- Univ. of Waterloo, Press SM, Study No.16, pp. 405-424 (1982).

Kauffman C.W., Yan Ch. and Nicholls J.A. Gaseous detonation in porous media/ 19th Symp. (Intern.) on Combustion - The Comb. Inst., pp. 521-597 (1982).

Knystautas R., Lee J.H., Moen I. and Wagner H.G. Direct initiation of spherical detonation by a hot turbulent gas jet. 17th Symp. (Intern) on Combustion, pp. 1235-1245 (1978).

Knystautas R., Lee J.H., Peraldi O. and Chan C.K. Transmission of a flame from a rough to a smooth-walled tube/ Progress in Astronaut. and Aeronaut., v.106, pp. 37-52 (1985).

Lee J.H., Knystautas R. and Freiman A. High speed turbulent deflagration and transition to detonation in H_2-air mixtures. Comb. and Flame, 56, pp. 227-239 (1984).

Lyamin G.A., Pinaev A.V. On gas combustion regimes in inert porous medium., Fiz. Goreniya Vzryva, 22, No.5, p. 64 (1986).

Lyamin G.A., Mitrofanov V.V., Pinaev A.V., Subbotin V.A. Gaseous and heterogeneous detonation in porous media./ Detonatsiya i Udarnye Volny, Chernogolovka, pp. 52-56 (1986).

Lyamin G.A., Pinaev A.V. On fast subsonic gas combustion regime in inert porous medium with a gradual increase of pressure in a wave., Fiz. Goreniya Vzryva, 23, No.4, p. 27 (1987).

Mamontov G.M., Mitrofanov V.V., Subbotin V.A. Gaseous mixture detonation regimes in rigid porous media./ Detonatsiya, Chernogolovka, pp. 106-110 (1980).

Moen I.O. The influence of turbulence on flame propagation in obstacle environmentals./ Fuel-Air Explosions - Univ. of Waterloo Press, pp. 101-135 (1982).

Nikolaevsky V.N., Basniev I.S. et al. Mechanics of saturated porous media. - Moscow, Nedra Publishers (1970).

Pinaev A.V. and Lyamin G.A. Fundamental laws governing subsonic and detonation gas combustion in inert porous media., Combustion, Explosion and Shock Waves (Fiz. Goreniya Vzryva), 25, No.4, p. 448 (1989).

Popov O.E., Kogarko S.M., Fotenkov V.A. On fast combustion of gaseous mixture in high-porosity media., Dokl. Akad. Nauk SSSR, v.219, No.3, pp. 592-595 (1974).

Rybanin S.S. On the theory of detonation in rough tubes. Fiz. Goreniya Vzryva, 5, No.3, p. 395 (1969).
Shchelkin K.I. Fast combustion and spin detonation of gases. Moscow, Voenizdat (1949).
Shchelkin K.I., Troshin Ya.K. Combustion gasdynamics. Moscow, Izdatel'stvo Akad. Nauk SSSR (1963).
Strizhevsky I.I., Zakaznov V.F. Commercial flame obstacles. Moscow, Khimiya (1974).
Subbotin V.A., Kuznetsova A.Ya. The combustion regimes of gaseous explosive mixtures in channels with variable cross-section. Dinamika Sploshnoi Sredy 68, Novosibirsk, pp. 124-131 (1984).
Vasiliev A.A., Mitrofanov V.V., Topchiyan M.E. Detonation waves in gases. Fiz. Goreniya Vzryva, 23, No.5, p. 109 (1987).
Voitsekhovsky B.V., Mitrofanov V.V., Topchiyan M.E. Detonation front structure in gases. Novosibirsk, Izdatel'stvo Sibirskogo Otdelenia Akad. Nauk SSSR (1963).
Wagner H.G. Some experiments about flame acceleration./ Fuel-Air Explosions.- Univ. of Waterloo Press, pp. 77-99 (1982).
Zel'dovich Ya.B., Kompaneets A.S. Theory of combustion and detonation of gases. Moscow, Izdatel'stvo Akad. Nauk SSSR (1944).
Zel'dovich Ya.B., Borisov A.A. et al. Low-speed quasi-detonation regimes of combustion of fuel-air mixtures in rough tubes. Dokl. Akad. Nauk SSSR, v.279, p. 1359 (1984).
Zel'dovich Ya.B., Gel'fand B.E. et al. Reaction zone at low-speed gaseous detonation in rough tubes. Khimicheskaya Fizika, No.2, pp. 279-286 (1985).
Zel'dovich Ya.B., Librovich V.B. et al. On appearance of detonation in nonuniformly heated gas. Prikladnaya Mekhanika i Tekhnicheskaya Fizika, No.2, pp. 76-84 (1970).

Physical Model of Dynamic Structure of the Surface of Detonation Wave

Anatoly A. Borissov and Oleg V. Sharypov

Institute of Thermophysics, Siberian Branch of the USSR Academy of Sciences, Novosibirsk 630090, USSR

The theoretical analysis of the planar detonation front stability with respect to three-dimensional perturbations is presented. It is shown that infinitesimal perturbations with certain wavelength can grow and form a complex spatial configuration. Nonlinear evolution equation for the front surface is developed. Various physical mechanisms are identified. Spontaneously appearing pulsating cellular structure of the front belongs to the class of auto waves which are observed in many non-equilibrium physical and chemical systems. On the basis of the developed non-linear evolution model it is possible to simulate the forming of two- and three-dimensional shock front structure.

The results of numerical calculations shown graphically and analytical conclusions agree with the known experimental data. The developed model is able to shed a light to the problem of physical explanation of critical tube diameter value and change of cellular structure regularity.

1. Introduction

The classical Chapman-Jouguet theory seeks the unique solution of the one-dimensional conservation equations across the detonation front in which the flow behind the wave is sonic. It does not require a knowledge of the structure of the wave itself, and involves only an equilibrium thermodynamic calculation for the detonation states (i.e.

A. A. Borissov (ed.), Dynamic Structure of Detonation in Gaseous and Dispersed Media, 77–108.

the detonation velocity, pressure, temperature and density ratios across the wave and the equilibrium composition of the product gases). The great number of experiments show that detonation wave fronts in gas mixtures as a rule have a non-steady three-dimensional structure (Voytsekhovsky,et.al. 1963). Planar stationary (according to the model of Ya. B. Zel'dovich, J. von Neumann and W. Döring (Zel'dovich, Kompaneyets, 1955) detonation wave appears to be unstable to transverse perturbations. The equilibrium or "static" detonation states obtained from the classical Chapman-Jouguet theory agree surprisingly well with experimental observations even in near-limit conditions when the flow structure near front is highly three-dimensional (Lee 1984).

In practice it is very important to be able to predict the mean size of the detonation front inhomogeneities. The ratio of the front inhomogeneity size per tube diameter determines self-sustaining or attenuating regime of detonation wave propagation. This information is essential for safety control of various technological devices. When solving the linearized gas-dynamic equations not allowing for the molecular dissipative processes (Zaydel' 1961), the detonation wave was represented in the form of compression and rarefaction shocks separated by induction zone. The equation of chemical kinetics was used for determination of induction zone size only but wasn't considered jointly with the linearized gas-dynamic equations. It was shown that the planar front of Chapman-Jouguet detonations is unstable practically in all gas mixtures. The instability appears for perturbations with the characteristic wavelength of the order of chemical transformation zone size in Ya. B. Zel'dovich's model l_z. The short- and long-wavelength perturbations attenuate under the influence of gas-dynamic factors without the assistance of viscosity, thermal conductivity or diffusion.

More exact mathematical formulations of the problem (Erpenbeck, 1962; Pukhnachyov, 1963) result in the analogous conclusions. The calculations predict the stability of the planar front in the tube with certain diameter when the detonation is overdriven.

The heuristic equation for detonation front evolution (Malomed 1984) qualitatively simulates main features of pulsating front but needs in more rigorous mathematical substantiation and doesn't allow to analyze the effects of specific chemical or gas-dynamic factors.

The direct numerical solution of the whole system of chemical kinetic and gas-dynamic equations (Taki, Fujiwara 1978; Markov 1981; Kailasanath, Oran, et.al. 1985; Schöffel, 1989) simulates the spontaneous development of pulsations of two-dimensional detonation front and successfully reproduces known experimental data. However the limitations of this method become more and more evident now. It is very difficult (or impossible) to make some conclusions on the nature of phenomena on the basis of numerical calculations only like as on the basis of experiments only. Moreover the numerical calculations have not any significant advantage to experiments neither in the cost nor in the reliability of results. Even the great successes achieved recently in the field of modelling of three-dimensional detonations (Fujiwara, et.al.,1989) can not satisfy the need in clear physical model of dynamic detonation structure.

The previous theoretical investigations of unsteady detonations maintain a number of unexplained problems. Particularly the substantiated and acceptable evolution models of two- and three-dimensional shapes of the front are not developed, the problem of critical tube diameter or propagation of detonation in channel with elastic walls exists, the question about the regularity of the cellular structure isn't trivial, the dependence of cell size a on l_z isn't a simple function and so on. At present the quantitative theory for the determination of dynamic detonation characteristics isn't developed yet, so the construction of simple theoretical model of non-stationary detonation front is of current concern.

The detonation front instability can be described as a finite-amplitude wave evolution in chemically non-equilibrium medium. The absence of conservation essentially complicates the investigation of processes in non-equilibrium systems relatively to the reversible phenomena when one succeed to convert to uniform description on the basis of Hamilton or Lagrange formalism. The spontaneous loss of stability results in the change of the state of system. The stability of new state is provided by energy exchange processes. As a rule, the structures generated in non-equilibrium systems exist due to permanent dissipation of energy and nonlinear mechanism of energy redistribution in the spectrum. They are named "dissipative structures". The surface of gas detonation shock front is the example of dissipative structure.

Nonlinear evolution equations describing wave-kinetic interaction in non-equilibrium media have been obtained for the first time by (Nakoryakov, Borissov 1976; Clarke 1978). The limits in which the amplification of one-dimensional perturbations have presented have been considered by (Abouseif, Toong 1981). The authors of this paper have obtained original model on the basis of which the evolution of gas detonation front have been modeled numerically (Borissov, Sharypov 1988; 1989-a; 1989-b; 1989-c; 1990-a; 1990-b).The developed model allows one not only to achieve certain correspondence of numerical solution to the phenomenon, but also to answer some questions connected with physical essence of detonations.

2. Governing Equation

The initial formulation of the mathematical problem involves the equations of perfect gas dynamics (conservation laws) with two-component one-step chemical reaction described by Arrhenius kinetics. Diffusion effects are neglected in the conservation equations here.

$$\frac{d\rho}{dt} = -\rho\,\mathbf{div}\,\mathbf{u}$$

$$\mathbf{u} = \mathbf{x}\,u + \mathbf{y}\,v + \mathbf{z}\,w$$

$$p = \rho R T$$

$$\frac{dp}{dt} = c^2 \frac{d\rho}{dt} - Q(\gamma - 1)\rho \frac{d\beta}{dt}$$

$$\frac{d\beta}{dt} = -n_1 \beta^{m_1} p^{m_1 - 1} \exp\left(-\frac{E}{RT}\right)$$

Here p is the pressure, ρ is the density, **u** is the velocity, T is the temperature, R is the universal gas constant, c is the sound speed, γ is the adiabatic exponent of the gas mixture (γ=const), β is the mass fraction of reactants, E is the activation energy, Q is the specific thermal release of the reaction, m_1 is the reaction order, n_1 is the positive pre-exponential factor.

One can write the known balance expression for the enthalpy and the expression for the velocity of detonations in Chapman-Jouguet regime ($D=const$, $p_o \gg p_-$, here subscripts "o" and "-" denote the values of parameters behind and ahead of the shock respectively).

$$\begin{cases} \dfrac{\gamma}{\gamma-1}\dfrac{p}{\rho} + Q\ (\beta-1) = \dfrac{p}{2\ \rho}\left(\dfrac{\rho}{\rho_-} + 1\right) \\ D^2 = \dfrac{p}{\rho_-}\dfrac{\rho}{\rho-\rho_-} \end{cases}$$

or in other form :

$$\frac{p}{\rho_-} = \frac{D^2}{(\gamma+1)}\left(1 \mp \sqrt{1-\frac{2\ Q(\gamma^2-1)(1-\beta)}{D^2}}\right)$$

$$\frac{\rho_-}{\rho} = \frac{1}{(\gamma+1)}\left(\gamma \mp \sqrt{1-\frac{2\ Q(\gamma^2-1)(1-\beta)}{D^2}}\right)$$

here the upper signs correspond to overdriven detonation wave and the lower ones - to under driven waves. Using the perfect gas state equation one can derive the dependence of temperature on the reactant concentration :

$$\frac{d\,T}{d\,\beta} = \frac{1}{R}\left[\frac{\rho_-}{\rho}\frac{d(\,p/\rho_-\,)}{d\,\beta} + \frac{p}{\rho_-}\frac{d(\,\rho/\rho_-\,)}{d\,\beta}\right] =$$

$$= \frac{Q\;(\gamma-1)\;(\;1\;-\;{}^{2\rho_-}/_{\rho}\;)}{R\left(\;(\gamma+1)\;{}^{\rho_-}/_{\rho}\;-\;\gamma\right)}$$

$$\left(\frac{d\,T}{d\,\beta}\right)_{x\to x_o} = \left(\frac{d\,T}{d\,\beta}\right)_{\beta\to 1} \longrightarrow \frac{Q\;(\gamma-1)\;(3-\gamma)}{R\;(\gamma+1)}$$

here x_o is the coordinate of the plane shock front. The factor $\left({}^{-E}/_{RT}\right)$ varies significantly quicker than other terms of the reaction rate equation, so when $\beta \longrightarrow \beta_o = 1$ and $T \longrightarrow T_o$ the kinetic equation may be approximately rewritten as :

$$\frac{d\,\beta}{d\,t} \approx -\,n_1\,\beta_o^{m_1}\,p_o^{m_1-1}\,exp\left({}^{-E}/_{RT_o}\right)\,exp\left(\,{}^{-E\,(T_o-T)}/_{RT_o^{\,2}}\right)$$

$$\left(\frac{T_o-T}{T_o}\right)_{T\to T_o} = \frac{1}{T_o}\left(\frac{d\,T}{d\,\beta}\right)_{\beta\to 1}\left(\;1\;-\;\beta\;\right) \longrightarrow -\,\frac{Q\;(1-\beta)(\gamma-1)(3-\gamma)}{R\;T_o\;(\gamma+1)}$$

Finally the kinetic equation in the vicinity of the front is of the form :

$$\left(\frac{d\,\beta}{d\,t}\right)_{x\to x_o} \approx -\,N\,exp\left(\,A\;(1-\beta)\right)$$

where

$$N = \;-\,n_1\,\beta_o^{m_1}\,p_o^{m_1-1}\,exp\left({}^{-E}/_{RT_o}\right) = const,$$

$$A = \frac{(3-\gamma)(\gamma-1)}{(\gamma+1)} \; \frac{E}{R\,T_o} \; \frac{Q}{R\,T_o} = const$$

Let us represent detonations in the form of superposition of a planar stationary wave (the model of Ya.B.Zel'dovich, J.von Neumann and W.Döring) and small three-dimensional unsteady perturbations :

$$H = \bar{H}(x) + H'(t,x,y,z), \qquad H'/\bar{H} \sim \varepsilon << 1$$

where $H = \begin{pmatrix} p \\ \rho \\ \mathbf{u} \\ \beta \end{pmatrix}$, ε– is the small parameter.

3. Stationary One-Dimensional Problem

The unperturbed one-dimensional detonation wave is immovable in the frame of coordinates: $t' = t$; $x' = x + D\,t$ (Figure 1). If the wave is self-sustaining than D is equal to the velocity of detonation in Chapman-Jouguet regime.

Under the conditions that the wave is planar and stationary in the vicinity of the shock front one can obtain the system of ordinary differential equations for $\bar{H}$:

$$\begin{cases} \bar{u}\,\dfrac{d\,\bar{\rho}}{d\,x} + \bar{\rho}\,\dfrac{d\,\bar{u}}{d\,x} = 0 \\[2ex] N\,exp\Big(A\,(1-\bar{\beta}) \Big) = -\,\bar{u}\,\dfrac{d\,\bar{\beta}}{d\,x} \\[2ex] \dfrac{d\,\bar{p}}{d\,x} = c^2\,\dfrac{d\,\bar{\rho}}{d\,x} - Q\,(\gamma-1)\,\bar{\rho}\,\dfrac{d\,\bar{\beta}}{d\,x} \\[2ex] \bar{\rho}\;\bar{u}\,\dfrac{d\,\bar{u}}{d\,x} = -\,\dfrac{d\,\bar{p}}{d\,x} \end{cases}$$

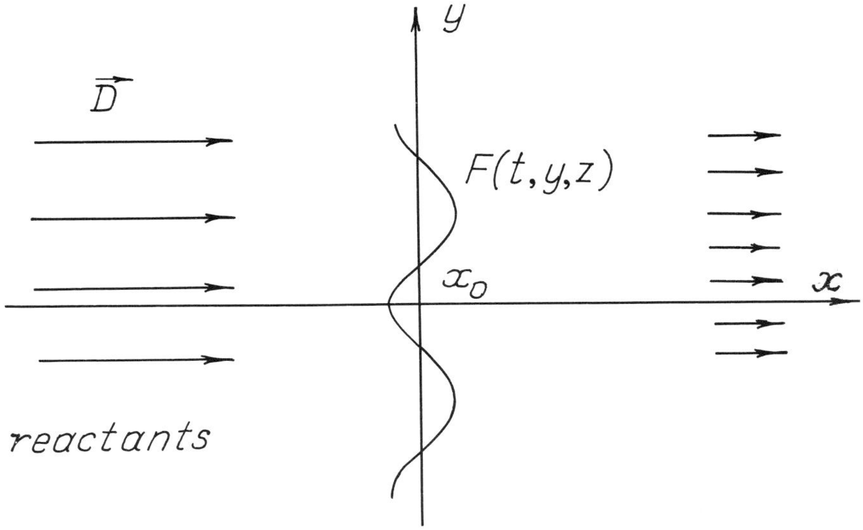

Figure 1. The general scheme of the problem.

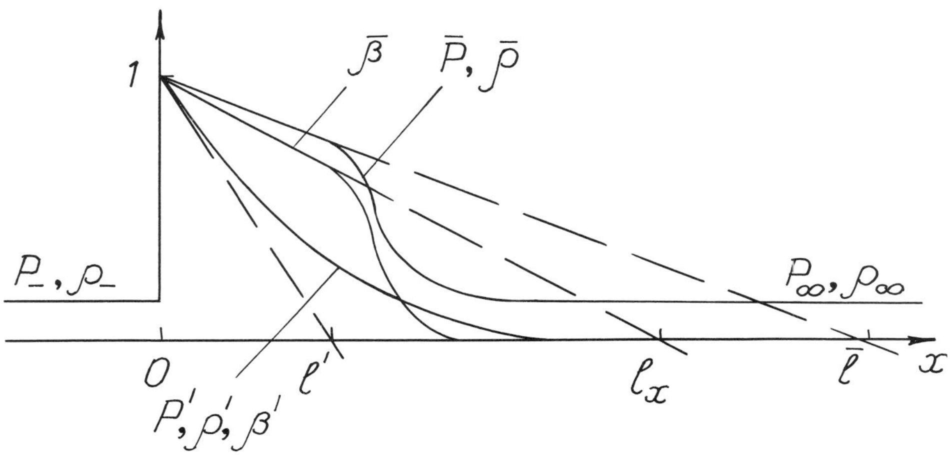

Figure 2. Scheme of detonation wave parameter variation.

The pressure and the density can be expressed in terms of the velocity in the equation of energy balance :

$$\bar{u}\,\frac{d\bar{u}}{dx}\left(\frac{c^2}{\bar{u}^2}-1\right) = -\,Q\,(\gamma-1)\,\frac{d\bar{\beta}}{dx}$$

For the Chapman-Jouguet detonation regime:

$$\frac{c^2}{\bar{u}^2} \longrightarrow \frac{2\gamma}{\gamma-1} >> 1 \;, \qquad \text{as} \quad x \longrightarrow x_o$$

So one can approximately write

$$\int \frac{d\bar{u}}{\bar{u}} \approx -\int \frac{Q\,(\gamma-1)}{c^2}\,d\bar{\beta}$$

$$\bar{u} = u_o\,exp\left(\frac{Q\,(\gamma-1)}{c^2}\,(1-\bar{\beta})\right)$$

Substituting the resulting expression to the kinetic equation one obtains:

$$\int exp\left(-A^{*}(1-\bar{\beta})\right)\,d\bar{\beta} = \int -\frac{N}{u_o}\,dx$$

$$\bar{\beta}(x)=1+\frac{1}{A^{*}}\,ln\left\{1-\frac{A^{*}N}{u_o}(x-x_o)\right\}\approx 1-\frac{x-x_o}{l_{ch}}$$

where $A^{*} = A - \dfrac{Q\,(\gamma-1)}{c^2}$, $l_{ch} = {}^{u_o}/_{N}$

It isn't difficult to write the expressions for $\bar{u}(x)$, $\bar{p}(x)$, $\bar{\rho}(x)$,

$\bar{T}(x)$, $\bar{v}(x)$, $\bar{w}(x)$ in the vicinity of the shock front :

$$\bar{u}(x) = u_o \exp\left((x-x_o)/\bar{l} \right) \approx u_o \left(1 + (x-x_o)/\bar{l} \right)$$

$$\bar{p}(x) = p_o + \rho_o u_o^2 \left(1-\exp\left((x-x_o)/\bar{l}\right)\right) \approx p_o - \rho_o u_o^2 (x-x_o)/\bar{l}$$

$$\bar{\rho}(x) = \rho_o \exp\left(- (x-x_o)/\bar{l} \right) \approx \rho_o \left(1 - (x-x_o)/\bar{l} \right)$$

$$\bar{T}(x) = T_o \left[1 - \frac{Q}{R\,T_o} \frac{(3-\gamma)(\gamma-1)}{A^*(\gamma+1)} \ln\left\{ 1- \frac{N\,A^*}{u_o} (x-x_o)\right\} \right] \approx$$

$$\approx T_o \left(1 + \gamma \frac{(3-\gamma)}{(\gamma+1)} \frac{x-x_o}{\bar{l}} \right)$$

$$\bar{v}(x) = \bar{w}(x) = 0$$

where $$\bar{l} = l_{ch} \frac{c^2}{Q\,(\gamma-1)}$$

l_{ch} is the reaction zone length defined by linear extrapolation of stationary dependence $\bar{\beta}(x)$ in the neighborhood of the front :

$$\bar{\beta}(x) = 0 \quad , \qquad \text{as} \qquad (x-x_o) \longrightarrow l_{ch}$$

The amplitude of stationary solution for gas dynamic parameters in the vicinity of the front changes by the factor e over a distance $\bar{l}$. The analogous distance for pulsations approximately equals to $l' = u_o/NA$ as it is shown below. Note that l' coincides with the size of chemical transformation zone in Zel'dovich's model l_Z (Zel'dovich, Kompaneyets 1955). Figure 2 shows schematically the dependencies (on the longitudinal coordinate x) of parameters of planar stationary Zel'dovich's detonation wave and perturbations (all the dependencies are made dimensionless with the help of their values at the point x_o).

Dashed lines in the Figure 2 show the asymptotes of solutions at $x \longrightarrow x_o$.

4. Unsteady Linearized Problem

Conserving only the terms of the 1st order infinitesimal, we'll seek the solution of linear problem in the form of harmonic superposition:

$$H'(t,x,y,z) = \delta H_{nm}(x)\ exp(\Omega_{nm} t + ik_n y + ig_m z)$$

$$\delta H(x) = \delta H_j exp\ (\lambda_j\ x)$$

$Re(H')$ corresponds to the physical parameters. The subscript "j" denotes different wave modes. For gas-dynamic modes we have the condition of perturbation decreasing as $x \longrightarrow +\infty$ ($Re\ (\lambda_j) < 0$) (Dyakov, 1954) . To substantiate this condition one could say the following. As $E/RT_o \longrightarrow \infty$ the solution of shock front stability problem must reduce to gas-dynamic limit case first investigated by (Dyakov, 1954). In this case the values of stationary solution's parameters ($\bar{p}_+ ;\ \bar{\rho}_+$) behind shock front are constant (don't depend on x) and perturbations attenuate as x grows. The last means that one examines the situation when the source of instability is the shock front itself, - perturbations "don't come to the front from infinity". Therefore

$$\left(\frac{\partial\ p'}{\partial\ x} >> \frac{d\ \bar{p}}{d\ x} \right)_{x \ge x_0} \quad ; \quad \left(\frac{\partial\ \rho'}{\partial\ x} >> \frac{d\ \bar{\rho}}{d\ x} \right)_{x \ge x_0}$$

In detonations the value of activation energy is finite but significantly more than 1. In the vicinity of the detonation front where heat released due to chemical transformation is small comparatively with energy of shock wave, i.e. in so-called induction zone, in equations taking place in gas-dynamic shock wave are approximately true. Consequently, when writing non-stationary equations, in the vicinity of the front one can suppose:

$$\left(\frac{d\,\bar{H}(x)}{d\,x}\right)_{x\to x_o} \sim \varepsilon^s, \quad s \geq 2\ ; \qquad \left(\frac{\partial\ H'}{\partial\ x}\right)_{x\to x_o} \sim \varepsilon$$

This assumption ($E/RT_o >> 1$) allows to separate the steady and perturbed parts of general system of equations and to solve linearized problem independently on the function $\bar{H}(x)$ .Using more exact approach (for example, Erpenbeck, 1962) one couldn't obtain productive theoretical model.

For the front surface we can write the similar equation:

$$x-x_o = F(t,y,z) = f_{nm}\ exp\ (\Omega_{nm} t + ik_n y + ig_m z) \sim \varepsilon$$

$Re(F)$ corresponds to the coordinate of physical surface of detonation front. The front will remain planar in the case if all the harmonics will be stable, i.e. $Re(\Omega_{mn}) \leq 0$. Non-zero imaginary part of Ω will result in pulsating detonation front.

The reactant concentration is close to zero at the distance l_z behind the front (according to Zel'dovich's model), therefore at $(x-x_o) \geq l_z$ linear approximation is unacceptable, l_z (or l') designates a virtual chemical transformation zone length, τ is the relevant reaction time ($\tau=1/(NA)$) . After the transformation of equations to the dimensionless form (using T_o , ρ_o, l_z, τ) one can write solutions for perturbations at $(x_o=0)$:

$$\delta\beta(x) = A_1\ S\ e^{\lambda_1 x} \qquad (1\text{-}a)$$

$$\delta p(x) = A_1\ e^{\lambda_1 x} - A_2\ (\Omega + \lambda_2)\ e^{\lambda_2 x} \qquad (1\text{-}b)$$

$$\delta\rho(x) = A_1(\lambda_1^2 - \kappa)\ e^{\lambda_1 x} + A_2\,\frac{\kappa^2 - \lambda_2^2}{\Omega + \lambda_2}\ e^{\lambda_2 x} + A_5\ e^{\lambda_3 x} \qquad (1\text{-}c)$$

$$\delta u(x) = A_1\lambda_1 e^{\lambda_1 x} + A_2 \lambda_2 e^{\lambda_2 x} + \frac{A_3 k^2 + A_4 g^2}{\lambda_3} e^{\lambda_3 x} \quad (1\text{-}d)$$

$$\delta v(x) = ik\ (A_1 e^{\lambda_1 x} + A_2 e^{\lambda_2 x} + A_3 e^{\lambda_3 x}) \quad (1\text{-}e)$$

$$\delta w(x) = ig\ (A_1 e^{\lambda_1 x} + A_2 e^{\lambda_2 x} + A_4 e^{\lambda_3 x}) \quad (1\text{-}f)$$

where λ_1 corresponds to the "chemical" wave mode, λ_2, λ_3 correspond to the gas-dynamic modes, κ is the wave number of two- dimensional perturbations, A_j and f are indefinite coefficients,

$$\lambda_1 = 1 - \Omega \ , \qquad (\lambda_2 + \Omega)^2 = c^2(\lambda_2^2 - \kappa^2) \ , \qquad \lambda_3 = - \Omega \ ,$$

$$\kappa = \sqrt{k^2 + g^2} \ \geq 0 \ , \qquad S = \frac{\gamma\ R\ T_o(\lambda_1^2 - \kappa^2) - 1}{(\gamma - 1)\ Q}$$

One can see that $Re(\lambda_3)$ corresponds to the gas-dynamic wave mode whose amplitude doesn't attenuate along the trajectory. So, following the assumption made above we have to require this mode indefinite coefficients to be equal to zero. $Im(\lambda_3)$ determines a gas-dynamic oscillatory mode.

The following connection between gas-dynamic component of pressure and density pulsations takes place in the vicinity of the shock front where the reaction doesn't begin yet :

$$(\delta p)_{gas} = \left(\frac{\partial p}{\partial \rho} \right) (\delta \rho)_{gas} = \left(\frac{\partial p}{\partial \rho} \right)_{x=0} (\delta \rho)_{gas} + \varepsilon^2 \approx c^2\ (\delta \rho)_{gas}$$

Then at $x=x_o$:

$$A_5 = A_2 \left(\frac{\lambda_2^2 - \kappa^2}{\Omega + \lambda_2} - \frac{\Omega + \lambda_2}{c^2} \right) = 0$$

For the other indefinite coefficients f , A_j one can write linear conservation conditions at the perturbed shock.

Transverse perturbations of the plane shock front are supposed to have long wave-lengths, i.e. their amplitudes are considerably less than their lengths. Consequently the difference between velocities of propagation along x-axis in the cases of planar and curved detonation front is pronounced only in the 2nd-order terms (non-linear effects, neglected here). This assumption corresponds to experimental data (Manzhalei, Mitrofanov, Subbotin, 1974) on the shape of front inhomogeneity and isn't too strong.

a) conservation of the reactant concentration

$$\beta(x=F-0) = 1 = \beta(x=F+0) = \beta(0) \left(\frac{\partial\beta}{\partial x} \right)_{x=0} F + \varepsilon^2$$

$$\delta\beta = f \left(- \frac{d\bar{\beta}}{dx} \right)_{x=0}$$

so in dimensionless form :

$$\delta\beta = {}^{f}/_{A} \tag{2}$$

b) mass flux conservation

$$\left[\rho \left(u - \frac{\partial F}{\partial t} \right) \right]_{x=F} = 0$$

$$f\,\Omega\,r = (\delta\rho + \delta u)_{x=0} \tag{3}$$

c) conservation of normal and tangential momentum flux components

$$\left[\left(p\, \delta_{jk} + \rho\, u_j\, u_k \right) n_k \right]_{x=F} = 0 \qquad (j,k = 1,2,3)$$

$$(\delta p + \delta \rho + 2\, \delta u)_{x=0} = 0 \tag{4}$$

$$i\, k\, f\, (D - 1) = (\delta v)_{x=0} \tag{5}$$

$$i\, g\, f\, (D - 1) = (\delta w)_{x=0} \tag{6}$$

where n_k is the component of front surface normal vector, r is the coefficient determining the intensity of the shock wave :

$$r = 1 - {}^{\rho_-}/\rho_o$$

The system of equations (1-a)-(1-f) has a unique non-trivial solution if the determinant of the system (2)-(6) written for indefinite coefficients equals to zero. This condition leads to the dispersion relation $\Omega(\kappa)$:

$$det \begin{pmatrix} \Omega + \dfrac{\kappa^2}{\lambda_3} & -r\,\Omega - \dfrac{1+\lambda_1 - {}^{\kappa^2}/\lambda_3}{A\, S} - r\, D\, \kappa^2/\lambda_3 \\ (\Omega + \lambda_2)(1 - {}^{1}/{}_{c^2}) & -2\, r\, \Omega + \dfrac{-1 - \kappa^2 + \lambda_1^2}{A\, S} \end{pmatrix} = 0$$

If $E/RT_o \longrightarrow \infty$ we obtain the gas-dynamic limit of this dependence :

$$2\, \Omega\, (\Omega^2 - \kappa^2) + (\Omega + \lambda_2)(1 - {}^{1}/{}_{c^2})(D\, \kappa^2 - \Omega^2) = 0$$

This equation was obtained for shock waves by (Dyakov, 1954). When

$c^2 \gg 1$ (strong shock wave) the front is stable :

$$Re(\Omega) = -\kappa \le 0 \tag{7}$$

$$Im(\Omega) = C_1 \kappa \tag{8}$$

where $C_1 = \pm \frac{1}{2} \left\{ \sqrt[3]{b + \sqrt{dis}} - \sqrt[3]{b - \sqrt{dis}} \right\}$,

$b = (18D - 10)\, 3^{-3/2}$, $dis = D^3 + 5D^2 + 3D - 9$,

The relation (8) corresponds to well-known usual wave equation

$$F_{tt} = C_1^{\,2} (F_{yy} + F_{zz}) ,$$

C_1 denotes the velocity of transverse waves travelling in the front plane (when $\gamma = 1.4$, then $D = 6$ and $|C_1| \approx 0.42\, D$).

From the physical point of view the criterion of planar shock wave stability can be explained by the following mechanism. Gas flow behind the concave section of slightly curved shock front appears to be convergent, i.e. stream tubes are constricted. The situation is opposite for the front sections which are convex in the direction of the wave propagation. This occurs due to a definite change in directions of streamlines crossing the front of oblique shock wave. The constriction of flow behind the wave front in the case of subsonic flow results in the increasing of the flow velocity, therefore concave sections will overtake the convex ones and front curvature will decrease. If flow behind the front is supersonic ($c^2 < 1$) then perturbations will increase at least until non-linear stabilizing factors will become significant.

As it was discussed, the mode $Re(\lambda_3)$ shouldn't influence on the stability of the solution, so the expression (7) for $Re(\Omega)$ could be obtained without noting the mode λ_3 . This mode (its imaginary part) determines only the oscillatory features ($Im(\Omega)$) . Really, let us put $A_3 = A_4 = 0$. In this case the equation for Ω is of the form :

$$\Omega \left(2 r \Omega + \frac{1}{A S} (1 + \kappa^2 - \lambda_1^2) \right) = \tag{9}$$

$$= (\Omega + \lambda_2) (1 - {}^1/_{c^2}) \left(r \Omega + \frac{1}{A S} (1 + \lambda_1) \right)$$

If $p_o >> p_-$ then $c^2 \approx \frac{2 \gamma}{\gamma - 1} >> 1$ and (9) has more simple form:

$$(\Omega - \lambda_2) + \frac{1}{\alpha} \frac{\left(1 + \kappa^2 + \lambda_2 + \Omega - \lambda_1^2 \right)}{(\lambda_1^2 - \kappa^2)} = 0 \tag{10}$$

where $\alpha = \frac{3-\gamma}{\gamma+1} r \frac{E}{R T_o}$

α is very important coefficient depending on the intensity of the shock wave, adiabatic exponent γ and also on the chemical features of gas mixture.

As $\kappa \longrightarrow 0$, $\kappa \longrightarrow \infty$ or $\alpha \longrightarrow \infty$ the dependence $\Omega(\kappa)$ is determined by the gas-dynamic limit (7). Only in the vicinity of $\kappa \approx 1$ the "chemical features" of equation (10) solution predominate and the unstable harmonic region takes place :

$Re(\Omega(\kappa)) \geq 0$, $Re(\lambda_{2,3}) \leq 0$

Noting that $Re(\Omega) \longrightarrow 0$ as $\kappa \longrightarrow 1$, in the neighborhood of $\kappa = 1$ one can approximately write, eliminating higher powers of $Re(\Omega)$:

$$Re(\Omega) = - \kappa + \frac{\kappa^2}{\alpha \kappa^2 - \alpha + 1}$$

or expanding the denominator in a series in powers of $(\kappa^2-1) \sim \varepsilon$, one obtains the expression :

$$Re(\Omega) = - \kappa + (1 + \alpha) \kappa^2 - \alpha \kappa^4 \tag{11}$$

In the two-dimensional case (detonation wave in a narrow plane channel without viscosity and thermal conductivity losses) the linear differential equation (12) written for the function F corresponds to the relation (11).

$$\frac{\partial F}{\partial t} = i \frac{\partial F}{\partial y} - (1 + \alpha) \frac{\partial^2 F}{\partial y^2} - \alpha \frac{\partial^4 F}{\partial y^4} \qquad (12)$$

According to (11) in the one-dimensional case (i.e. $\kappa \longrightarrow 0$) increment $Re(\Omega)$ tends to zero. So there exists the asymptotic transition to Zel'dovich-Neumann-Döring model for stationary planar detonations.

Linear equation (12) predicts amplification of harmonics (see Figure 3) with the wave-lengths λ_{un} ($\lambda_{max} > \lambda_{un} > \lambda_{min}$),

where $\lambda_{max} = \dfrac{4 \pi l_z \alpha}{\sqrt{1 + 4 \alpha} + 1}$, $\lambda_{min} = 2 \pi l_z$

5. Non-Linear Evolution Equation

If one considers each element of the surface of weakly curved shock wave front as a planar wave propagating along the normal at the velocity D then the velocities of different front sections in the direction of x-axis are different and curvature decreases (see Figure 4). This mechanism is described by non-linear terms of the evolution equations (Zel'dovich, 1966). Setting down the surface equation

$$\Psi (t,x,y,z) = F (t,y,z) - x = 0 ,$$

one defines the normal deformation rate v_n of this surface at each point (x', y', z') :

$$v_n = \mathbf{n} \frac{d \mathbf{X}}{d t} , \text{ where } \mathbf{X} = \{ x', y', z' \} , \quad \mathbf{n} = \frac{\nabla \Psi}{|\nabla \Psi|}$$

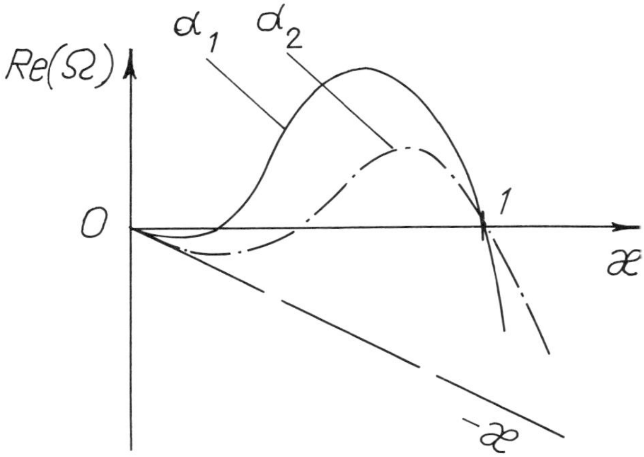

Figure 3. The dependence of infinitesimal perturbation's increment on wave number and parameter α ($\alpha_1 > \alpha_2$) (corresponding to equation (11)).

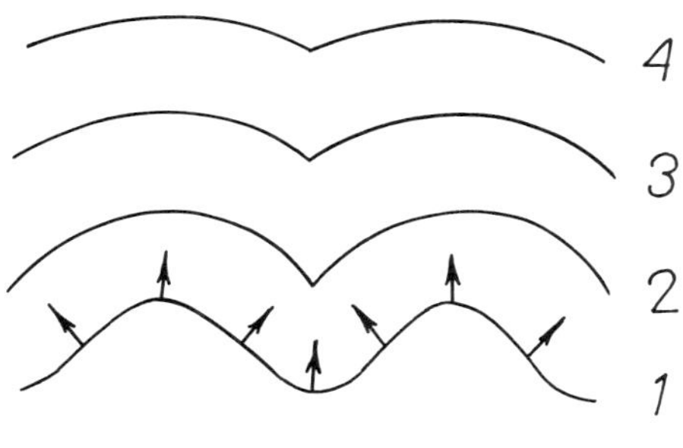

Figure 4. The stabilizing effect of weak non-linear kinematic mechanism.

Since
$$\frac{d\Psi}{dt} = \frac{\partial \Psi}{\partial t} + \nabla \Psi \frac{d\mathbf{X}}{dt} = 0 \quad ,$$

then
$$v_n = - \frac{\partial \Psi}{\partial t} \left(|\nabla \Psi| \right)^{-1}$$

On the other hand $v_n = \mathbf{n}\mathbf{v}_1 - v_2$, where $\mathbf{v}_1$ is the vector ofthe medium velocity , v_2 is the module of the normal velocity relatively to medium. As a result one arrives to the kinematic condition

$$\frac{\partial \Psi}{\partial t} = v_2 |\nabla\Psi| - \mathbf{v}_1 \nabla\Psi$$

If the medium is at rest then $\mathbf{v}_1 = 0$ and v_2 is the velocity of planar detonation wave $v_2 = - D$. Then

$$\frac{\partial \Psi}{\partial t} = - D \sqrt{\left(\frac{\partial \Psi}{\partial x}\right)^2 + \left(\frac{\partial \Psi}{\partial y}\right)^2 + \left(\frac{\partial \Psi}{\partial z}\right)^2} = - D \sqrt{1 + \left(\frac{\partial \Psi}{\partial y}\right)^2 + \left(\frac{\partial \Psi}{\partial z}\right)^2}$$

Since we solve the problem in the long-wavelength approximation:

$$\left(\frac{\partial F}{\partial y}\right)^2 << 1 \; , \qquad \left(\frac{\partial F}{\partial z}\right)^2 << 1 \; , \qquad \text{then}$$

$$\frac{\partial \Psi}{\partial t} = \frac{\partial F}{\partial t} \approx - D \left\{ 1 + \frac{1}{2} \left(\frac{\partial F}{\partial y}\right)^2 + \frac{1}{2} \left(\frac{\partial F}{\partial z}\right)^2 + \varepsilon^4 \right\}$$

The obtained relation completes the linear evolution equation for the front surface, i.e. the first term brings one to the immovable frame of coordinates and the second is related to the high order terms which are not taken into account in linear approximation.

$$\frac{\partial F}{\partial t} = i \frac{\partial F}{\partial y} - (1 + \alpha) \frac{\partial^2 F}{\partial y^2} - \alpha \frac{\partial^4 F}{\partial y^4} - \frac{D}{2} \left(\frac{\partial F}{\partial y} \right)^2 + \varepsilon^4 \qquad (13)$$

Non-linearity could stabilize the solution by redistribution the energy in the spectrum between the amplified and attenuated harmonics. Independent in linear case unstable harmonics interact now and generate high- and low-frequency secondary harmonics which are attenuated according to the linear dispersion relation. This mechanism could provide dynamic stability of the spectrum and lead to forming of cellular structure of solution. In its turn the imaginary part of Ω (being of gas-dynamic nature, equation (8)) causes solution's pulsations and propagation of transverse waves.

The developed equation (13) looks close to Sivashinsky's equation (Sivashinsky, 1977) for hydrodynamic instability in flames

$$F_t - \mu^2 \nabla^2 F + \frac{1}{2} \left(\nabla F \right)^2 = \eta^2 \int_{-\infty}^{\infty} |\mathbf{k}| \, e^{i\mathbf{k}(\mathbf{y}-\mathbf{z})} \, F(\mathbf{z}, t) \, d\mathbf{k} \, dz$$

But the very important difference is that long-wave linear perturbations according to Sivashinsky's equation are unstable. As the authors' experience confirms this evolution equation leads to the permanent increase of the inhomogeneity length, i.e. there is no steady solution in the infinite space. On contrary, equation (13) requires long-wave infinitesimal disturbances to attenuate and doesn't allow infinite growth of the solution's period. In this sense equation (13) is physically more substantiated (more realistic) than Sivashinsky's one.

The usual values of the parameter α for the gas detonations are about 10 . So the range of instability in the spectrum and the maximum value of $Re(\Omega)$ in the equation (13) aren't small, this fact restricts the possibilities of analytical investigation of its solutions.

6. Results of Numerical Modelling

The linear analysis indicates the instability of all possible finite-difference schemes of equation (13). This is the feature of mathematically incorrect equation. A spectral method is used for numerical modelling. Fast Fourier transform of the signal is performed at each time step ($\hat{\tau}=0.025$, $h=0.2$, $M=256$) and the spectrum (A^n_m) is recalculated according to the dispersion relation (11)

$$A^n_m = A^n_m \; exp \quad (\hat{\tau}\Omega_m), \qquad\qquad m=1, \; M/_2+1,$$

where $\Omega_m = \Omega\,(\kappa_m)$. On calculations $M/_2 + 1$ harmonics are taken into account. After reverse fast Fourier transform a correction due to non-linearity is introduced. It is calculated using the elementary explicit difference scheme

$$F_j^{\,n+1} = \left(F_j^{\,n+1} \right)_{linear} + \frac{\hat{\tau}}{4\,h^2}\,\frac{D}{2}\left(F^{\,n}_{j+1} - F^{\,n}_{j-1} \right)^2$$

Here $\hat{\tau}$ is the time step, h is the spatial step, M is the number of mesh points, n and j are number of time step and number of mesh point, respectively, m is the number of harmonics. Periodic boundary conditions are set, the calculation region exceeds significantly (approximately by six times) the characteristic scale of inhomogeneity.

Generally the realized algorithm is equal to calculations with the help of explicit finite-difference scheme because only the values from previous time layer are used for calculations. This algorithm turns out to be stable in the sense that physically absurd solutions are absent. The total computer time doesn't exceed six minutes. The amplified numerical computer's "noise" existing due to non-ideal exactness of computer operations is used as a weak random initial perturbation of trivial solution.

As a result of numerical modelling of the equation's (13) solution ($\alpha=3$, $D=6$) (see Figure 5) a quasi-stationary cellular structure (convex towards the wave propagation) is generated. The symmetry of initial trivial solution is broken. The velocity of the shock front displacement is approximately 1.1 times the equilibrium

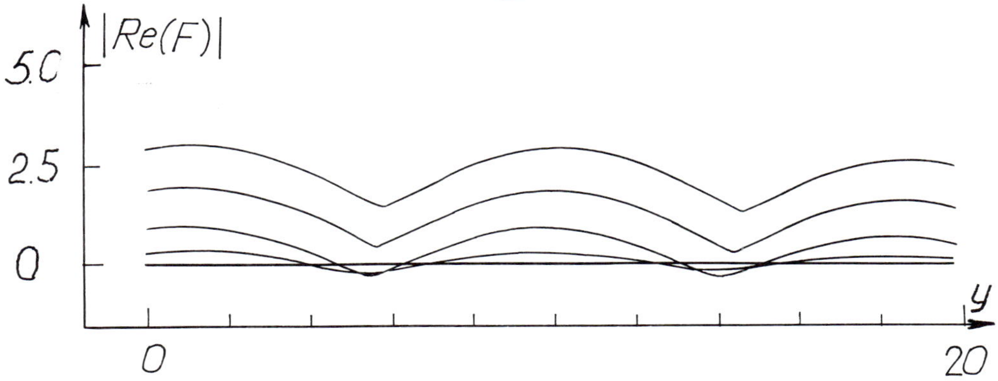

Figure 5-a. Formation of two-dimensional front cellular structure; dimensionless time: 0; 17.5; 20.; 22.5; 25.

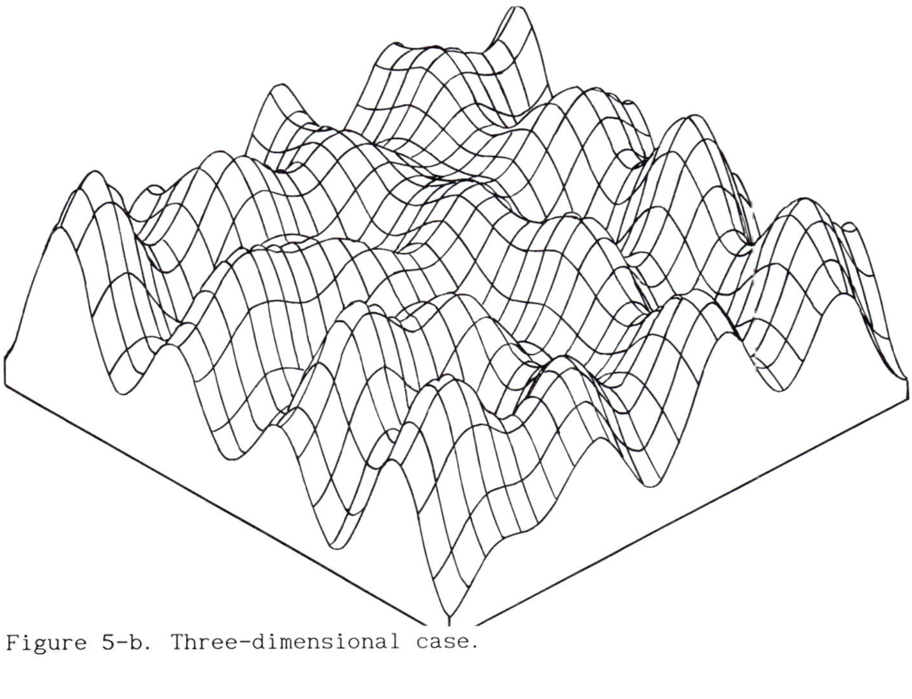

Figure 5-b. Three-dimensional case.

Chapman-Jouguet detonation velocity D , i.e. the mean solution of the non-linear equation turns out to be non-zero and increases. Linear part of equation (13) is written according to relation (11) which defines only the real part of Ω . Noting the imaginary part of Ω on can rewrite the dispersion relation in the general form: $\Omega = Re(\Omega) +$ $+ i\, Im(\Omega)$ or to eliminate indefinite sign in the expression (8) for $Im(\Omega)$,

$$(\Omega - Re(\Omega))^2 = Im^2(\Omega)$$

where $Re(\Omega)$ and $Im(\Omega)$ are defined by equations (11) and (8), respectively. This dispersion relation corresponds to the linear part of the equation which can be written as following :

$$L(F) = C_1^2 \frac{\partial^2 F}{\partial y^2}$$

where $L(F)$ is non-linear differential operator

$$L(F) = \left\{ \frac{\partial}{\partial t} - i \frac{\partial}{\partial y} + (1+\alpha) \frac{\partial^2}{\partial y^2} + \alpha \frac{\partial^4}{\partial y^4} + \frac{D}{2} \frac{\partial F}{\partial y} \frac{\partial}{\partial y} \right\}^2 F$$

This equation is too cumbersome so one could try to divide it into two parts to solve them successively. The first on is equation (13) and the second is the wave equation

$$\frac{\partial^2 F}{\partial t^2} = C_1^2 \frac{\partial^2 F}{\partial y^2}$$

These two equations are to be solved separately but at the each time step and the results must be combined. The solution obtained as a result of such procedure turns out to be essentially non- steady. Cyclic motion of transverse waves appears. The solution represents cellular pulsating structure (Figure 6). Quasi- stationary solution of equation (13) is used here as the initial condition.

If one follows the trajectories of the cell joints (or any other

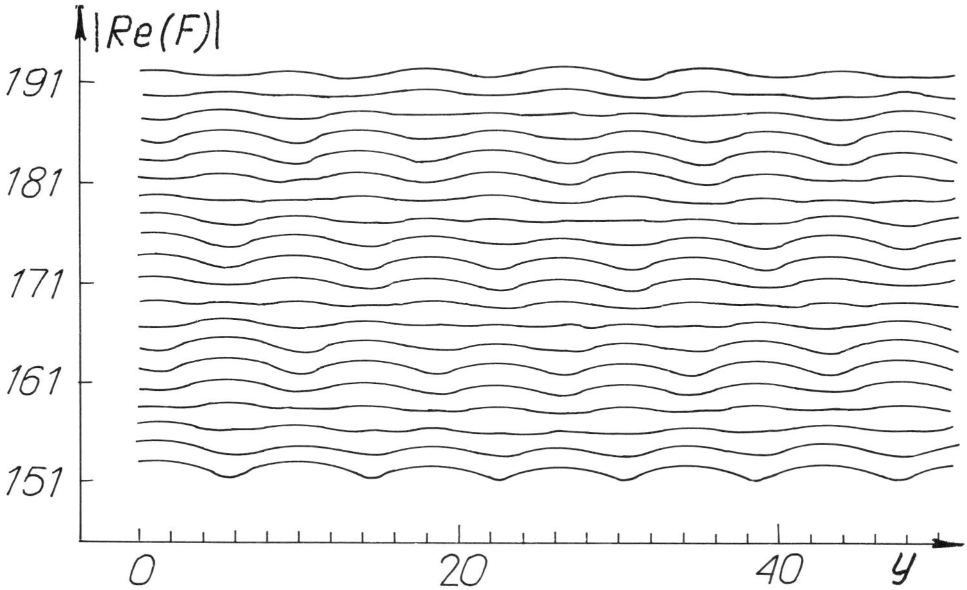

Figure 6. Unsteady cellular detonation front.

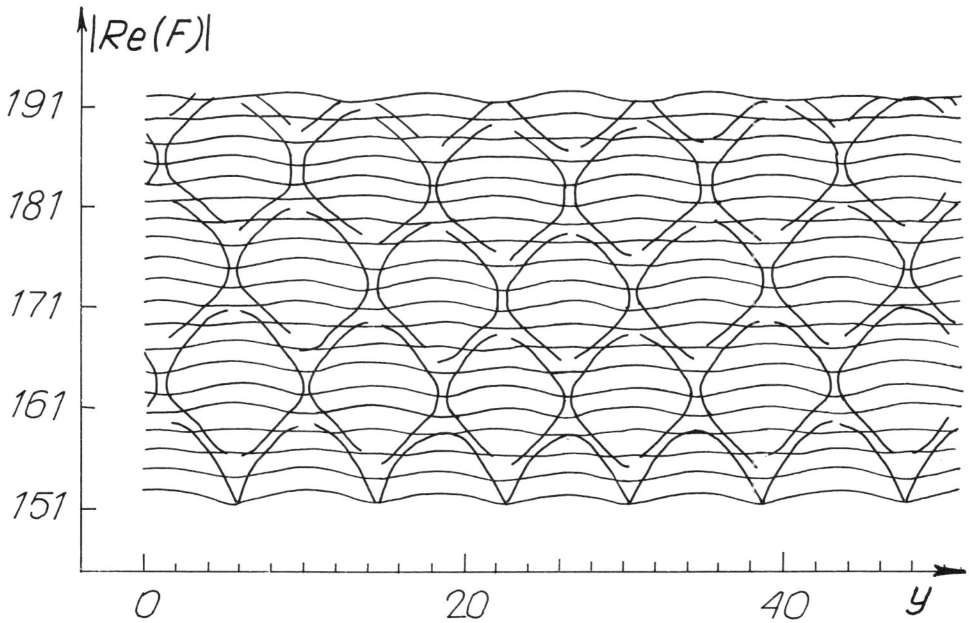

Figure 7. Pulsating detonation front. Trajectories of cell joints calculated with small systematic error are placed between the dotted lines ($\alpha = 3$); regular structure is observed.

phase of the surface) then the scheme reproducing detonation wave imprints on sooty channel wall is obtained (Figures 7,8).

7. Comparison with Experimental Data

First of all it needs to underline that direct comparison of the model's results with experimental measurements of cell size is difficult by two causes:

i. characteristic size of inhomogeneity of solution generally isn't determined by linear relations and doesn't correspond to the most unstable harmonics because of non-linearity of the evolution equation, its solution one can obtain only numerically;

ii. the main difficulty is indefinite values of kinetic parameters. In some experimental works the relation of cell size per length of induction zone l_{ind} is reported but l_{ind} is determined in different ways by different authors and can differ significantly from the scale l_z used here. On the basis of this model one can write a general expression for the cell size which doesn't contradict to experiments:

$$a = 2\pi\, l_z\, \varphi(\alpha, D) \qquad (14)$$

where $$l_z = \frac{\gamma+1}{\gamma-1}\, \frac{p_o^{1-m_1}\, \beta_o^{-m_1}\, u_o}{(3-\gamma)}\, \frac{R\,T_o}{Q}\, \frac{R\,T_o}{E}\, \exp\,(\,E/RT_o\,)$$

Usually the proportionality between cell size a and l_{ind} is considered and dependence $a(Q)$ is not taken into account. But Q is necessary to be taken into account for investigation of overdriven detonation front dynamics when thermal release can significantly decrease and even change the sign (Nikolayev, Topchiyan, 1977).

The expression for cell size a shows that its dependence on E/RT_o has minimum at $E/RT_o=1$ and tends to infinity as $E/RT_o \longrightarrow \infty$ or $E/RT_o \longrightarrow 0$. But experimentally observed is only the branch where $E/RT_o > 1$. So the expression (14) is in accordance with empirical data (Vasilyev, Mitrofanov, Topchiyan, 1987) : the cell size in overdriven regime is less than in Chapman-Jouguet one.

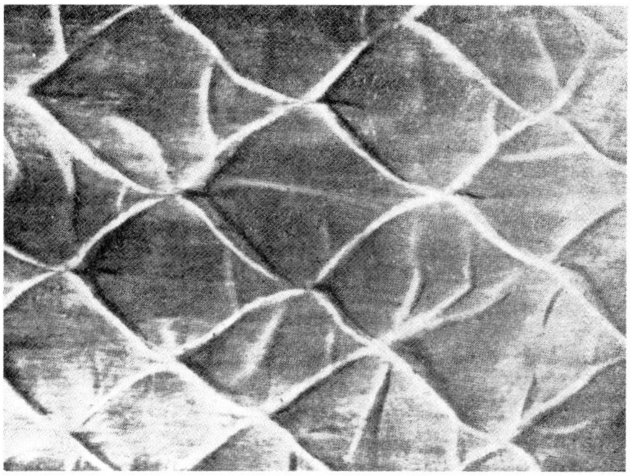

Figure 8. The detonation wave imprint on a soot-coated inside wall of narrow co-axial channel (fish-scale pattern) (Voytsekhovsky, Mitrofanov, Topchiyan, 1963).

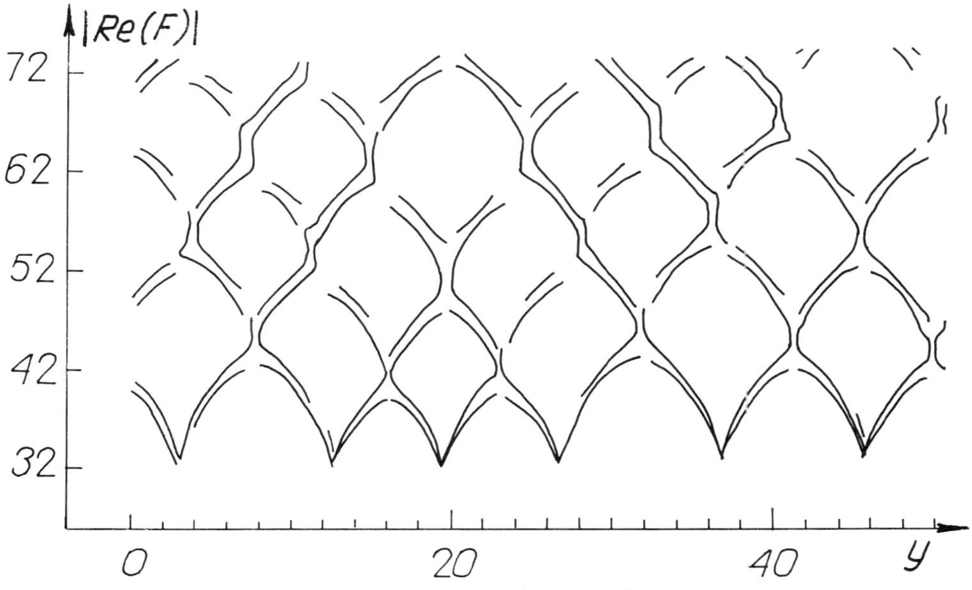

Figure 9. Pulsating detonation front ($\alpha = 5$); quasi-regular structure is observed.

On the basis of equation (14) the well-known experimental dependence $a(\phi)$ (ϕ is equivalence ratio can be explained not only traditionally by the connection $E(\phi)$ but also by thermal release function $Q(\phi)$: $Q = Q_{max}$, as $\phi \approx 1$, so according to (14) $a = a_{min}$, as $\varphi \approx 1$.

According to expression (14) there is no universal connection between cell size a and chemical transformation zone length l_z. The coefficient of proportionality $2\pi\ \varphi(\alpha, D)$ depends on dimensionless activation energy, shock wave intensity and adiabatic exponent γ. The function $\varphi(\alpha, D)$ can not be determined in explicit form, - it denotes non-linear features of evolution equation.

The value of a increases as the pressure p_- of unperturbed medium decreases (due to reduction of the temperature T_o behind the front). This agrees with experimental data of (Voytsekhovsky, Mitrofanov, Topchiyan, 1963). The relation a/l_z doesn't depend on p_- (equation (14)) according to (Westbrook, Urtiew, 1983).

May be the most important is the agreement with experiments in the velocity of transverse waves ($|C_1| \approx 0.4\ D$) and pulsations of longitudinal velocity of front propagation ($D_{max} \approx 1.5\ D$; $D_{min} \approx 0.7\ D$). The correspondence of front inhomogeneity's shape is also good: the ratio of amplitude per length in calculations is about 0.17 (Figure 5) and in experiments the value between 0.1 and 0.2 is measured by (Manzhalei, Mitrofanov, Subbotin, 1974).

8. Discussion and Conclusions

The comparison of theoretical results with experimental data shows that this simple quasi-linear evolution model is able to predict certain reliable information. In spite of the existence of large gas parameters' variations behind the natural pulsating detonation front relatively to the stationary values the used quasi-linear approach turns out to be admissible. Apparently the cause is that the deviation of the state of all gas mass from the state in planar stationary detonation wave is in general sufficiently weak. It would be interesting to try to answer with the help of the model some questions which are not clear yet.

The first one: The experiments (Vasilyev, Mitrofanov, Topchiyan, 1987) show that regular cellular structure of detonations is destroyed as the activation energy increases. This is not quantitative effect but one can observe cells of quite different sizes. The regularity also depends on the intensity of detonation wave and adiabatic exponent γ. There is no any theoretical model now to explain this dependence. But it is easily seen from the equation (11) for real part of Ω or from Figure 3 that the width of the unstable harmonics region in the spectrum is a function of parameter α. So the larger α, the larger quantity of unstable harmonics which present in solution, therefore the regularity of the cellular structure disappears. If one would remind the expression for α he would see that it is proportional to activation energy and depends on the wave intensity and γ.

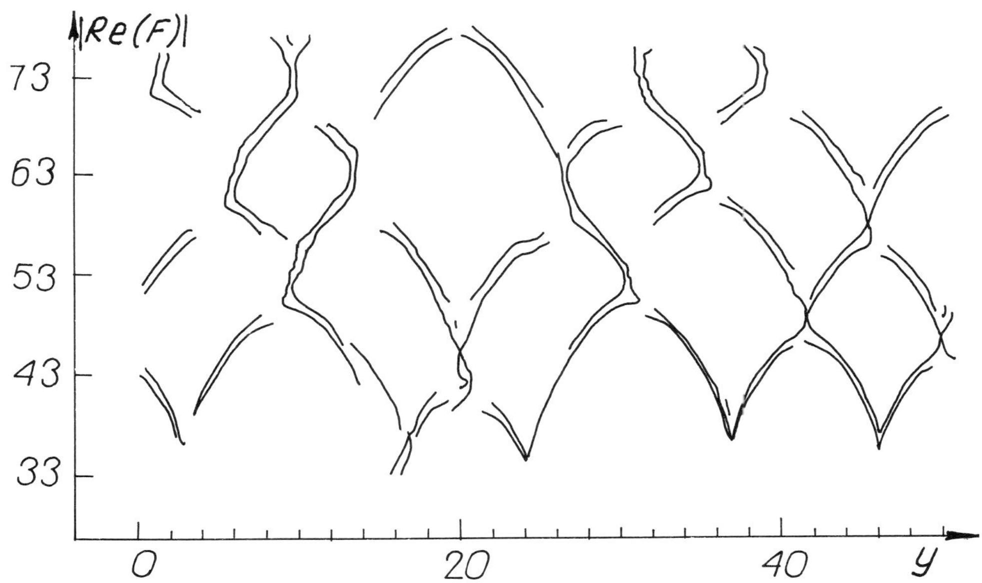

Figure 10. Pulsating detonation front ($\alpha = 7$); irregular structure is observed.

This is very clear physical explanation for the interesting experimental observation. The results of calculations as α = 3, 5, 7 are represented in Figures 7,9,10 to illustrate this effect.

The second problem is connected with so-called "13 λ --correlation" of critical tube diameter d_c . When the planar detonation wave emerges into a volume or when the detonations propagates in the channel with elastic walls then rarefaction waves can transform it to the deflagration regime. There is the qualitative criterion which states that critical radius of the tube equals to effective thickness of detonation wave Δ_H . This thickness is the distance from the shock front to the equilibrium Chapman-Jouguet plane. But the problem of Chapman-Jouguet plane position is very complicative and isn't solved yet, so in practice this criterion turns out to be useless.

On the basis of the model presented here one could try to formulate a new criterion. First of all there is a certain characteristic time of the process which could be included in new criterion. Namely, it is the time of forming of cellular structure t_f which must be not more than time of rarefaction wave propagation to the tube axis $t_c = d_c/2C$. This criterion notes the competition of two opposite physical mechanisms. The rarefaction waves tend the front to become smooth, but the sharp folds are the origins of fast reaction and without them detonations transforms to deflagration. So the critical condition requires the time of fold forming t_f to be equal to t_c . In other words the time of forming of cell structure must be equal to the time of its destruction. May be there is a certain physical connection between this new criterion and the traditional one.

The experiments (Vasilyev, Mitrofanov, Topchiyan, 1987) show that critical tube diameter can be two times more or less than 13 cell sizes, the criterion $d_c = 13\lambda$ is not universal and depends on the kinetics of reaction. Except this the critical diameter is approximately inversely proportional to the wave intensity. The represented model could explain these facts. In calculations the characteristic time of forming of cellular structure t_f grows as α decreases. This is due to proportionality of increment $Re(\Omega)$ to α . The larger real part of Ω the less time t_f and the less relation of critical tube diameter per cell size. For $\alpha = 3$ and $\alpha = 5$ the relations d_c/λ differ approximately two times.

Thus the developed approximate mathematical model for the evolution of gas detonation front helps us to explain the physical essence of certain phenomena. Particularly it shows the great significance of parameter α (proportional to activation energy, γ and intensity of shock wave) for description of the dynamics of gas detcnations.

References

Abouseif G.E., Toong T.Y., 1981 J. Fluid Mech., **103**, pp. 1-22.

Borissov A.A., Sharypov O.V., 1988 European Colloquium on Mechanics 241, Nonlinear Waves in Active Media, Tallinn, USSR, September 27-30, 1988.

Borissov A.A., Sharypov O.V., 1989-a Intern. Forum on Mathematical Modelling and Computer Simulation of Processes in Energy Systems, Sarayevo, Yugoslavia, March 20-24, 1989.

Borissov A.A., Sharypov O.V., 1989-b Izvestiya Sibirskogo Otdel. Akad. Nauk SSSR, Ser. tekhn. nauk, 1989,No.2, pp.50-55 (in Russian) (transl.: Soviet Journal of Applied Physics).

Borissov A.A., Sharypov O.V., 1989-c. 3-d Intern. Seminar on Flame Structure, Alma-Ata, USSR, September 18-22, 1989 (to be published by Nauka, USSR, in 1991).

Borissov A.A., Sharypov O.V.,1990-a. 1st Asian-Pacific Intern. Symposium on Combustion and Energy Utilization, Beijing, China, October 15-18, 1990.

Borissov A.A., Sharypov O.V., 1990-b. 3-d Intern. Conference "Lavrentyev Readings in Mathematics, Mechanics and Physics", Novosibirsk, USSR, September 10-14, 1990.

Clarke J.F., 1978 J. Fluid Mech., **89**, pp. 343-355.

Dyakov S.P., 1954 Zh. Eksperim. Toeret. Phys., **27**, pp. 288-295 (in Russian).

Erpenbeck J.J., 1962 Physics of Fluids, **5**, No 5.

Fujiwara T., Reddy K.V., 1989. 3-d Intern. Seminar on Flame Structure, Alma-Ata, USSR, September 18-22, 1989 (to be published by Nauka, USSR, in 1991).

Kailasanath K., Oran E.S., Boris J.P. and Young T.R., 1985 Combustion and Flame, **61**, pp. 199-209.

Lee J.H.S., 1984 Ann. Rev. Fluid Mech., **16**, pp. 311-336.

Malomed B.A., 1984 Zh. Techn. Phys., **54**, pp. 233-240.

Manzhalei V.I., Mitrofanov V.V., Subbotin V.A., 1974 Phys. Goreniya Vzryva, **10**, No 1, pp. 102-110 (in Russian) (transl.: Combustion, Explosion and Shock Waves).

Markov V.V., 1981 Doklady Akad. Nauk SSSR, **258**, pp. 314-317 (in Russian).

Nakoryakov V.E., Borissov A.A., 1976. Phys. Goreniya Vzryva, 1976, **12**, No 3, pp. 414-422 (in Russian) (transl.: Combustion, Explosion and

Shock Waves).

Nikolayev Yu.A., Topchiyan M.E., 1977 Phys. Goreniya Vzryva, **13**, No 3 (in Russian) (transl.: Combustion, Explosion and Shock Waves).

Pukhnachyov V.V., 1963. Doklady Akad. Nauk SSSR, **149**, pp. 798-801 (in Russian).

Schöffel S.U., 1989 in the book: Finite Approximations in Fluid Mechanics, ed. by E.N.Hirschel.- Verlag, Braunschweig, pp. 351-365.

Sivashinsky G.I., 1977 Acta Astronautica, **4**, pp. 1177-1206.

Taki S., Fudjiwara T., 1978 AIAA Journal, **16**, pp. 73-77.Vasilyev A.A., Mitrofanov V.V., Topchiyan M.E., 1987 Phys.Goreniya Vzryva, **23**, No 5, pp. 109-131 (in Russian) (transl.:Combustion, Explosion and Shock Waves).

Voytsekhovsky B.V., Mitrofanov V.V., Topchiyan M.E., 1963 The Structure of Gas Detonation Front.- Novosibirsk, USSR, Nauka (in Russian).

Westbrook C.K., Urtiew P., 1983 Phys. Goreniya Vzryva, **19**, No 6, pp. 65-76 (in Russian) (transl.: Combustion, Explosion and Shock Waves).

Zaydel' R.M., 1961 Doklady Akad. Nauk SSSR, **136**, pp. 1142-1145 (in Russian).

Zel'dovich Ya.B., Kompaneyets A.S., 1955 The Theory of Detonations.- Moscow, USSR, Gostekhteorizdat (in Russian).

Zel'dovich Ya.B., 1966 Zh. Prikladn. Mekhan. Tekhn. Phys., No 1, pp. 102-104 (in Russian).

Theoretical Description of Direct Initiation of Detonation for One-Step Chemistry*

J.W. Dold † A.K. Kapila ‡ and M. Short †

† *School of Mathematics, University of Bristol, Bristol BS8 1TW, United Kingdom*

‡ *Department of Mathematical Sciences, Rensselaer Polytechnic Institute, Troy, New York 12180-3590, U.S.A.*

Abstract. In this article, a theoretical picture is presented of a sequence of events that ensue when a homogeneous reactive medium is subjected to conditions that make it detonate. For example, it may be struck by a strong shock wave or it may be subjected to a suddenly increased (initial) temperature. It is assumed that the chemistry is thermally sensitive and that the temperatures involved are high enough to generate a significant degree of reactivity. Typically one finds that a form of reaction-wave is generated. If initial temperatures are sufficiently widespread (as is automatically the case after shock impingement) this reaction wave travels supersonically with respect to the sound speed at every point of its structure. It therefore assumes the asymptotic form of a quasi-steady 'weak' detonation—involving no shock wave in its local structure. Only as this wave slows down sufficiently to produce a slightly subsonic flow relative to the wave at some point do characteristics begin to travel faster than the wave, quickly giving rise to a shock-wave. As this shock progressively moves through the wave, it transforms the structure from a weak to a 'strong' Zel'dovich, Von Neuman, Döring detonation wave.

The details of such an evolution are revealed by asymptotic arguments based on a large chemical activation energy in a one-dimensional evolution. Nevertheless, many aspects of the evolution may be expected to apply more generally, at least wherever these conditions prevail in an approximate sense.

* Much of what is presented in this paper may found in a similar contemporary review article [3] which we extend significantly in these pages.

A. A. Borissov (ed.), Dynamic Structure of Detonation in Gaseous and Dispersed Media, 109–141.

1. Introduction

The properties and structure of detonation waves are so rich that, despite the fact that many fundamental concepts are now very old, there are many more questions remaining to be answered. Particularly in the last few decades, there has been a wealth of new and intriguing experimental findings (such as those presented as topical issues by other authors in this book) that await some adequate theoretical description.

A large part of the unfilled gap in theoretical understanding lies in the field of unsteady behaviour [1]. Cellular detonation provides a good example of this, as do problems of deflagration to detonation transition. One of the main difficulties is the large amplitude of changes that occur. Combined with strongly nonlinear responses, typical observed detonation phenomena become very complicated indeed. In the same evolution, it may be necessary to combine weakly nonlinear (or semi-linear) responses with strongly nonlinear structures, moving multiple shock-wave and contact-surface boundaries in more than one dimension, transonic converging and diverging flows, wave focussing and shock collision, instabilities, inhomogeneities in the medium and possibly turbulence. The details of cellular detonation propagation may well involve chaotic interactions. Faced with such complexity it is little wonder that theoretical studies have not yet been able to make substantial progress in describing such behaviour in detail.

Apart from being very difficult to perform accurately, numerical simulations of such phenomena are typically no more helpful than experiments in achieving any significant understanding. They help only to pose the basic theoretical questions which themselves may be very broadly based. The most fruitful form of analysis that can reasonably be expected at this stage is the theoretical examination of particular aspects of any unsteady evolution under conveniently simplifying conditions.

A useful way of theoretically separating some of the most complex interactions is through the use of asymptotic techniques. Where small or large parameters arise in a system of equations (as they often do in combustible systems [2]) this can help to isolate a variety of boundary-layers, or regions in which simplified mathematical sub-problems may arise that can be tackled with greater generality to yield a deeper understanding. Solutions of this sort often shed some further qualitative understanding on more general related problems.

In this spirit, we outline some recent investigations of detonation initiation in a one-dimensional geometry, that have been studied using asymptotic approaches based on a large chemical activation energy in an essentially one-step reaction. This paper adds significantly to, and reproduces much of what has been presented in, an earlier review article [3]. In addition, it draws from a related review [4] and presents some new, as yet unpublished, research ideas. The emphasis in the discussion is on physical interpretation and conceptual ideas so that mathematical details are reduced to a minimum. A number of aspects of detonation phenomena are discussed with the basic purpose of illustrating the way in which a chemically reacting medium can ignite to produce a detonation wave. The exposition is not comprehensive, serving mainly to outline connected ideas rather than to deal with all possible realisations of them. The reader should always bear in mind the more general aspects of the particulars discussed briefly in this paper. They suggest, and often are, subjects for further research.

At the heart of the discussion lies the self-evident notion that detonations arise as a result of chemical and compressibility interactions in a reacting medium. For the purpose of mathematical modelling, it is easiest to think of this medium as consisting of a mixture of ideal gases. Of course other media sustain detonations [2], ranging from non-ideal gases to condensed (liquid or solid) phases of explosive mixtures and even multiphase or amorphous mixtures, but for a brief, qualitatively correct exposition, ideal gases provide a useful basis for understanding many of the phenomena and their mathematical representations.

After introducing an ideal gas model for reacting mixtures, some steadily travelling solution properties are outlined for a chosen (simplest) case where the total number of gas molecules is unchanged by the chemistry. The analysis for changes in the number of molecules is a little more complicated, but would still lead to essentially analogous results. Depending on the ability of the chemistry to proceed at a significant rate, it is shown that reaction waves (or 'fast flames' [5, 6]) of many different speeds are possible.

These may or may not involve shock waves (representing either strong or weak detonations respectively), and may indeed travel at unbounded Mach numbers for suitable behaviour of the chemistry. The word 'propagate' needs to be interpreted carefully in such a context. Over a wide range of wave-speeds, it is shown that solutions describing steady reaction waves are not necessarily unique, having a continuum of possible shock locations separating weak (supersonic) from strong (subsonic) parts of the structure. The analysis of the reaction waves presented here is tied in with more classical steady detonation analyses based on Rankine–Hugoniot relations.

By then focussing attention on a simple one-step model for the chemistry, with a large activation temperature, unsteady behaviour is examined in more detail. It is found that ignition is possible for appropriate ranges of initial and boundary situations, and that the interaction between chemistry and compressibility of the medium during an induction process leads naturally to the emergence of quasi-steadily travelling weak-detonation structures [7]. The movement of these reaction waves arises through spatial variation in the induction process rather than any mechanism for propagation. The paths of such 'induction flames' can be calculated numerically using a fairly simple procedure applied to an asymptotic model [8].

These waves provide a more accurate and more realistic generalisation of the 'spontaneous flame' conjecture of Zel'dovoch [9]. This approach, which is based on finding an induction time at constant volume for a given initial setup, is also generalised in this paper in another very simple way that reveals a picture that is qualitatively more correct. By dropping the insistence on a *constant volume* induction process in defining Zel'dovich's 'spontanteous flame' one finds an alternative definition that approximates the 'induction flames' more correctly by also allowing for the same compressible interaction as one would find in a locally steady or quasi-steady reaction wave.

A natural limitation on the possible mass flux through any weak detonation shows that the reaction wave that emerges in this way cannot normally remain shockless—as the induction flame slows down below the speed at which a part of the flow through it becomes subsonic (the Chapman–Jouget speed), a shock wave forms to transform the basic nature of the reaction wave from a weak to a strong, genuinely self-propagating, detonation structure [10]. In this way, a Zeldovich–von Neumann–Döring detonation [11] is born.

Through such theoretical discussions one is able to observe a full sequence of asymptotically distinct processes that lead to an overall description of detonation initiation for a thermally-sensitive, one-step reacting system. Different forms of chemical progress may involve qualitatively different sequences of events [12].

2. Basic Modelling

2.1 Diffusive effects

A useful point of departure is to take note of an essential property of all realistic gaseous chemical reactions. They invariably involve the rearrangement of atoms between molecules. As such, some appropriate kind of contact between molecules is necessary before any change can take place. Weak molecular collisions are normally not enough—a relatively large amount of energy being required for a collision to tear apart strong molecular bonds in order for the atoms to be free to recombine in a different molecular arrangement, usually accompanied by an overall release of energy.

Since only rare, relatively energetic, thermal interactions between the right sorts of molecules can trigger a chemical change, it must be true that most molecular collisions in gaseous reactants at moderate combustion temperatures (say, about 1500 °K) result in no chemical change at all. It follows that the time taken for a medium to react chemically t_c must, at least, be many times as long as the mean molecular collision time t_m. It follows that the Von Kárman number [13], which measures the ratio of t_c to t_m, must invariably be large,

$$K\acute{a} = \frac{t_c}{t_m} \gg 1. \tag{1}$$

Typical values for $K\acute{a}$ would be of the order of 10^5 or greater. If one now considers any diffusive process (say the average movement of a labelled gas molecule), this is akin to a random walk for which the average distance moved is proportional to the square root of the number of steps or molecular collisions. Thus the velocity of a diffusive process decreases as the inverse square root of this number. After only one or two collisions, the velocity is still of the order of magnitude of the average molecular speed which, in turn, is of the same order of magnitude as the speed of sound.

It follows that a typical value of the Mach number of a diffusive process that takes place over N molecular collisions is simply $M_N^d = 1/N^{1/2}$. For a diffusive process taking place over the chemical time-scale t_c, this Mach number becomes

$$M_{K\acute{a}}^d = 1 \Big/ \sqrt{K\acute{a}} \ll 1. \tag{2}$$

On the other hand, velocities and changes of velocity in detonations are typically of the order of the sound speed or greater. This elementary argument makes it clear that where such changes in velocity take place due to chemical effects, diffusive effects can only play a very small (or negligible) role, and where large velocity changes are strongly influenced by diffusive effects (such as viscosity and thermal conductivity, within the internal structure of a shock wave) too few molecular collisions would be involved for significant chemical changes to occur. Similar arguments can also be applied to non-gaseous mixtures whenever the chemical time-scale is much longer than the time-scale for interaction between neighbouring molecules.

It is clear therefore that it is perfectly realistic to neglect all diffusive effects in examining detonation phenomena, except possibly in the internal structure of shock waves—relative errors being only of the order of $1/K\acute{a}$. Shock waves are adequately treated as discontinuities across which Rankine–Hugoniot jump conditions can be applied at constant chemical composition. These are discussed later. Moreover, while any other part of a detonating structure (apart from a shock-wave) may be in a non-equilibrium state chemically, it will nevertheless be very close to thermodynamic equilibrium.

2.2 Partial inertial confinement

Of key importance for studies of detonation, is the basic principle that interactions between chemistry and compressibility must take place. In order to understand this in a simple way, it is helpful to introduce two spatial scales. Let us imagine a chemical reaction releasing energy in a region that happens to be of the spatial size ℓ_c. This energy tends to disturb the medium around it, with disturbances propagating away at a typical (chemically frozen) speed of sound $a_c = (\gamma P_c/\rho_c)^{1/2}$, where P_c and ρ_c simply represent typical values of pressure P and density ρ. In the chemical time t_c these disturbances would propagate a distance of the order of $\ell_a = t_c a_c$.

There are two possible extremes in comparing the values of the length scales ℓ_c and ℓ_a. If the acoustic length scale ℓ_a is much greater than the chemical length scale ℓ_c, then disturbances are rapidly propagated away from the reacting region. The chemically reacting medium is able to expand easily, pushing aside the surrounding less-rapidly reacting medium with little overall change in pressure. On the other hand, if ℓ_a is much smaller than ℓ_c, then the medium is practically unable to move as the reaction proceeds. The velocity and density of the medium remain unchanged while the pressure increases. This represents a limit of inertial confinement where the medium is unable to move significantly during the time available.

In a typical detonation, significant changes take place in both pressure and density, as well as velocity and temperature, so that one should naturally expect to find a situation between these two extremes. Thus there must, at least, be a partial degree of inertial confinement at some stage with $\ell_a \approx \ell_c$. Only then is it possible to develop a substantial level of interaction between the effects of exothermic chemistry and the compressibility of the medium.

2.3 Ideal gas equations

In general for an ideal gas $\alpha \in \mathcal{A}$ ($\mathcal{A}$ is the set of all chemical species present in a reacting mixture of gases), kinetic theory shows that the difference between the specific heat at constant pressure of α, $C_{P\alpha}$, and the specific heat at constant volume, $C_{V\alpha}$, is

$$C_{P\alpha} - C_{V\alpha} = \frac{\gamma_\alpha - 1}{\gamma_\alpha} C_{P\alpha} = \frac{R}{W_\alpha} \tag{3}$$

where γ_α is the ratio of specific heats, $C_{P\alpha}/C_{V\alpha}$; R is the universal gas constant; and W_α is the molecular weight of α in mass units per mole. We can thus write the useful expression for predicting the value of the specific heat

$$C_{P\alpha} = \frac{\gamma_\alpha}{\gamma_\alpha - 1} \frac{R}{W_\alpha} \tag{4}$$

which shows that $C_{P\alpha}$ generally increases for decreased values of both γ_α and W_α. In this, the classical principle of equipartition of energy may be used to estimate γ_α as being equal to the constant $(f_\alpha + 2)/f_\alpha$, where f_α is the number of degrees of freedom of motion of a molecule of α. For a monatomic, linear or nonlinear gas molecule, f_α is 3, 5 or 6, giving $\gamma_\alpha = \frac{5}{3}$, $\frac{7}{5}$ or $\frac{4}{3}$, respectively. Better estimates, based on quantum or statistical mechanics, are available [14] to show how energy distribution in rotational, vibrational and electronic modes inject some temperature dependence into $C_{P\alpha}$ and γ_α in many real substances.

For a given set of values of the mass fractions y_α $(\alpha \in \mathcal{A})$, one can identify the effective specific heat at constant pressure of the mixture C_P, the number of moles per unit mass n, the effective molecular weight W and the effective ratio of specific heats γ as follows:

$$C_P = \sum_{\alpha\in\mathcal{A}} y_\alpha C_{P\alpha} = R \sum_{\alpha\in\mathcal{A}} \frac{\gamma_\alpha}{\gamma_\alpha - 1} \frac{y_\alpha}{W_\alpha}$$

$$n = \frac{1}{W} = \sum_{\alpha\in\mathcal{A}} \frac{y_\alpha}{W_\alpha} = \frac{\gamma - 1}{\gamma} \frac{C_P}{R} \tag{5}$$

$$\gamma = \frac{C_P}{C_P - nR} = \left(\sum_{\alpha\in\mathcal{A}} \gamma_\alpha \frac{y_\alpha / W_\alpha}{\gamma_\alpha - 1} \right) \Big/ \left(\sum_{\alpha\in\mathcal{A}} \frac{y_\alpha / W_\alpha}{\gamma_\alpha - 1} \right).$$

The values of C_P and γ are now functions of the mixture proportions, and so are not necessarily constant when these change, essentially through chemical reaction. For reactions that preserve the total number of gas molecules, n and W remain unchanged.

The set of conservation equations governing the evolution of pressure P, density ρ, temperature T, velocity $\vec{u}$, specific enthalpy h and the species mass-fractions y_α can be written as follows:

$$\begin{aligned}
&\text{Continuity of Mass:} && \dot{\rho} + \rho \nabla \cdot \vec{u} = 0 \\
&\text{Momentum:} && \rho \dot{\vec{u}} + \nabla P = 0 \\
&\text{Energy:} && \rho \dot{h} = \dot{P} \\
&\text{Ideal Gases:} && P = nR\rho T \\
&\text{Species Production:} && \rho \dot{y}_\alpha = K_\alpha
\end{aligned} \tag{6}$$

where K_α is the rate of chemical production of α measured in terms of mass per unit volume per unit time. The superscript 'dot' notation represents a particle-following or Lagrangian time derivative, e.g. $\dot{\rho} = (\partial/\partial t + \boldsymbol{u} \cdot \nabla)\rho$.

The specific enthalpy h is composed of the specific enthalpies of the component species of the mixture in the manner

$$h = \sum_{\alpha\in\mathcal{A}} y_\alpha \left(h_{0\alpha} + \int_{T_0}^{T} C_{P\alpha}\, dT \right) \tag{7}$$

where $h_{0\alpha}$ is the energy of formation at the reference temperature T_0 and $\int_{T_0}^{T} C_{P\alpha}\, dT$ is the thermal energy of the species α per unit mass. Differentiation leads to the form of the energy equation,

ENERGY: $$C_P\dot{T} = \dot{P}/\rho - \dot{q}$$

where $$\dot{q} = -\sum_{\alpha\in\mathcal{A}} \dot{y}_\alpha \left(h_{0\alpha} + \int_{T_0}^{T} C_{P\alpha}\, dT \right). \tag{8}$$

The 'chemical energy' q that may be released by changes in chemical composition can thus be seen to depend on temperature if $C_{P\alpha}$ is not the same for all species. This is because one species may contain a different amount of thermal energy from another species at the same temperature. However if the overall specific heat C_P remains constant, then the energy per unit mass released by the chemistry, $-[q]$, depends only on changes in the overall energy of formation, $\sum y_\alpha h_{0\alpha}$, of the mixture per unit mass.

It is useful to think of the chemical energy q as being normalised to have a 'base value' of zero at a suitable burnt reference state in chemical equilibrium. Thus the value of q at any stage broadly represents the amount of chemical energy remaining to be released, and $q \to 0$ as the reaction approaches its final completion ($\dot{y}_\alpha \to 0,\ \forall\, \alpha \in \mathcal{A}$). In general, this picture has to be qualified by recognising that q need not change monotonically, and that the state of chemical equilibrium may change for different values of the thermodynamic variables T, P and ρ.

Differentiation of the ideal gas law gives $\dot{P} = nR\rho T(\dot{\rho}/\rho + \dot{T}/T + \dot{n}/n)$. Thus, using the continuity equation and the energy equation, it follows that

$$\frac{C_P}{nR}\dot{P}/\rho + C_P T \nabla\cdot\vec{u} = C_P\dot{T} + C_P T\frac{\dot{n}}{n} = \dot{P}/\rho + C_P T\frac{\dot{n}}{n} - \dot{q} \tag{9}$$

from which one can obtain an equation relating pressure and velocity changes,

COMPRESSIBILITY: $$\frac{1}{\gamma - 1}\dot{P}/\rho + C_P T \nabla\cdot\vec{u} = C_P T\frac{\dot{n}}{n} - \dot{q}. \tag{10}$$

The terms on the right hand side of this equation represent forcing terms for either volumetric expansion, $\nabla\cdot\vec{u}$, or an increase in pressure, $\dot{P}$. The balance of chemical activity that goes into either expansion or pressure must depend upon the particular gasdynamical situation.

It may be of interest to bear in mind that non-chemical effects could equally well provide a similar forcing in equation (10). Thus, for example, radiant heating of a compressible medium (say, interstellar dust or a laser-heated material) or the constriction of a flow in a channel (such as a wind tunnel) could be modelled in a similar way. Some of the phenomena discussed below might thus have relevance to quite different situations.

2.4 Reaction Models

In order to close the system of equations, a set of chemical rate-laws providing expressions for the species production rates K_α is needed. For most practical combustion reactions, this is in fact an entire science in itself. Starting with a set of reactants $\mathcal{F}$, a typical pathway for the progress of the chemistry towards a set of product species $\mathcal{P}$ takes place via a set of intermediate chemical 'radicals' $\mathcal{R}$. The reaction steps $\mathcal{F} \rightleftharpoons \mathcal{R} \rightleftharpoons \mathcal{P}$ sketch this type of overall behaviour, with the arrows $\rightleftharpoons$ typically involving many alternative (often reversible) connections. Although the overall process should release a significant amount of internal energy in a combustible system, some reaction steps may be endothermic and there is no guarantee that the release of chemical energy would proceed monotonically. Likewise, the number of moles per unit mass n may also change non-monotonically as some reactions act to increase, while others decrease, the total number of molecules.

To model all of this in a simple, yet general, way is not a straightforward task. One approach is to assume that the reaction may be characterised by a single, suitably normalised, chemical-state parameter λ (possibly, but not necessarily, restricted to the range $\lambda \in [0, 1]$). This reduces the total number of variables y_α that characterise the chemical state to the one variable λ by an assumption that $y_\alpha = \tilde{y}_\alpha(\lambda)$ under all circumstances. The chemical energy q and specific mole-number n are then (possibly non-monotonic) functions of λ :

$$q = \tilde{q}(\lambda) \quad \text{and} \quad n = \tilde{n}(\lambda). \tag{11}$$

With an approach of this sort, the rate of progress of the chemical reaction can typically be expressed in the the form $\dot{\lambda} = -f(\lambda, T, \rho)$.

If only a single chemical reaction-step $F \rightarrow P$ were to take place, converting reactant directly into product, then λ could (for example) be taken to represent a scaled value of the reactant mass-fraction y_F, decreasing from (say) 1 to 0 as the reaction proceeds. In this case, the rate of chemical reaction would be described by an Arrhenius law, typically of the form

$$\dot{\lambda} = -f(\lambda, T, \rho) = -\frac{\lambda}{t_0} \exp\left(-T_A/T\right) \tag{12}$$

where the constants T_A and t_0 are the activation temperature of the reaction and a time that reflects the rate of chemical change at very high temperatures, respectively. This time is generally longer than the mean molecular collision time t_m. A fuller reaction scheme normally involves rate laws for each reaction step that are generalisations of equation (12).

A fairly simple model for the chemistry can now be posed by using equation (12) as a *representative* rate-law for the chemical-state parameter λ, without necessarily assuming that the chemistry proceeds in only one step (for which the functions $\tilde{q}$ and $\tilde{n}$ would be simple linear functions of λ). This model will not, however, be used explicitly until unsteady effects are considered in section 4. For equation (12) to reflect the general properties of the progress of most more complex chemical schemes it is necessary for the activation temperature T_A to be large enough to make $\dot{\lambda}$ a sensitive function of temperature. This is achieved by insisting that

$$T_A \gg T \tag{13}$$

for all temperatures of interest. This property can provide the basis for an asymptotic analysis of equations (4)–(12).

It remains only to specify the functions $\tilde{q}$ and $\tilde{n}$ of equation (11). Since these terms play a similar role in the compressibility equation (10), all qualitative features can be modelled using only $\tilde{q}$. For this reason, we assume that $\dot{n} \equiv 0$, and for a similar reason we take C_P and γ to remain constant. Different variations of q with λ can lead to different detonation structures and evolutions, so that it is best at this stage to leave $\tilde{q}(\lambda)$ unspecified. While being a good representative model for a detonable medium, the equations presented above may easily be generalised. It should be noted, however, that it is not always possible to identify a single reaction-progress parameter λ for unsteady systems, and that different forms of evolution may arise for different chemical models [12].

2.5 One dimensional Lagrangean coordinate

A 'stream-function' or mass-coordinate in one dimension, that remains fixed for any fluid element or particle, can be defined as

$$\psi = \int_{x_0(t)}^{x} \rho \, dx \tag{14}$$

where the position $x = x_0(t)$ moves with the local fluid velocity, and time is held fixed in the integration. It is also convenient to introduce the specific volume $V = 1/\rho$. Using subscript notation to represent partial derivatives, the model equations then become

$$\begin{aligned} V_t - u_\psi &= 0 \\ u_t + P_\psi &= 0 \\ VP_t + \gamma P u_\psi &= \gamma PV \frac{n_t}{n} - (\gamma - 1) q_t \\ T &= \frac{PV}{nR} \end{aligned} \tag{15}$$

and

$$\lambda_t = -f(\lambda, T, \rho) = -\frac{\lambda}{t_0} \exp(-T_A/T)$$

with

$$q = \tilde{q}(\lambda) \quad \text{and} \quad n_t = 0$$

the latter equation being modelled as shown for simplicity. Usefully, if q_t, n_t and γ are specified, then the first three of these equations become decoupled from the rest so that it is possible to express thermodynamic changes only in terms of changes in q, n and γ, without explicitly requiring a model for the chemistry.

3. Steady reaction waves or fast flames

This fact is particularly useful in seeking solutions representing steady waves. For this purpose, each of the thermodynamic and chemical variables can be considered to behave as a function of a phase coordinate Ξ in the manner

$$f = f(\Xi) \quad \text{with} \quad \Xi = \psi - mt \tag{16}$$

where m denotes the steady mass flux of material passing through a wave travelling from negative to positive values of ψ. The analysis of this type of reaction wave is not new [1], but is presented here for the primary purpose of stressing some aspects that are normally glossed over in more conventional presentations. The most important of these are the possibility of finding uniformly supersonic 'weak' detonation solutions, and the non-unique nature of solutions that may involve a shock-transition between 'weak' and 'strong' parts of the same steady wave. As quasi-steady reaction-waves, these structures are very important for understanding the later unsteady analysis of the initiation of detonation.

With the form of behaviour (16) (and with n and γ constant), the first four of equations (15) are easily solved to give the relations

$$\begin{aligned}
P - P_0 &= -m^2 (V - V_0) \\
u - u_0 &= \frac{P - P_0}{m} \\
(\gamma - 1)m^2(q_0 - q) &= \left(P_0 + V_0 m^2\right)(P - P_0) - \frac{\gamma + 1}{2}\left(P^2 - P_0^2\right) \\
T &= \frac{P}{nRm^2}\left(P_0 + V_0 m^2 - P\right) \\
\text{and} \qquad \mu^2 &= \frac{m^2 V}{\gamma P} = \frac{1}{\gamma}\left(\frac{P_0 + V_0 m^2}{P} - 1\right)
\end{aligned} \tag{17}$$

for any suitable reference values P_0, V_0, q_0 and u_0. The parameter μ represents the mach number of the flow through the wave at any point and is included for future reference.

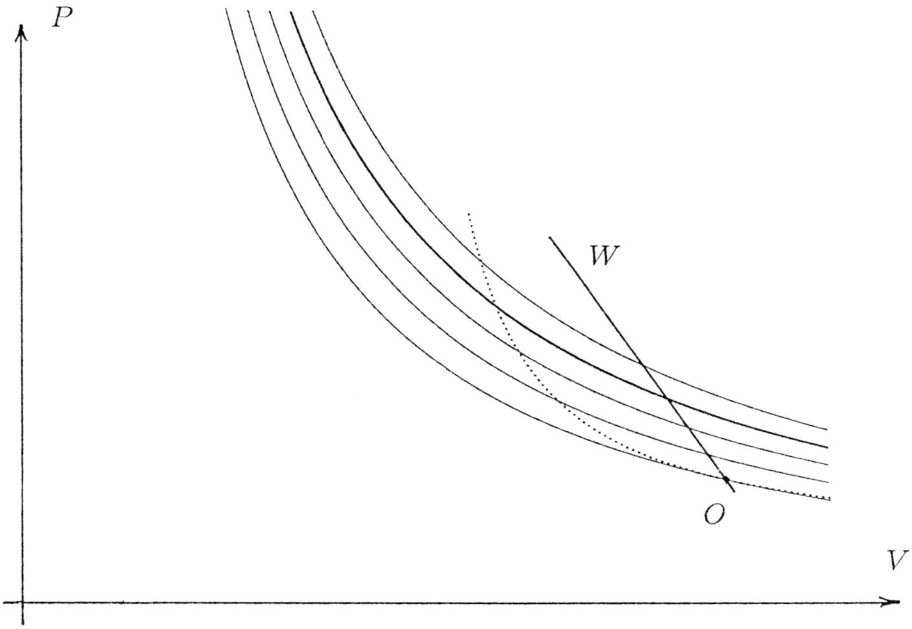

Figure 1: *Rankine–Hugoniot diagram based on equation (18) showing a burnt Hugoniot curve (thick curve) for* $q = 0$ *and partially reacted Hugoniot curves (thin curves) for different values of* q. *The dotted curve traces sonic points,* $\mu = 1$, *and the straight line* $O - W$ *represents a typical Rayleigh line.*

The first of these equations represents a Rayleigh line, while elimination of m^2 between the first and third equations leads to the Hugoniot relation

$$\frac{\gamma+1}{\gamma-1}(PV - P_0V_0) + P_0V - V_0P = 2(q_0 - q). \tag{18}$$

It follows that the thermodynamic variables in the wave are related throughout by Rankine-Hugoniot relations, as obtained for the appropriate value of the chemical energy q at any point—these are sketched in figure 1. This result emphasises the fact that, while being far from chemical equilibrium, we may consider chemical and gasdynamic interactions of this type to occur in a state of local thermodynamic equilibrium. Exactly the same principle would apply for non-constant values of n and γ.

It may be noted that the chemical energy q and temperature T are given in the relations (17) by simple (but different) quadratics in pressure. A useful way of redrafting these relations

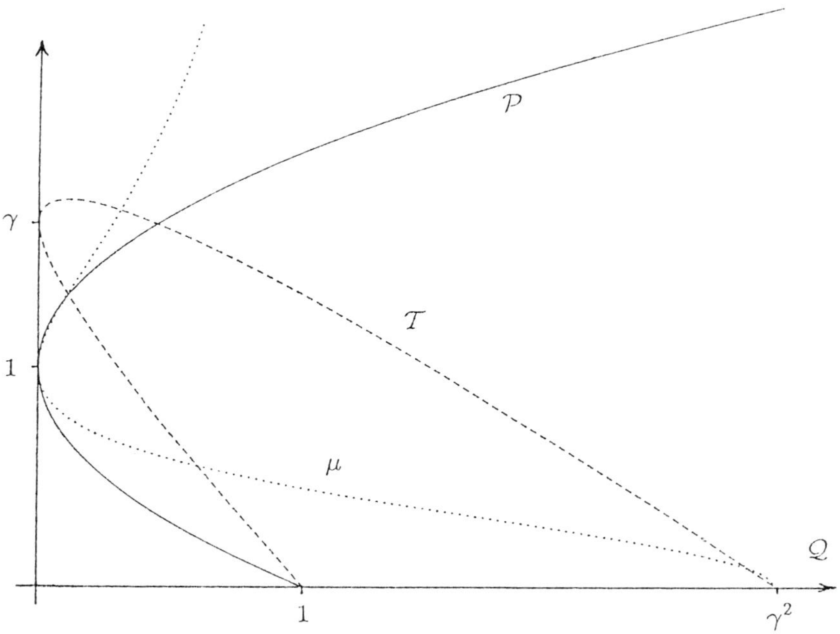

Figure 2: *Temperature (dashed curve), pressure (solid curve) and Mach number (dotted curve) variations with chemical energy through a steady reaction wave.*

is found by defining the dimensionless parameters

$$\mathcal{P} = P \Big/ \frac{P_0\,(1+\gamma\mu_0^2)}{\gamma+1}, \qquad \mathcal{T} = T \Big/ \frac{P_0^2\,(1+\gamma\mu_0^2)^2}{(\gamma+1)^2 nRm^2}$$

and

$$\mathcal{Q} = \mathcal{Q}_b + (q - q_b) \Big/ \frac{P_0^2\,(1+\gamma\mu_0^2)^2}{2(\gamma^2-1)m^2} \tag{19}$$

where $\mathcal{Q}_b = 1 + \dfrac{\gamma+1}{(1+\gamma\mu_0^2)^2}\left(\gamma - 1 - 2\gamma\mu_0^2\right) - \dfrac{2(\gamma^2-1)}{(1+\gamma\mu_0^2)^2}\,\dfrac{(q_0-q_b)m^2}{P_0^2}$,

μ_0 being the flow Mach number through the wave at the reference values, and q_b is the final burnt value of the chemical energy q. The relations (17) now become simply

$$\begin{aligned} \mathcal{T} &= \mathcal{P}\,(\gamma + 1 - \mathcal{P}) \\ \mathcal{Q} &= (\mathcal{P} - 1)^2 \\ \mu^2 &= \frac{\gamma + 1 - \mathcal{P}}{\gamma\mathcal{P}}. \end{aligned} \tag{20}$$

Values for $\mathcal{P}$, $\mathcal{T}$ and μ^2 are plotted against values of $\mathcal{Q}$ in figure 2. It can be seen that two branches of solution exist. The subsonic upper (deflagration or 'strong' detonation

[2]) branch, with $\mu^2 < 1$, involves higher pressures and temperatures for any given value of $\mathcal{Q}$, while the supersonic lower (or 'weak' detonation) branch, with $\mu^2 > 1$, involves lower pressures and temperatures. Reaction wave solutions would generally follow one or other of these branches as $\mathcal{Q}$ changes through the wave towards a final stable chemical equilibrium value. In equations (19) this is defined as $\mathcal{Q}_b$, a final value that is, in general, a function of the wave speed or mass flux m through the reaction-wave. Thus different speeds generally lead to different final values. Clearly no speed is possible that would make $\mathcal{Q}_b < 0$, since real solutions only exist for $\mathcal{Q} \geq 0$.

A variety of solutions is then possible, provided only that a chemical reaction can actually proceed to drive $\mathcal{Q}$ towards $\mathcal{Q}_b \geq 0$, not necessarily monotonically, in the range $\mathcal{Q} \geq 0$. A chemical change that releases energy would cause an overall decrease in $\mathcal{Q}$ even though local increases may occur in some parts of the wave. It is interesting to note that temperatures on the strong detonation or deflagration branch do not increase monotonically as $\mathcal{Q}$ decreases. A point of maximum temperature occurs just as the flow through the wave reaches the isothermal sound speed, $\mu^2 = 1/\gamma$. Any subsequent release of chemical energy is converted into kinetic rather than thermal energy. Moreover, it causes some of the existing thermal energy also to be converted into kinetic energy.

Figure 2, together with the scalings (19), provides a more detailed unravelling of possible Rayleigh line variations occuring in a more conventional Rankine–Hugoniot diagram (as illustrated by the line $O - W$ in figure 1).

There are two ways in which solutions can swop between the deflagration and weak detonation branches. If $\min\{\mathcal{Q}\} = 0$ with $\mathcal{Q}_b > 0$, then as $\mathcal{Q}$ decreases to zero and increases again, solutions can move smoothly from one branch to the other, passing through a sonic point at $\mathcal{Q} = 0$. This exothermic–endothermic change between branches can generally only happen for a unique 'eigenvalue' mass flux, which we denote by m_S (in order to emphasise the presence of a sonic point in the flow).

The other means of changing from one branch to the other is via a shock-wave (across which $\mathcal{Q}$ remains constant). This, of course, is only able to *increase* pressures so that it can only cause a change from the weak detonation to the deflagration or strong detonation branch. Thus a solution can start by moving along the lower (weak) branch until, at some value of $\mathcal{Q}$, the solution jumps to the upper branch where the solution continues to evolve as $\mathcal{Q}$ continues to change. Such solutions would remain on the deflagration branch for all mass fluxes that give $\min\{\mathcal{Q}\} > 0$. The condition $\min\{\mathcal{Q}\} = 0$ (or $m = m_S$) would identify the lower bound for such mass fluxes. Curiously, if $\mathcal{Q}_b > 0$, an exothermic–endothermic change could then, potentially, return the solution once again to the lower branch from which yet another shock wave increase could not be ruled out.

Unless the reaction-rate actually is zero at some unburnt stable equilibrium point on the weak detonation branch, all such solutions involving a shock-jump from the weak branch to the strong branch are non-unique. There is no criterion that fixes the value of Q at which the shock-wave would lie, so that a continuum of steady reaction-wave structures involving both weak detonation and deflagration parts is possible for all mass fluxes in the range $m \geq m_S$.

According to these arguments, solutions describing both weak and strong detonations are possible for all mass fluxes bounded below by m_S provided only that chemical changes can realistically take place at 'weak' temperatures and pressures. Moreover, mixed reaction waves are also possible with shock-waves separating weak from strong (or deflagration) parts of the structure. We make no attempt here to discuss the stability of such structures (or any more exotic variations that may be possible for particular exothermic–endothermic chemical changes). Rather, we find in the next section that these solutions can arise in an unsteady or quasi-steady evolution in which they may have some transient existence for which questions of stability are not particularly relevant.

Propagation: At this point it is worth commenting on the role of mechanisms for propagation in these waves. In the strict interpretation of the word, 'propagation' does not simply denote movement, but requires some action on the part of the object in question that brings about a reduplication of the object at some other point and time. Since information in this compressible system cannot travel faster than the local speed of sound, this kind of action is only possible for the subsonic deflagration or strong detonation reaction waves. These 'fast flames' do genuinely propagate.

On the other hand, no information can travel forwards through weak detonation waves. Being uniformly supersonic, these very-fast flames do not really propagate. Their movement is entirely attributable to preset conditions in the form of (say) initial data and/or upstream boundary conditions in the leading parts of the wave (e.g. as $\psi \rightarrow \infty$). It is helpful to imagine such a fast flame as representing either a convected explosion [6] or a spatially non-uniform explosion process [9]. An example of the latter is considered in the next section.

4. Unsteady behaviour

4.1 Induction

We will now invoke the reaction-rate model (12), requiring only that it should hold true for relatively small temperature changes, of the order of $T_i^2/T_A \ll T_i$, about some reference temperature T_i. This and the other reference values, P_i, V_i, u_i, q_i and $\lambda_i = 1$, may be taken to typify the state of a medium behind an initiating shock-wave that triggers a chemical reaction into life, or it may represent peak initial values in the conditions produced by bulk compression or radiant heating of an explosive medium, either in a container or in an unbounded region. At this state, the medium would explode in a time of order t_i (to be specified below).

A dimensionless asymptotic form of the chemical and compressible model is now usefully expressed in terms of the dimensionless time and mass-coordinates

$$\tau = \frac{t}{t_i} \qquad \text{and} \qquad \chi = \frac{\psi}{t_i\sqrt{\gamma P_i/V_i}}. \tag{21}$$

With these as independent variables, the thermodynamic parameters may be considered to satisfy the following asymptotic relations

$$\begin{aligned} \frac{T}{T_i} &\sim 1 + \frac{T_i}{T_A}\phi(\tau,\chi) \qquad & u &\sim u_i + \sqrt{\gamma P_i V_i}\,\frac{T_i}{T_A}v(\tau,\chi) \\ \frac{P}{P_i} &\sim 1 + \gamma\frac{T_i}{T_A}p(\tau,\chi) \qquad & \lambda &\sim 1 - \frac{C_P T_i}{\tilde{q}'(1)}\frac{T_i}{T_A}\ell(\tau,\chi). \end{aligned} \tag{22}$$

After substituting these rescalings into equations (15) and neglecting lower order terms, one obtains the set of equations

$$\begin{aligned} \phi_\tau &= (\gamma-1)p_\tau + \ell_\tau \\ p_\tau + v_\chi &= \ell_\tau \\ v_\tau + p_\chi &= 0 \\ \ell_\tau &= e^\phi \end{aligned} \tag{23}$$

after selecting the following definition for t_i

$$t_i = t_0\,\frac{C_P T_i}{\tilde{q}'(1)}\frac{T_i}{T_A}\exp(T_A/T_i). \tag{24}$$

The perturbation model (23) describes the small-scale behaviour of thermodynamic variables in the presence of (at least) a partial inertial confinement.

In general, these equations would need to be solved subject to an appropriate specification of initial and boundary conditions. There are clearly many possible forms for this, representing different physically imposed conditions as suggested earlier. For the present discussion, it is simplest to suppose only that initial conditions

$$\begin{aligned} \phi(0,\chi) &= \phi_0(\chi), & u(0,\chi) &= u_0(\chi), \\ p(0,\chi) &= p_0(\chi) \quad \text{and} & \ell(0,\chi) &= \ell_0(\chi) \end{aligned} \tag{25}$$

are imposed over an unbounded domain $\chi \in (-\infty, \infty)$.

Assuming appropriate differentiability, equations (23) can be condensed into a single differential equation for ϕ, namely Clarke's equation [15]

$$\left(\phi_\tau - \gamma e^{\phi}\right)_{\tau\tau} = \left(\phi_\tau - e^{\phi}\right)_{\chi\chi}. \tag{26}$$

By associating different terms in this equation, one can observe some of the rich structure of possible interactions and processes taking place as the medium reacts. These include the presence of the adiabatic sound speed of unity; the isothermal sound speed of $\gamma^{-1/2}$; constant pressure blowup if $\partial_\chi \gg \partial_\tau$; constant density blowup if $\partial_\tau \gg \partial_\chi$; and possible transition from the former to the latter [16].

Integration along characteristics τ and $\chi \pm \tau$ leads to the integral forms of equations (23),

$$\begin{aligned} p + v &= R_0(\chi - \tau) + \int_0^\tau e^{\phi(s,\chi-\tau+s)}\,ds \\ p - v &= L_0(\chi + \tau) + \int_0^\tau e^{\phi(s,\chi+\tau-s)}\,ds \\ \phi &= (\gamma - 1)p + S_0(\chi) + \int_0^\tau e^{\phi(s,\chi)}\,ds \end{aligned} \tag{27}$$

where $\qquad R_0 = p_0 + v_0, \quad L_0 = p_0 - v_0,$

$\qquad$ and $\quad S_0 = \phi_0 - (\gamma - 1)p_0.$

The functions R_0, L_0 and S_0 represent right travelling, left-travelling and stationary (with respect to any fluid element) thermodynamic waves, respectively. By eliminating p and v, equations (23) can now also be condensed into a single integral equation for ϕ,

$$\begin{aligned} \phi \;=\; & \int_0^\tau \left[e^{\phi(s,\chi)} + \frac{\gamma - 1}{2}\left(e^{\phi(s,\chi-\tau+s)} + e^{\phi(s,\chi+\tau-s)} \right) \right] ds \\ & + S_0(\chi) + \frac{\gamma - 1}{2}\left[R_0(\chi - \tau) + L_0(\chi + \tau) \right]. \end{aligned} \tag{28}$$

Under conditions where p_0, v_0 and ϕ_0 are uniformly bounded from below within a suitable range of characteristics, this equation shows that ϕ would remain bounded from below by the singular function $\widehat{\phi} = -\ln(\widehat{\tau}_I - \tau)$ for some appropriate constant $\widehat{\tau}_I < \infty$. It follows that ϕ would become unbounded, $\phi(\tau_I, \chi_I) \to \infty$, at some location χ_I in the finite time $\tau \to \tau_I < \widehat{\tau}_I$.

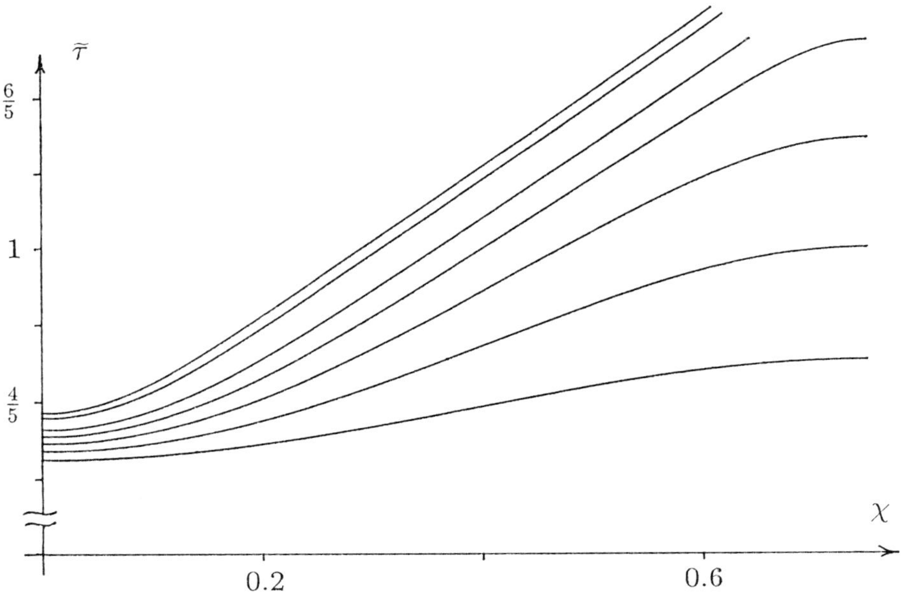

Figure 3: *Calculated curves of* $\tilde{\tau}(\chi)$ *for* $\gamma = 1.4$ *with the initial conditions* $p_0 \equiv v_0 \equiv \ell_0 \equiv 0$ *and* $\phi_0(\chi) = \frac{\kappa}{2}\left[\cos\left(\pi\chi/\frac{3}{4}\right) - 1\right]$ *for* $\kappa = 0.2,\ 0.4,\ 0.6,\ 0.8\ 1.0,\ 1.4$ *and* 1.6.

This is highly significant for the emergence of detonation in the context of the present discussion. However, it is equally important to note that ϕ does not only become unbounded at an isolated point; it also grows to infinity at other locations. In fact, so long as the integrations in (28) are suitably well-defined, there must exist a path or continuum of singularity times $\tau = \tilde{\tau}(\chi)$ at which ϕ becomes unbounded:

$$\phi(\tau, \chi) \to \infty \quad \text{as} \quad \tau \to \tilde{\tau}(\chi). \tag{29}$$

The time τ_I simply represents the first time of singularity, $\tau_I = \min\{\tilde{\tau}(\chi)\}$. Some numerical results for the flame-path $\tilde{\tau}$, are reproduced in figure 3 for a sample set of periodic initial conditions.

This singularity path has certain important properties. Firstly, since the integrals in (28) can only increase along any characteristic, $\chi = \chi_0 \pm (\tau - \tau_0)$, it follows that the speed of the singularty path $1/\tilde{\tau}'$ must be supersonic, that is greater than unity:

$$\tilde{\tau}'(\chi) < 1. \tag{30}$$

Also, towards the first point of singularity, the gradient $\tilde{\tau}'(\chi)$ tends to zero, so that the path of singularity first appears with *infinite* speed. Using results found for the shape of the function ϕ as $\tau \to \tau_I$ [16, 17], and the local solutions (32) below, the singularity path can typically behave in one of at least the following two ways near its initial point:

$$\begin{aligned} &\tilde{\tau} \sim \tau_I + \kappa_p \chi^{(2\gamma-1)/\gamma} \\ \text{or} \quad &\tilde{\tau} \sim \tau_I + \kappa_i (\chi - \chi_I)^2 \end{aligned} \tag{31}$$

for some constants κ_p and κ_i. The former behaviour arises when the singularity evolves from an initial temperature maximum at a rigid boundary with non-zero gradient (as typically found if the reaction is triggered by a strong piston-driven or reflected shcck wave). The latter behaviour arises from a temperature distribution with zero gradient at its peak value.

Because e^{ϕ} becomes very large close to $\tau = \tilde{\tau}$, the behaviour close to the singularity is locally quasi-steady. As a result, the equations are readily solved in this region to yield the asymptotic structures of ϕ, v, p and ℓ close to the singularity

$$\begin{aligned} \phi &\sim -\ln\left(\frac{\gamma - \tilde{\tau}'^2}{1 - \tilde{\tau}'^2}\varpi\right) + F\varpi \ln \varpi \\ \phi &\sim \left(\gamma - \tilde{\tau}'^2\right) p + a + A\varpi \ln \varpi \\ \ell &\sim \left(1 - \tilde{\tau}'^2\right) p + b + B\varpi \ln \varpi \\ v &\sim \tilde{\tau}' p + c + C\varpi \ln \varpi \end{aligned} \tag{32}$$

where $\varpi = \tilde{\tau} - \tau$, and

$$\begin{aligned} A = B &= \frac{\gamma + 3\tilde{\tau}'^2}{(\gamma - \tilde{\tau}'^2)^2}\tilde{\tau}'' \\ C &= -\frac{2\tilde{\tau}'\tilde{\tau}''}{(\gamma - \tilde{\tau}'^2)^2} \\ F &= -\frac{\gamma - 1}{2}\frac{A}{1 - \tilde{\tau}'^2}. \end{aligned} \tag{33}$$

The functions $a(\chi)$, $b(\chi)$ and $c(\chi)$, as well as the function $\tilde{\tau}(\chi)$, must depend on the set of boundary and initial conditions that are relevant to the particular situation being modelled.

One means of numerically calculating the singularity path as well as these functions is to introduce a parametric coordinate $\nu(\chi, \tau)$ in such a way [8] that $\phi \to \infty$ as $\nu \to \infty$.

Expressed in terms of χ and ν as independent coordinates (using subscripts to represent partial derivatives), the model (23) can be written in the form

$$\begin{aligned} p_\nu &= \frac{\tau_\nu}{1-\tau_\chi^2}\left(e^\phi - \tau_\chi p_\chi - v_\chi\right) \\ v_\nu &= \frac{\tau_\nu}{1-\tau_\chi^2}\left(\tau_\chi e^\phi - \tau_\chi p_\chi - v_\chi\right) \\ \ell_\nu &= \tau_\nu e^\phi \\ \phi_\nu &= \ell_\nu + (\gamma-1)p_\nu. \end{aligned} \tag{34}$$

Integration with respect to ν can then be carried out once a suitable definition is found for τ_ν. There are many ways in which this can be done in order that the singularity path is approached slowly as $\nu \to \infty$. In particular one must have $\tau_\nu \to 0$ in this limit. The numerical results shown in figure 3 were obtained in this way.

4.2 Generalising Zel'dovich's 'Spontaneous Flame' Concept

It is easily seen from the scalings (21), (22) and (24) and the set of equations (23) and (25) that a uniform mixture, in which no spatial (or material) gradients appear, will produce a singular growth in ϕ uniformly in the finite time

$$\tilde{\tau} = \gamma^{-1}\exp(-\phi_0). \tag{35}$$

Without loss of generality, if T_i is selected to represent the initial temperature then $\phi_0 = 0$, so that one can identify a large activation temperature asymptotic estimate for the constant volume explosion time as

$$\tilde{t} = \frac{t_0}{\gamma}\frac{C_P T_i}{\tilde{q}'(1)}\frac{T_i}{T_A}\exp(T_A/T_i). \tag{36}$$

For substances that are able to react after being provided with some sufficiently high initial temperature, Zel'dovich [9] proposed that one could generalise this formula to apply to situations in which the initial temperature is not spatially constant $T(0,\psi) = T_i(\psi)$. In this case, $\tilde{t}(\psi)$ would represent a fairly simple approximation to the path in space and time of a 'spontaneous' reaction wave. By differentiating equation (36) one finds Zel'dovich's expression for the 'spontaneous' flame-speed or more correctly, in this case, mass-flux

$$\tilde{m}(\psi) = \left(\frac{d\tilde{t}}{d\psi}\right)^{-1} = \frac{\gamma\tilde{q}'(1)}{t_0 C_P}\frac{T_A}{2T_i - T_A}\exp(-T_A/T_i)\times\left(\frac{dT_i}{d\psi}\right)^{-1}. \tag{37}$$

Given such an expression for the speed of a flame, based only on initial conditions, Zeldovich was able to conjecture that a Zeldovich–von Neumann–Döring detonation would be formed close to the point where this speed falls below the speed that a Chapman–Jouget detonation would have under the same local conditions. We will be examining this conjectured transition in greater detail later.

There is a very clear parallel between the 'spontaneous' path (36) and the singularity path $\tilde{\tau}$ which arises from fully solving the reactive–acoustic perturbation equations (23). Because they arise from a fuller model without assuming local constant volume behaviour, such a solution is a more accurate representation of the actual path of any reaction wave that emerges. Also, equations (23) may be solved not only for a set of initial conditions, but under such moving boundary conditions as may (for example) represent the initiation of a reaction through shock-heating where it is not immediately possible to identify an *initial* temperature distribution $T_i(\psi)$. Moreover, as pointed out above, the singularity path $\tilde{\tau}$ is always supersonic for any reasonable boundary and initial conditions while, on the other hand, the 'spontaneous' speed (37) can have any value for a suitably chosen initial temperature so that it fails to reproduce this significant feature.

However, it is a simple matter to obtain a generalisation of Zel'dovich's approach that does reproduce this feature. Examining equations (23) or the asymptotic solutions (32) it is again easy to note that a factor of $(\tilde{m}^2 - P_i/V_i)/(\tilde{m}^2/\gamma - P_i/V_i)$ enters into the relationship between ϕ_τ and ℓ_τ for any reaction-wave of constant mass-flux $\tilde{m}$. A natural way of extending equations (36) and (37) is therefore to define

$$\tilde{t} = \frac{\tilde{m}^2 - \gamma P_i/V_i}{\tilde{m}^2 - P_i/V_i} \times \frac{t_0}{\gamma}\frac{C_P T_i}{\tilde{q}'(1)}\frac{T_i}{T_A}\exp(T_A/T_i)$$

$$\text{with} \qquad \tilde{m} = \left(\frac{d\tilde{t}}{d\psi}\right)^{-1} \tag{38}$$

The explosion time $\tilde{t}(\psi)$ now depends on the spontaneous mass-flux $\tilde{m}(\psi)$ in such a way that $\tilde{t}$ is prevented from growing too rapidly (thereby decreasing $\tilde{m}$) when the mass-flux $\tilde{m}$ decreases towards the local acoustic impedance $(\gamma P_i/V_i)^{1/2}$. Thus, with this generalisation, it can be seen that the earlier constant-volume estimate (36) is reproduced for strongly supersonic reaction waves, $\tilde{m}^2 \gg \gamma P_i/V_i$, but that some accounting for compressible interactions prevents the wave from slowing down below sonic speeds. This generalisation is therefore in better qualitative agreement with the results of fuller induction-period examinations.

It is worth noting that the system (38) is equivalent to a nonlinear first order ordinary differential equation for $\tilde{t}(\psi)$. It is therefore not as simple as Zel'dovich's formula (36) for obtaining explicit expressions, although any results should be more realistic. It is also simpler than the partial differential equation system (23). Another feature of note is the exclusion of mass-fluxes for waves with Mach numbers between the adiabatic and isothermal levels when $\gamma^{-1} \leq \tilde{m}^2/(\gamma P_i/V_i) \leq 1$. Also, for initial conditions that would produce strongly subsonic waves, $\tilde{m}^2 \ll P_i/V_i$, the formula selects constant pressure induction times, which are again more realistic for such slow reaction waves.

4.3 Induction flame; weak detonation

Naturally, as the perturbation quantities ϕ, p, v and ℓ become too large, they lose their physical meaning. However, the quasi-steady behaviour (32) near the path of singularity remains significant in that it shows the emergence of a chemical reaction wave involving large-scale variations in thermodynamic and chemical variables. Provided only that the reaction rate $f(\lambda, T, \rho)$ remains much larger than its value during the induction stage of the evolution, the relations (17) describe the quasi-steady variation of temperature, pressure etc. through the wave. Because this strongly active chemical wave and its movement arise as a direct result of a spatially varying induction process, it is natural to describe it as an 'induction flame.'

Selecting $P_0 = P_i$, appropriate values for the reference parameters V_0, q_0 and u_0, and the mass flux m, that determine the thermodynamic and chemical variation through the flame, are obtained by asymptotic matching with the local singularity structure (32) in the limit $T_i/T_A \to 0$. This gives

$$\begin{aligned} P_0 &= P_i, & m(\psi) &\sim \sqrt{\gamma P_i/V_i}\Big/\tilde{\tau}'(\chi), \\ V_0(\psi) &= \frac{nRT_0}{P_i} \quad \text{where} & T_0(\psi) &\sim T_i\left(1 + \frac{T_i}{T_A}a(\chi)\right), \\ q_0(\psi) &\sim q_i - C_P T_i \frac{T_i}{T_A} b(\chi) \quad \text{and} & u_0(\psi) &\sim u_i + \sqrt{\gamma P_i V_i}\,\frac{T_i}{T_A} c(\chi) \end{aligned} \tag{39}$$

in which m would generally vary significantly as a function of ψ. The remaining functions, V_0, T_0, q_0 and u_0, are in fact constant to leading order as $T_i/T_A \to 0$.

Since the mass flux through the induction flame, m, starts off being infinite (where $\tilde{\tau}'$ is zero), and therefore must be very large at least for a short while, the induction flame initially has the structure of a weak detonation. In figure 2, it would be described by thermodynamic variations along the lower curves. On the Rankine–Hugoniot diagram, figure 1, it would be described by changes that take place along a continuum of Rayleigh–lines, such as $O - W$, that gradually decrease in slope after initially being vertical. Although steady solutions that begin on the (supersonic) weak detonation branch can, in principle, be lifted by a shock-wave to the (subsonic) deflagration branch, there is no mechanism in the early progress of the induction flame that would create such a shock.

Thus, although thermodynamic changes generally increase the local sound speed as the combustion proceeds, the induction flame passes through the medium at a rate that is uniformly supersonic. In the sense discussed earlier, it does not strictly 'propagate' itself forwards since information cannot travel faster than the flame. The only way in which compressibility and chemistry can interact to determine the speed and structure of the induction flame is during the induction process, before the flame has actually appeared at any location.

One vital feature that is worth highlighting before proceeding further is the thermodynamic state of the medium at the point at which the compressible forcing terms in equation (10) approach zero (possibly by changing sign). Since we are modelling n as remaining constant, the chemical energy q approaches a minimum value $\widehat{q}$ at this stage. For a reaction in which q decreases monotonically, $\widehat{q}$ would be the same as the final burnt value q_b of q (which may, possibly, be normalised to zero). Provided the chemical reaction approaches this point much more rapidly than the rate of reaction during the induction process, the thermodynamic state is determined by the intersection between the appropriate Rayleigh line and the weak detonation part of the Hugoniot curve corresponding to $\widehat{q}$ in figure 1. From equations (17), it is determined by the lower root of the equation determining the chemical energy q, as q approaches $\widehat{q}$. This gives

$$P = \widehat{P} = \frac{P_0 + V_0 m^2}{\gamma + 1}\left(1 - \sqrt{1 + \frac{(\gamma^2 - 1)P_0^2 - 2(\gamma + 1)m^2(\widehat{Q} + P_0 V_0)}{(P_0 + V_0 m^2)^2}}\right) \tag{40}$$

where $$\widehat{Q} = (\gamma - 1)(q_0 - \widehat{q}\,).$$

Other thermodynamic quantities are then determined by the remainder of equations (17).

As the mass flux m decreases, one finds that $\widehat{P}$ increases until the point is reached where the two roots become equal—there are no real roots for lower values of m. Equation (40) and the last of equations (17) make it clear that this happens precisely at a sonic point ($\widehat{\mu}^2 \to 1$). As in section 3, we use the subscript S to denote the state at this sonic critical point, obtaining values for P_S, V_S, u_S, T_S and m_S.

4.4 Shock formation

At this point, the induction flame is supersonic with respect to the sound speed in the medium ahead of it, where an unsteady induction process still prevails. Thus the critical mass flux m_S can be attained for some perfectly realistic value of $\widetilde{\tau}\,' = \widetilde{\tau}_S{}' < 1$. In general, it is very likely that an induction flame would slow down below this critical mass-flux at some stage.

When this happens, the steady solution descriptions presented in section 3 are no longer tenable over the entire range of chemical energy q. Once $m < m_S$, the steady wave description is only feasible, in terms of the scaled chemical energy $\mathcal{Q}$ (in equations (19), (20) and figure 2), within the domain where real solutions can be found, namely where $\mathcal{Q} \geq 0$. Something else must happen wherever $\mathcal{Q} < 0$, at the least.

This breakdown in the quasi-steady weak detonation description of the induction flame can be analysed by focussing attention on the point where a sonic flow first appears. For simplicity, we will consider cases in which the chemical energy q decreases monotonically so that the sonic point only appears towards the very end of the chemical reaction. It is also convenient to describe the reaction-rate at this point using the simple model formula

$$\lambda_t = -\frac{\lambda}{t_c} \tag{41}$$

where the 'fast' chemical time t_c is very much shorter than the induction time t_i identified in equation (24). In terms of the reaction-rate model (12), one should expect to find $t_c = t_0 \exp(T_A/T_S) \ll t_i$, although a different chemical mechanism and model from that governing the induction stage might well describe the final stage of the reaction.

By identifying the 'acoustic impedance' $\alpha = \sqrt{\gamma P/V}$, the momentum and compressibility equations from (15) can be combined to describe evolutions along the characteristics $\partial\psi/\partial t = \pm\alpha$:

$$P_t \pm \alpha P_\psi \pm \alpha(u_t \pm \alpha u_\psi) = \gamma P \frac{n_t}{n} - (\gamma - 1)\frac{q_t}{V}. \tag{42}$$

For an induction flame that slows down through a critical mass flux $m_S > 0$, it is worth noting that only the equation that follows forward-travelling characteristics is likely to reveal significant unsteady behaviour. Also, with $t_c \ll t_i$, all of the remaining equations can be shown still to lead to good quasi-steady descriptions in coordinates that shadow the induction flame. It follows that only the third of equations (17) runs into difficulty near $m = m_S$, all of the other relations remaining useable. Moreover, the appearance of the square-root singularity of P in equation (41) indicates that changes in pressure dominate over changes in mass flux in these relations. All of these assertions can be verified using formal asymptotic techniques [7, 10].

In particular one obtains

$$\alpha \sim \sqrt{\frac{\gamma P}{V_S}\left(1 + \frac{P - P_S}{V_S m_S^2}\right)} \sim m_S + \frac{\gamma + 1}{2}\frac{P - P_S}{V_S m_S}$$
$$\text{and} \qquad u_t \pm \alpha u_\psi \sim \frac{P_t \pm \alpha P_\psi}{m_S} \tag{43}$$

after using the fact that $\mu_S = 1$. The resulting form of equation (42) that is relevant for following the induction flame path over small ranges of $P - P_S$, is a chemically-forced Burgers' equation

$$P_t + \frac{\gamma+1}{2}\frac{P-P_S}{V_S m_S}P_{\psi-\widetilde{\psi}} \sim \frac{m_S}{2}\left(\frac{\gamma-1}{V_S}q_{\psi-\widetilde{\psi}} - \frac{\gamma P_S}{n_S}n_{\psi-\widetilde{\psi}}\right) \tag{44}$$

in which the time differentiation is taken along paths of constant $\psi - \widetilde{\psi}$, where the induction flame path is represented by $\psi = \widetilde{\psi}(t)$. It may be convenient to think of $\widetilde{\psi}(t)$ as tracing the path of some fixed 'small' value of λ, or perhaps the path of maximum chemical activity. To a large degree, the exact definition of $\widetilde{\psi}(t)$ is a matter of choice.

Extending equations (40) by simply dropping the superscript '^' throughout, the behaviour of pressure close to P_S, takes on the asymptotic form

$$\frac{\gamma+1}{2}(P-P_S)^2 \sim [(P_S - P_0)V_0 - Q)]\,2m_S(m-m_S) + (\gamma-1)m_S^2(q-\widehat{q}) \tag{45}$$

for small enough values of $m - m_S$ and $q - \widehat{q}$. This expression describes variations in pressure over those parts of the domain where unsteady effects governed by equation (44) are still unimportant. It therefore provides a matching or initial/boundary condition for equation (44) once the variations in mass flux $m - m_S$ and chemical energy $q - \widehat{q}$ are known.

Near the critical point, the mass flux through the induction flame, $m = \widetilde{\psi}'(t)$, is given by

$$\widetilde{\psi}'(t) - m_S \sim -m_S\frac{t-t_S}{t_i}\frac{\widetilde{\tau}_S{}''}{\widetilde{\tau}_S{}'^2} + \cdots \tag{46}$$

which changes significantly over the induction time-scale t_i. The time t_S represents the moment at which $\widetilde{\psi}'(t_S) = m_S$. Also, close to the critical point which we take to exist at the end of the chemical reaction, the model rate equation (41) has the quasi-steady solution

$$\lambda = \widetilde{\lambda}\exp\left(\frac{\psi-\widetilde{\psi}}{m\,t_c}\right) \tag{47}$$

for a suitable constant $\widetilde{\lambda}$ that depends on the definition of $\widetilde{\psi}(t)$.

The unsteady development of pressure can now be cast into an asymptotic form of analysis by identifying the rescalings

$$\begin{gathered}\psi - \widetilde{\psi} = m_S t_c(\xi - \xi_0) \quad \text{and} \quad t - t_S = \widehat{t}_S\zeta \\ \text{with} \qquad P - P_S \sim \widehat{P}_S\,\theta(\zeta,\xi)\end{gathered} \tag{48}$$

in which

$$\begin{gathered}\widehat{P}_S^3 = V_S m_S^4\left[(P_S - P_0)V_0 - \widehat{Q}\right]\frac{8}{\gamma+1}\frac{\widetilde{\tau}_S{}''}{\widetilde{\tau}_S{}'^2}\frac{t_c}{t_i} = O\left(\frac{t_c}{t_i}P_S^3\right), \\ \widehat{t}_S = \frac{2}{\gamma+1}\frac{V_S m_S^2}{\widehat{P}_S}t_c \ll t_i \quad \text{and} \quad \xi_0 = \ln\left((\gamma-1)\frac{\widehat{t}_S}{t_c}\frac{\widetilde{q}'(0)\widetilde{\lambda}}{\widehat{P}_S V_S}\right).\end{gathered} \tag{49}$$

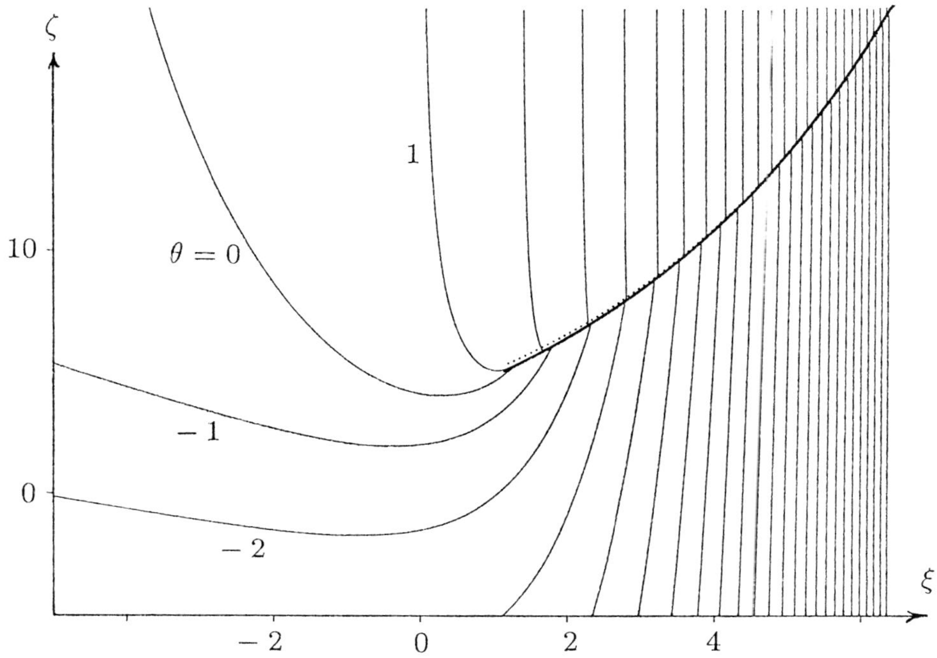

Figure 4: *Pressure perturbation contours* θ, *showing the formation and movement of a shock wave (thick curve). Contours are plotted at unit increments of* θ *from* -24 *to* $+24$.

The problem can now be expressed in the simplified form

$$\theta_\zeta + \theta\theta_\xi = \tfrac{1}{2}e^\xi$$

$$\text{with} \qquad \theta \sim -\sqrt{e^\xi - \zeta} \quad \text{as} \quad \begin{cases} \zeta \to -\infty \\ \quad \text{or} \\ \xi \to \infty. \end{cases} \tag{50}$$

A more general form of this problem can be produced by replacing the first occurrence of e^ξ by $f'(\xi)$ and the second by $f(\xi)$, for any differentiable function f that ultimately depends on the reaction-rate model and the behaviour of n and q. For example, the simple case $f = \xi^2$ would model a situation in which the compressibility forcing terms of equation (10) change sign before the end of the chemical reaction.

By defining a characteristic η such that

$$\zeta - \eta + \int_\xi^\infty \frac{d\xi}{\theta} = 0 \tag{51}$$

it becomes a straightforward task to solve equations (50) to give

$$e^{\xi} - \theta^2 = \eta = e^{\xi} \sin^2 \left(\frac{(\zeta - \eta)\sqrt{\eta}}{2} \right) \tag{52}$$

for positive values of η. When η is negative, $\sqrt{\eta}$ can be replaced by $\sqrt{-\eta}$, with $\sin^2$ being replaced by $-\sinh^2$. Equivalently, the characteristic variable η can be eliminated to give

$$1 - \theta^2 e^{-\xi} = \sin^2 \left(\frac{(\zeta + \theta^2 - e^{\xi})\sqrt{e^{\xi} - \theta^2}}{2} \right) \tag{53}$$

in which the sign of the contents of the square root can be changed as before. Solutions must be selected such that θ is negative and satisfies the asymptotic boundary condition in (50) for all sufficiently large positive values of ξ and/or $-\zeta$. Resulting pressure profiles at different times ζ are plotted in figure 4.

An examination of the solution (53) reveals that an infinite pressure gradient first arises when

$$\zeta = \zeta_* \approx 5.076205, \quad \xi = \xi_* \approx 1.142412 \quad \text{and} \quad \theta = \theta_* = 1/\sqrt[3]{3}. \tag{54}$$

This identifies the moment of shock-formation. Subsequently, if the path of the shock-wave is denoted by $\xi = X(\zeta)$, then the shock-wave propagates ahead at the speed [18]

$$X'(\zeta) = \frac{\theta(\zeta, X^+) + \theta(\zeta, X^-)}{2} \tag{55}$$

the initial value of which is $X'(\zeta_*) = \theta_* = 1/\sqrt[3]{3}$, with $X(\zeta_*) = \xi_*$. The superscripts $^+$ and $^-$ denote evaluation of θ immediately ahead of and behind the shock discontinuity respectively. As the shock moves forwards, it stabilises asymptotically about the path

$$X(\zeta) \sim 4 \ln(\zeta/4) \tag{56}$$

shown as a dotted line in figure 4. Thus it can be seen that the shock wave slows down dramatically as it moves forwards, approaching the shock-speed

$$X'(\zeta) \sim 4/\zeta. \tag{57}$$

4.5 Development of a strong-detonation

The existence of this shock wave (or indeed any positive value of θ) marks a fundamental change in the nature of the reaction wave. It shows that a part of the flow through the wave has become subsonic, and that genuine propagation mechanisms can at last play a role in its evolution. As will be seen, these mechanisms ultimately dominate and lead to a truly 'propagating' detonation wave.

The physical significance of the behaviour (57) of the shock wave becomes clearer as soon as one refers back to dimensional coordinates. If $\psi = \Psi(t)$ represents the path of the shock-wave in terms of ψ and t, then the mass flux through the shock is given by

$$\Psi'(t) - \widetilde{\psi}'(t) \sim \frac{\gamma+1}{2} \frac{\widehat{P}_S}{V_S m_S^2} m_S X'(\zeta) = O\left((t_c/t_i)^{1/3} m_S X'(\zeta)\right) \tag{58}$$

from equations (48) and (49). This should be compared with the difference between the induction flame mass flux and the critical (sonic) mass flux m_S, which can be seen from equations (46), (48) and (49) to be

$$\widetilde{\psi}'(t) - m_S \sim -\frac{2}{\gamma+1} \frac{V_S m_S^2}{\widehat{P}_S} \frac{t_c}{t_i} \frac{\widetilde{\tau}_S''}{\widetilde{\tau}_S'^2} m_S \zeta = O\left((t_c/t_i)^{2/3} m_S \zeta\right). \tag{59}$$

Since $(t_c/t_i)^{1/3} \gg (t_c/t_i)^{2/3}$ it follows that the initial unsteady shock speed (relative to the critical sonic speed) is much faster than the imbalace between the induction flame-speed and the sonic speed. However, as the time-perturbation ζ increases this imbalance decreases, until for

$$\zeta = O\left((t_i/t_c)^{1/6}\right) \quad \text{or} \quad t - t_S = O\left((t_i/t_c)^{1/2} t_c\right) \tag{60}$$

the two speeds become similar but still relatively small, at the order of $(t_c/t_i)^{1/2}$ compared with the sonic speed. At this point, the subsonic part of the reaction wave that lies behind the shock-wave involves a pressure excursion of the order of $\widehat{P}_S\sqrt{\zeta}$ or $P_S(t_c/t_i)^{1/4}$. This is generally small compared with $P_S(T_S/T_A)$ which confirms, in retrospect, that the model reaction rate (12) does indeed reduce to the simple linear form (41) for the purpose of analysing the early shock behaviour.

During the time-scale identified by (60), it is no longer justifiable to neglect changes in the convective mass-flux in equations (43) and (44). Without entering into the details (see reference [10]), analysis of this stage reveals that two connected quasi-steady evolutions emerge. Ahead of the shock-wave, the induction flame continues to behave as previously described. It takes the form of a quasi-steady weak detonation travelling forwards at the relatively slowly-varying mass flux m.

However, the evolution behind the shock-wave also becomes quasi-steady [10] with thermodynamic changes taking place in the manner of a quasi-steady strong detonation as described in section 3. There is, however, one crucial feature of this wave. Examining the Rankine–Hugoniot diagram in figure 1, it can be seen that as the induction flame slows down along

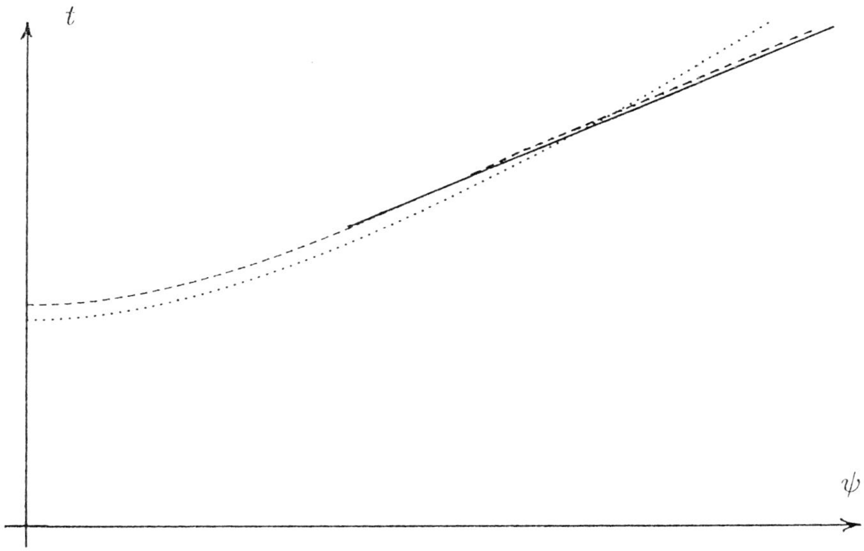

Figure 5: *The movement of an induction flame, between the dotted perturbation singularity path and the dashed path of maximum chemical activity, until it is transformed into a strong detonation following the solid shock path.*

the weak detonation part of Rayleigh lines (such as $O-W$) of decreasing slope, the pressure that remains behind the wave increases. In the subsequent unsteady, now chemically inert, evolution that takes place it follows that an expansion wave must follow behind the chemical reaction wave. This ensures that a sonic flow is always found very close to the point at which the chemical energy q reaches its minimum value. It thus, effectively, fixes the mass flux traversed by the strong detonation part of the reaction wave to be very close to the critical sonic mass flux m_S.

It may be noted that an analogous pair of quasi-steady structures would be found in the more general case where the number of moles per unit mass n is able to change. The sonic point then arises when the chemical forcing term on the right hand side of the compressibility equation (10) approaches zero. Again this may happen either at the end of the chemical reaction or, if the chemical forcing is generated by non-monotonic or competing changes, before the reaction is complete.

Since $m_S > m$ at this stage, the strong detonation structure actually travels faster than the induction flame and simply overtakes it as sketched in figure 5. In this, the induction

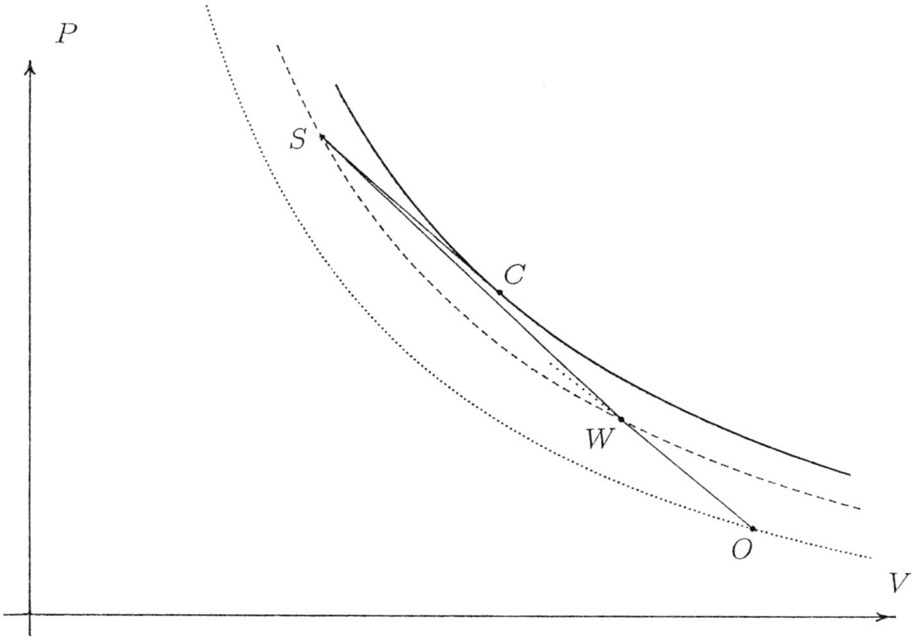

Figure 6: *Rankine–Hugoniot diagram showing weak detonation, shock and strong detonation Rayleigh lines. The solid curve, dashed curve and dotted curve represent fully burnt, partially burnt and fully unburnt Hugoniot curves respectively.*

flame path $\psi = \tilde{\psi}(t)$ (that may, most usefully, be thought of as representing a path of maximum chemical activity) is asymptotically close to the singularity path $\tau = \tilde{\tau}$. Once the induction flame slows down below the critical sonic mass flux m_S, the quasi-steady shock and deflagration or strong detonation complex (formed after the brief unsteady evolution over the time-scale (60)) continues to move at a mass flux that remains very close to m_S. It therefore *propagates* steadily forward through the slowing induction flame.

As it does so, the shock wave passes into less and less burned parts of the induction flame. Correspondingly, the value of the chemical energy q at the shock location generally increases (although not necessarily monotonically). In terms of figure 2, this would correspond to a shock location (joining weak and strong detonation branches of solution) that gradually moves into increasing values of $\mathcal{Q}$. It should be noted that, because the scalings (19) that define $\mathcal{Q}$, $\mathcal{P}$ and $\mathcal{T}$ depend on the mass flux m which is slightly different for the two wave structures, the value of $\mathcal{Q}$ also changes very slightly through the shock wave. This does not, however, significantly alter the overall picture.

On a Rankine–Hugoniot diagram, the variation is not quite so clearly demonstrated. Since the difference between m and m_S is relatively small, the evolution all lies close to a single Rayleigh line. The difference between the Rayleigh lines illustrated in figure 6 is exaggerated for greater clarity. Starting at the point O, the induction flame increases along a Rayleigh line $O - W$ that has too small a slope to be able to reach a fully burnt Hugoniot curve. At some point W where the reaction has only partially proceeded, a shock wave raises the pressure along the Rayleigh line $W - S$ to the point S on a suitable Hugoniot curve for the same fixed chemical state. Finally, the pressure decreases again along the deflagration part of another Rayleigh line $S - C$ that reaches the corresponding fully burnt Hugoniot curve at the tangent point C. This corresponds to the sonic point or Chapman–Jouget point that is found at the end of the chemical reaction. Since the shock wave travels slightly faster than the deflagration that follows it (in order to be able to progress ahead into less reacted regions), the slope of the line $W - S$ is slightly greater than $S - C$. As the shock wave advances, the Hugoniot curve containing the points W and S gradually shifts so as to correspond to less advanced chemical change.

The strong detonation ultimately propagates ahead of the induction flame into the region where unsteady induction processes still prevail. It then has the typical form of a Zeldovich–von Neumann–Döring detonation wave [1], propagating into a weakly nonuniform medium.

5. Conclusions

Using relatively mild assumptions about the nature of a compressible reacting medium and the way in which it undergoes an exothermic reaction, this article has demonstrated a number of features of both steady and unsteady behaviour. Steady combustion waves of many different speeds and configurations are possible. Weak detonations (that do not involve a shock wave) are found to fall within this spectrum, and should only really be excluded if chemical changes are unable to proceed significantly at the appropriate 'weak' values of the thermodynamic parameters.

Indeed they are found to describe the structure of 'induction flames' that emerge in a natural way from a wide class of unsteady ignition processes in compressible media. Since the flow through such a weak detonation is uniformly supersonic, these waves do not genuinely self-propagate but rely on conditions for reaction runaway that are set by events ahead of the waves. Importantly, a transition from such a weak detonation to a strong detonation takes place automatically once the induction flame slows down below a speed at which a sonic point first appears in the flow.

An unsteady evolution then describes the formation of a shock wave and the development of a strong detonation. The flow through this particular reaction wave is subsonic, so that information can be transmitted ahead and thus contribute to a genuine propagation of the wave. As a result, the strong detonation is able to move ahead and completely engulf the induction flame.

No attempt has been made to be comprehensive in describing these events. For example, questions about the stability of steady reaction wave structures have been left completely aside. However, the reader is encouraged to bear in mind more general questions that are hardly touched on in this discussion. Most particularly, it illustrates how some of the difficult theoretical issues involved in examining detonation behaviour can be reduced to meaningful proportions by using asymptotic techniques.

Acknowledgements

Many people contributed substantially towards formulating the arguments that are presented in this article. John Clarke was most instrumental in helping to develop some of the ideas concerning unsteady behaviour, while the document also benefitted from many other useful discussions, especially with Amable Liñán, Paul Clavin and Sergei Frolov. Collaboration between Bill Dold and Ashwani Kapila was facilitated by a NATO collaborative research grant. Bill Dold is supported by a Science and Engineering Research Council fellowship and Mark Short holds a S.E.R.C. studentship, and Ashwani Kapila is supported by the Los Alamos national Laboratory and by the U.S. Army Research Office.

Above all, this article arose from a visit to the Instite of Thermophysics in the Academic Town of Novosibirsk, arranged by Anatoly Borisov, whose warm hospitality (and that of many others in Academgorodok) is gratefully acknowledged by the first author.

References

1. W. Fickett and W.C. Davis, 'Detonation,' University of California Press, Berkeley, 1979.
2. F.A. Williams, 'Combustion Theory,' 2nd ed., Benjamin/Cummings, California, 1985.
3. J.W. Dold, *Emergence of a Detonation within a Reacting Medium*, in 'Fluid Dynamical Aspects of Combustion Theory,' A. Tesei and M. Onofri (Eds.), Longman, U.K., to appear.

4. A. K. KAPILA, *An Introduction to Detonation Theory*, in 'Fluid Dynamical Aspects of Combustion Theory,' A. Tesei and M. Onofri (Eds.), Longman, U. K., to appear.

5. D. R. KASSOY, *Mathematical Modelling for Planar, Steady, Subsonic Combustion Waves,* Ann. Rev. Fluid Mech., **17** (1985) 267–287.

6. J. F. CLARKE, *Fast flames, waves and detonation*, Prog. in Energy and Combustion Science **15** (1989) 241–271.

7. J. W. DOLD and A. K. KAPILA, *Asymptotic analysis of detonation initiation for one-step chemistry: I–emergence of a weak detonation,* submitted.

8. J. W. DOLD, *Induction period generation of a supersonic flame,* in 'Numerical Combustion,' A. Dervieux and B. Larrouturou, Eds., Lecture Notes in Physics **351**, Springer Verlag, 1989, pp. 245–256.

9. YA. B. ZELDOVICH, *Regime classification of an exothermic reaction with nonuniform initial conditions,* Combustion and Flame, **39** (1980) 211–214.

10. J. W. DOLD and A. K. KAPILA, *Asymptotic analysis of detonation initiation for one-step chemistry: II–from a weak structure to ZND,* submitted.

11. YA. B. ZELDOVICH and A. S. KOMPANEETS, 'Theory of Detonation,' Academic Press, 1960.

12. J. W. DOLD and A. K. KAPILA, *Comparison Between Shock Initiations of Detonation using Thermally-Sensitive and Chain-Branching Chemical Models*, Combustion and Flame, to appear.

13. T. VON KÁRMAN, *Models in thermogasdynamics*, in 'I Modelli nella Technica,' Atti del Convegno di Venezia, **1** (1955) 643–651.

14. M. W. CHASE, C. A. DAVIES, J. R. DOWNEY, D. J. FRURIP, R. A. MCDONALD and A. N. SYVERUD, *JANAF Thermochemical Tables, Third Edition*, J. Phys. Chem. Ref. Data **14**, Suppl. 1 (1985) 17–18.

15. J. F. CLARKE, *Finite amplitude waves in combustible gases*, in 'The Mathematics of Combustion,' J. D. Buckmaster, Ed., SIAM Publications, Philadelphia, 1985, pp. 183–245.

16. J. W. DOLD, *Dynamic transition of a self-igniting region,* in 'Mathematical Modelling of Combustion and Related Topics,' C-M. Brauner and C. Schmidt-Laine, Eds., Martinus Nijhoff, Dordrecht, 1988, pp. 461–470.

17. T. L. JACKSON, A. K. KAPILA and D. S. STEWART, *Evolution of a reaction center in an explosive material*, SIAM J. Appl. Math., **49** (1989) 432–458.

18. G. B. WHITHAM, 'Linear and Nonlinear Waves,' Wiley, New York, 1974.

GROUP METHODS FOR THE STUDY OF DETONATION WAVE ANALYSIS

M.TORRISI

Dipartimento di Matematica - Università di Catania
Viale A.Doria, 6 - 95125 CATANIA

Summary

After a brief outline on the algorithm for the determination of Lie symmetries of a PDEs system we find some classes of similarity solutions by a double group transformation.

1. Introduction

Many authors looked for similarity solutions in detonation theory as had been done in fluid dynamics; one can see, e.g., [1] and the well known monographs of L.Sedov [2] and V.P.Korobeinikov [3].

Only recently the group analysis approach has been used in detonation to obtain the similarity solutions and, if they exist, also other invariant solutions. In [4], J.D.Logan and J.J.Perez, starting from a planar model and working in Lagrangian coordinates, performed the possible Lie symmetries of the system, by requiring also the invariance of Rankine-Hugoniot relations associated with the governing equations as boundary conditions. Successively, Logan and others again considered the problem of obtaining detonation waves via group analysis [5] [6].

Here the Author taking some results of some of his previous works [7] [8] as a starting point, wishes to present a simple procedure based on the application of some ideas of group analysis which allows us to obtain exact solutions when we assume some special classes of rate functions and, as boundary conditions, the Rankine-Hugoniot relations in the strong shock approximation. After introducing in sect. 2 the

A. A. Borissov (ed.), Dynamic Structure of Detonation in Gaseous and Dispersed Media, 143–156.

governing equations, in sect. 3 we give a brief outline of group analysis methods which transform the PDEs system of governing equations in a ODEs system and after by introducing the associated group [9] and in agreement with a procedure presented in [10], we transform the previous ODEs system in a new autonomous system of ODEs. In sect. 4 we integrate this system in some special cases.

2. Governing equations

It is usual, when we consider detonation phenomena, to assume as governing system the equations which describe the evolution of a binary exotherming reacting mixture when we neglect the transport effects which in the monodimensional case and in conservative form can be written [11], [12] as:

$$\begin{aligned}
&\rho_t+(\rho u)_x= -\frac{\alpha u\rho}{x}\\
&(\rho u)_t+(\rho u^2+p)_x= -\frac{\alpha u^2\rho}{x}\\
&\left(\left(\frac{1}{2}u^2+e\right)\rho\right)_t+\left(u\left(\rho\left(\frac{1}{2}u^2+e\right)+p\right)\right)_x=-\frac{\alpha u}{x}\left(p+\rho\left(\frac{1}{2}u^2+e\right)\right)\\
&(\rho\lambda)_t+(\rho u\lambda)_x=\rho w(\rho,e,\lambda)-\frac{\alpha\rho u}{x}\lambda
\end{aligned} \tag{2.1}$$

where x and t are the independent variables of space and time, α=0,1,2 the geometric factor, while with u,ρ,p,e,λ we denote respectively the particle velocity, the density, the internal energy, the pressure and the progress variable of reaction defined as mass fraction of the product of reaction ($0\leq\lambda\leq1$). The constitutive function $w=w(\rho,e,\lambda)$ is the rate function characterized by the chemical kinetics. Assuming, for simplicity's sake, the reactant and product are both polytropic with the same γ-gas law, we again add the constitutive relation to the system (2.1)

$$e=\frac{p}{\rho(\gamma-1)}+(1-\lambda)q \tag{2.2}$$

where q is heat of reaction.

By using the constitutive relation (2.2) we can write the system (2.1) as follows:

$$\begin{aligned}
&\rho_t+\rho_x u+\rho u_x = - \frac{\alpha u\rho}{x} \\
&u_t+uu_x+\frac{p_x}{\rho}=0 \\
&p_t+\gamma p u_x+u p_x=q(\gamma-1)\rho w-\alpha\gamma\frac{pu}{x} \\
&\lambda_t+u\lambda_x=w(p,\rho,\lambda)\ .
\end{aligned} \tag{2.3}$$

Here we assume that reaction is begun by a purely fluidynamic strong shock progressing in quiescent gas ($p_0=u_0=0$) so that we take, as auxiliary conditions, the Rankine-Hugoinot relations:

$$\begin{aligned}
&u_1=\frac{2}{\gamma+1}D \qquad \rho_1=\frac{\gamma+1}{\gamma-1}\rho_0 \\
&p_1=\frac{2}{\gamma+1}\rho_0 D^2 \qquad \lambda_1=0
\end{aligned} \tag{2.4}$$

where D is the shock speed and the subscripts 0 and 1 refer respectively to the quantities ahead and behind the shock front. We must remark that in general

$$\rho_0=\hat{\rho}x^{-\Omega} \tag{2.5}$$

with $\Omega\geq 0$ in agreement with classical cases studied in explosion theory (see e.g. [2] [3]); while the condition (2.4.IV) means that there is no chemical reaction at the shock front [12].

3. On the group analysis approach

The main aim of the group analysis approach to find special solutions of a PDEs system of the kind

$$H_j(x_\alpha, v_i, v_{s,\alpha}) = 0 \qquad \begin{matrix} \alpha=0,1 \\ i,j,s=1,2,3,4 \end{matrix} \tag{3.1}$$

with

$$v_{s,\alpha} := \frac{\partial v_s}{\partial x_\alpha}, \tag{3.2}$$

concerns the finding of solutions which are invariant with respect to any special group of infinitesimal transformations

$$\begin{aligned} x'_\alpha &= x_\alpha + \epsilon X_\alpha(x_\beta, v_i) \qquad & \alpha,\beta=0,1 \\ & & i,j=1,\ldots,4 \\ v'_j &= v_j + \epsilon \eta_j(x_\beta, v_i) \end{aligned} \tag{3.3}$$

which leaves invariant the form of the PDEs system and the possible auxiliary conditions (initial, boundary ...) when they exist.

In order to give a brief outline of the standard procedure used we rewrite the system (2.1) in the form (3.1) by putting.

$$x_0 := t \qquad x_1 := x \qquad (v_1, v_2, v_3, v_4) := (\rho, u, p, \lambda) \ . \tag{3.4}$$

We distinguish two steps:

a) *The determination of infinitesimals* $X_{,\alpha}$ *and* η_j *which leave the system invariant.*

Following the classical monographs [13-18] we can write the invariance conditions briefly as:

$$LH_i = 0 \tag{3.4}$$

where the operator

$$L=X_{\alpha}\frac{\partial}{\partial x_{\alpha}}+\eta_{j}\frac{\partial}{\partial v_{j}}+\eta_{j\alpha}\frac{\partial}{v_{j,\alpha}} \tag{3.5}$$

is the Lie derivative in the direction of the extended field vector, and

$$\eta_{j\alpha}=\eta_{j,\alpha}+\eta_{j,v_s}v_{s,\alpha}-X_{\beta,\alpha}v_{j,\beta}-X_{\beta,v_s}v_{s,\alpha}v_{j,\beta} \tag{3.6}$$

are the extended infinitesimals of (3.3).

We introduce the relations (3.6) in the equations (3.4) and then eliminate $v_{j,0}$ by means of (3.1) so that we obtain the equations as polynomial expressions in $v_{j,1}$ which must vanish.

The vanishing of coefficients of these expressions leads, in general, to an over determined system (called *determining system*) from which we get X_{α} and η_j and some constraints on the constitutive function which appear in the governing equations. In our case we get restriction on the rate function.

Here, bearing in mind some applications only, we report only the following special results concerning the X_{α} and η_j infinitesimals in the planar case (α=0), which, taking the (3.4) into account, can be written as [19]

$$X_0=at+c \qquad X_1=bx+f_1t+d$$

$$\eta_1=f_2\rho\,, \qquad \eta_2=(b-a)u+f_1\,, \qquad \eta_3=(2(b-a)+f_2)p \tag{3.7}$$

$$\eta_4=2(b-a)\lambda+f_3\,,$$

while the rate function must satisfy the following PDE:

$$(3a-2b)w=[2(a-b)\lambda+f_3]w_{,\lambda}+[2(a-b)+f_2]pw_{,p}+f_2\rho w_{,\rho} \tag{3.8}$$

In this article we do not consider the cases a=b or the case when some of arbitrary constants a,b,c,d could be zero.

If we further require the invariance of boundary conditions (2.4) then $f_1=0$, $f_2=-\Omega$ and $f_3=0$ arise.

In the following we will consider, for simplicity's sake, only the case $\Omega=0$, that is the case of a shock front progressing in a homogeneous gas at rest.

So the (3.7) become:

$$X_0=at+c \qquad X_1=bx+d \qquad \eta_1=0\ , \qquad \eta_2=(b-a)u$$
$$\eta_3=2(b-a) \qquad \eta_4=2(b-a)\lambda \tag{3.9}$$

and (3.8)

$$(3a-2b)w=2(a-b)\lambda w_{,\lambda}+2(a-b)pw_{,p} \tag{3.10}$$

which easily gives

$$w=p^{\beta}f\left(\frac{\lambda}{p}\ ,\rho\right) \tag{3.11}$$

with

$$\beta=\frac{3a-2b}{2(a-b)} \tag{3.12}$$

and f arbitrary function of its arguments.

b) *The calculations of invariant solutions.*

After having determined the transformations which leave our system invariant and its boundary conditions, we look for the invariant solutions by using the invariance condition (*invariant surface condition*) which can be written as

$$X_{\alpha}v_{j,\alpha}=\eta_j\ . \tag{3.13}$$

By taking into account the (3.9) we integrate (3.13) and obtain

$$
\begin{aligned}
\rho(t,x)&=R(\sigma)\\
u(t,x)&=(c_3t+1)^{c_2-1}U(\sigma)\\
p(t,x)&=(c_3t+1)^{2(c_2-1)}P(\sigma)\\
\lambda(t,x)&=(c_3t+1)^{2(c_2-1)}\Lambda(\sigma)
\end{aligned}
\tag{3.14}
$$

where

$$c_2:=\frac{b}{a}\ ,\qquad c_3:=\frac{a}{c}\ ,\qquad c_4:=\frac{b}{d}\ ,\qquad \sigma=\frac{c_4x+1}{(c_3t+1)^{c_2}}\ .$$

while R,U,P,Λ are functions which are to be determined by the system (2.3).

So, by substituting (3.9) in (2.3) we get the following ODEs system (*reduced system*)

$$
\begin{cases}
\left(c_4U-c_2c_3\sigma\right)R'+c_4RU'=0\\
\left(c_4U-c_2c_3\sigma\right)U'+c_4R^{-1}P'=(1-c_2)c_3U\\
\left(c_4U-c_2c_3\sigma\right)P'+\gamma c_4U'P=2(1-c_2)c_3P+(\gamma-1)qRF^{\beta}f\\
\left(c_4U-c_2c_3\sigma\right)\Lambda'=2(1-c_2)c_3\Lambda+P^{\beta}f
\end{cases}
\tag{3.15}
$$

Since the shock front must be a similarity curve passing through $t=0$, $x=0$, its equation is:

$$c_4x+1=(c_3t+1)^{c_2}$$

from which it follows

$$D(t)=D_0(c_3t+1)^{c_2-1}$$

where

$$D_0=\frac{c_2c_3}{c_4}=D(0) \qquad (3.15')$$

corresponds to a noted initial shock velocity and the R-H conditions become:

$$\begin{cases} R(1)=\dfrac{\gamma+1}{\gamma-1}\rho_0 & U(1)=\dfrac{2D_0}{\gamma+1} \\ P(1)=\dfrac{2}{\gamma+1}\rho_0D_0^2 & \Lambda(1)=0\ . \end{cases} \qquad (3.16)$$

So a boundary problem was transformed into a initial values problem for a ODEs system and this, in general, when we cannot find exact solutions, allows us to apply numerical methods easily . Nevertheless, as we have to finding some classes of exact solutions in mind, we note that the system (3.15) is again invariant with respect to the transformation group (*associated group*) [9]:

$$\begin{cases} \sigma^*=\mu\sigma \quad U^*=\mu U \quad P^*=\mu^{2+A}P \quad R^*=\mu^AR \\ \Lambda^*=\mu^2\Lambda \qquad\qquad A\in R-\{0,-2\} \end{cases} \qquad (3.17)$$

provided that f have the following functional form:

$$f\left(\frac{\Lambda}{P},R\right)=R^{2(\frac{1-\beta}{A})-\beta}\hat{f}\left(\frac{\Lambda R}{P}\right) \qquad (3.18)$$

where $\hat{f}$ is an arbitrary functions of its arguments.

So, by assuming f belongs to the class of function characterized by (3.18) and by applying the following change of variables [10]

$$\begin{cases} \eta=\ln\sigma \qquad V_1=\dfrac{R}{\sigma^A} \qquad V_2=\dfrac{U}{\sigma} \\ V_3=\dfrac{P}{\sigma^{2+A}} \qquad V_4=\dfrac{\Lambda}{\sigma^2} \end{cases} \tag{3.19}$$

to the system (3.15) and to the auxiliary conditions (3.16), we obtain

$$\begin{cases} \left(c_4V_2-c_2c_3\right)V_1'+c_4c_3V_1V_2' = -\left(c_4V_2-c_2c_3\right)AV_1+c_4V_1V_2 \\ \left(c_4V_2-c_2c_3\right)V_2'+c_4V_1^{-1}V_3' = -\left(c_4V_2-c_2c_3\right)V_2+ \\ \qquad -c_4(2+A)V_1^{-1}V_3+(1-c_2)c_3V_2 \\ \gamma c_4V_3V_2'+\left(c_4V_2-c_2c_3\right)V_3' = -(2+A)\left(c_4V_2-c_2c_3\right)V_3+ \\ \qquad -\gamma c_4V_2V_3+2(1-c_2)c_3V_3+(\gamma-1)qV_1^{(1-\beta)(1+\frac{2}{A})}V_3^{\beta}\cdot\hat{f}\left(\dfrac{V_1V_4}{V_3}\right) \\ \left(c_4V_2-c_2c_3\right)V_4'=-2\left(c_4V_2-c_2c_3\right)V_4+2(1-c_2)c_3V_4+ \\ \qquad +V_1^{2(\frac{1-\beta}{A})-\beta}V_3^{\beta}\cdot\hat{f}\left(\dfrac{V_1V_4}{V_3}\right) \end{cases} \tag{3.20}$$

with

$$V_1(0)=\frac{\gamma+1}{\gamma-1}\rho_0,\qquad V_2(0)=\frac{2D_0}{\gamma+1},$$
$$V_3(0)=\frac{2D_0^2}{\gamma+1}\rho_0,\qquad V_4(0)=0\ . \tag{3.21}$$

It is interesting to note that in (3.20) the independent variable does not appear so that this fact, as we will see in the next section, allows us to look for some classes of exact solution.

4. Some classes of exact solutions

Here we look for solutions of system (3.20), with auxiliary conditions (3.21), of the form [7,8]:

$$V_1=V_{10}=\frac{\gamma+1}{\gamma-1}\rho_0 \qquad V_2=V_{20}=\frac{2D_0}{\gamma+1}$$
$$V_3=V_{30}=\frac{2}{\gamma+1}\rho_0 D_0^2 \qquad V_4=V_4(\eta) \tag{4.1}$$

this implies

$$\begin{cases} -V_1V_2c_4(A+1)+c_2c_3AV_1=0 \\ -c_4V_2^2+V_2c_3-c_4(2+A)\dfrac{V_3}{V_1}=0 \\ -(2+A+\gamma)c_4V_2V_3+(c_2A+2)c_3V_3- \\ \qquad\qquad +(\gamma-1)q\hat{f}\,V_1^{(1-\beta)(1+\frac{2}{A})}V_3^{\beta}=0 \end{cases} \tag{4.2}$$

so we must assume:

$$\hat{f}=f_0=\text{cost}. \tag{4.3}$$

This means that we choose the rate function w of the form

$$w=f_0 p^{\beta} \rho^{2(\frac{1-\beta}{A})-\beta} \tag{4.4}$$

This form is analogous to the one used in [1] and [6] where the authors assumed that the reaction rate is proportional to the internal energy. As in those works, we can observe that in the rate function (4.4) the Arrhenius factor is absent.

This form of rate function is of interest in the detonation of condensed explosives, is a good approximation at the beginning of reaction and finally, provides a vehicle for the mathematical tractability of the model.

Therefore we can solve the system (4.2) for some of the constants which appear there.

We observe that we can consider f_0 known from the functional form of the rate law and q known from the chemical characteristics of the reactive medium, while we must determine c_2, c_3, c_4 and A from (3.15′) and (4.2).

From the (4.2), after some calculations we obtain

$$A=\frac{2}{\gamma-1} \tag{4.4}$$

$$c_2=\frac{1}{2} \tag{4.5}$$

so that

$$c_3=-2\left(\frac{\gamma-1}{\gamma+1}\right)^{1+\gamma} \rho_0^{1-\gamma} D_0^2 q f_0 \tag{4.6}$$

while from the initial shock velocity we can obtain

$$c_4=-\left(\frac{\gamma-1}{\gamma+1}\right)^{1+\gamma} \rho_0^{1-\gamma} D_0 q f_0 \ . \tag{4.7}$$

V_4, instead, is given by the solution of the problem

$$\begin{cases} \frac{1}{2}\left(\frac{1-\gamma}{\gamma+1}\right)V'_4=2\left(1-\frac{1}{\gamma+1}\right)V_4+f_0V_0^{2(\frac{1-\beta}{A})-\beta}V_{30}^{\beta} \\ V_4(0)=0 \end{cases} \tag{4.8}$$

Then after having integrated (4.8), by going back to the original variable t,x, we get.

$$\rho(t,x)=\frac{\gamma+1}{\gamma-1}\rho_0\left[\frac{c_4x+1}{(c_3t+1)^{1/2}}\right]^{\frac{2}{\gamma-1}} \tag{4.9}$$

$$u(t,x)=\frac{2D_0}{\gamma+1}\frac{c_4x+1}{c_3t+1} \tag{4.10}$$

$$p(t,x)=\frac{2\rho_0D_0^2}{\gamma+1}(1+c_3t)^{-1}\left[\frac{c_4x+1}{(c_3t+1)^{1/2}}\right]^{\frac{2\gamma}{\gamma-1}} \tag{4.11}$$

$$\lambda(t,x)=-\frac{D_0^2}{q(\gamma+1)\gamma}(1+c_3t)^{-1}\left[\left(\frac{c_4x+1}{(c_3t+1)^{1/2}}\right)^{-2(\frac{1+\gamma}{\gamma-1})}-\frac{(c_4x+1)^2}{(c_3t+1)}\right]. \tag{4.12}$$

It is worthwhile noticing that, taking into account (4.6) and (4.7) we may obtain some comments about these solutions [8].

Finally, specializing the previous results for $\gamma=3$ we obtain

$$\rho(t,x)=2\rho_0\frac{c_4x+1}{(c_3t+1)^{1/2}} \tag{4.13}$$

$$u(t,x)=\frac{1}{2}D_0\frac{c_4x+1}{c_3t+1} \tag{4.14}$$

$$p(t,x)= \frac{1}{2}\,\rho_0 D_0^2\, \frac{(c_4x+1)^3}{(c_3t+1)^{5/2}} \tag{4.15}$$

$$\lambda(t,x)=-\frac{1}{12}\,\frac{D_0^2}{q}\left[\frac{(c_3t+1)}{(c_4x+1)^4}-\left(\frac{c_4x+1}{(c_3t+1)}\right)^2\right] \tag{4.16}$$

where

$$c_3= -\frac{1}{8}\left(\frac{D_0}{\rho_0}\right)^2 qf_0 \qquad c_4=-\frac{1}{16}\,\frac{D_0}{\rho_0^2}\,qf_0\ . \tag{4.17}$$

Aknowledgements:

This research was supported by C.N.R. in accordance with the agreement for Scientific Collaboration between C.N.R. and the U.S.S.R. Acad. Sci. It was also supported by M.U.R.S.T. through Fondi 40% and Fondi 60%.

REFERENCES

[1] Stenberg H.M., 1970, *Similarity solutions for reactive shock waves.* Q. I. Mech. Appl. Math. 23, 77-99.

[2] Sedov L.I., 1959, *Similarity and dimensional methcds in mechanics,* Academic Press, New York.

[3] Korobeinkov V.P., 1976, *Problems in the theory of point explosion in gases,* Proceeding of the Steklov Institute of Mathematics, n.119 (1973), A.M.S. Providence, Rodhe Island.

[4] Logan, J.D., Perez J.J., 1980, *Similarity solutions for reactive shock hydrodynamics.* SIAM J. App. Math. 29, 512-27.

[5] Holm D.D., Logan J.D., 1983, *Self-similar detonation waves*, J.Phys A 16, 2035-47.

[6] Logan J.D., Woerner E.L., 1989, *Sensitivity of Self-similar* ZND

waves in condensed media, IMA J. of Appl. Math. 43, 167-184.

[7] Torrisi M., 1989, *Similarity solutions and growth af weak discontinuities in a non constant state for a reactive polytropic gas.* Int. J. Nonlinear Mechanics 24 (5), 441-449.

[8] Torrisi M., Lombardo S., 1991, *Similarity solutions of the wood-kirkwood model in detonation theory,* Int. J. Engng Sci. 24 (4) 523-532.

[9] Dresner L., *Similarity solutions of non linear partial differential equations,* RNM 88, Pitman, London (1987).

[10] Ames W.F. and Donato A., 1988, *On the evolution of weak discontinuities in state characterized by invariant solutions* Int. J.of Non-Linear Mech., 23 (2) 167-174.

[11] Williams F.A., 1985, *Combustion theory,* The Benjomin/Cummings Publishing Company Inc., Menlo Park California.

[12] Fickett W. and Davis W.C., 1979, Detonation. University of California Press.

[13] Ames W.F., 1972, *Non linear partial differential equations in Engineering*, Academic Press New York.

[14] G.V.Bluman and J.D.Cole, 1974, *Similarity method for differential equations,* Springer Verlag Berlin.

[15] Ovsiannikov L.O., 1982, *Group Analysis of differential equations,* Academic Press, New York.

[16] Ibragimov N.H., 1985, *Transformation group applied to Mathematical Physics,* D.Reidel Publishing Company, Dordrecht.

[17] Bluman G.W. and Kumei S., 1989, *Symmetries and differential equations,* Springer Verlag, New York.

[18] Ames W.F. and Rogers C., 1989, *Non linear Boundary Value Problems in Science and Enginenering,* Academic Press, New York.

[19] Lalicata M.V. and Torrisi M., 1991, *Group Analysis approach for a binary reacting mixture,* to appear.

Transition and Structure of Dust Detonations

F. Zhang and H. Grönig

Shock Wave Laboratory, RWTH Aachen
5100 Aachen, Federal Republic of Germany

Abstract. The present paper deals with the transition from deflagration to detonation and the detonation structure in organic dust fuel-gas oxidizer mixtures in tube. The complete process of detonation evolution can be distinguished in four stages: 1. initial partial ignition; 2. pressure wave amplification by coherent energy release; 3. unsteady reaction shock; and 4. spin detonation. By decreasing the initiation energy the monotonic acceleration in stage 2 turns into a multistage one. The three- dimensional structure of the detonation wave front forms within the stage of the "reaction shock" in which a shock front is closely coupled with the reaction zone behind it. In this stage, an abrupt violent onset of detonation, as observed in gaseous mixtures, in some cases occurs. The spin structure exists in both the periphery and the interior region of the cross section. The detailed description of the structure shows that transverse waves play a dominant role in stable detonation propagation in those heterogeneous mixtures.

1. INTRODUCTION

Detonation in dust mixtures can be divided in dust, hybrid and dusty detonations, corresponding to the mixtures of dust fuel- gas oxidizer, dust and gas fuel-gas oxidizer, and inert dust in reactive gas mixtures, respectively. During the last two decades there has been much interest on studies concerning dust detonations.[1-10] Among them little

A. A. Borissov (ed.), Dynamic Structure of Detonation in Gaseous and Dispersed Media, 157–213.

work has been reported for an unconfined mixture.[9-10] The majority of the experimental studies in this area were carried out in tubes, mostly of a few centimeters in hydraulic diameter. There, the measured propagation velocities were 20 - 30 % less than those predicted by the equilibrium Chapman-Jouguet theory. The observed transition process from deflagration to detonation exhibited mainly a smooth acceleration to the steady state, without a violent onset of detonation accompanied by a retonation wave. This phenomenon was referred to as "quasi-detonations".[11] So far it had not been definitively established if a dust fuel-gas oxidizer suspension can support a stable self-sustained detonation wave. Further, in case such a stable dust detonation existed, it was unknown whether transverse waves are mainly responsible for its propagation.

It is of interest that in a tube of 0.6 m in diameter and 42 m long, Gardner et al.[7] showed the transition process in coal dust-air mixtures, where, at the end of the tube, the velocity reached 2850 m/s and a peak pressure of 81.5 bar was observed. The wave there did not appear to reach a stable state. However, this study predicted that in dust-gas mixtures, a transition from deflagration to stable detonation is possible. Recently it was experimentally demonstrated, that in a tube of 141 mm diameter, in corn starch dust-oxygen mixtures a stable self-sustained detonation was formed, and it possessed a spin structure.[12-13] Thus, in these two-phase system, transverse waves are predominant in stable propagation of detonations, similar as in homogeneous mixtures.

On this basis, one can discuss the dust detonability concerning experimental tubes, similar to the gaseous detonation. First, owing to two-phase relaxations and the heterogeneous effects the induction zone length in a dust fuel-gas oxidizer mixture is much larger than those in homogeneous gaseous mixtures, as shown by studies on ignition delay times.[14-15] This leads to a long reaction zone, which gives rise to wall losses, which in turn increase the induction zone length. The transverse wave spacings or cell sizes of dust detonations must be considerably extended due to the large induction zone. Hence, the minimum tube diameter for propagation of a stable detonation must be of the order of magnitude of this cell size, or otherwise the development of detonations with cellular structure will be confined. Secondly, as observed in Ref. 13, in these two-phase system the preceding state of the detonation comprises of an accelerating shock wave which is closely coupled with the reaction zone behind it. The physically significant property of this stage is that during its acceleration the symmetric smooth front structure turns into a three dimensional one. This reaction shock complex could appear to be a detonation, but in fact represents an unsteady process. Thus, if the ratio of tube length and diameter is not sufficient, the flow can still remain in this prestage of the detonation.

In this paper, after introduction of the detonation tube for study of dust-gas mixtures at the Shock Wave Laboratory in Aachen, an estimate of initiation energy concerning planar wave and free jet initiation will be derived. Then, the observation of the stable dust detonation, its structure, and transition from deflagration to stable detonation will be discussed. At last the detonability in dust-air mixtures will be briefly concerned.

2. EXPERIMENTAL SETUP

The experiments were carried out in a horizontal tube of 141 mm inside diameter (Fig. 1). It consists of a 4 m initiation section, a 17.4 m test section, and a 2.36 m^3 expansion tank. To produce the initiation pulse, a mixture of 2 H_2 + O_2 in the initiation section is ignited at one end and subsequently injected into the test section at the other end, either as a planar wave over the whole cross section or as a free jet through an orifice 20 mm in diameter.

The difficulty in using a horizontal tube of large extension, arises in producing a homogeneous dust-gas suspension ahead of the wave. Thus, a method according to the Laval's pressure principle is applied to fulfill this requirement. A dispersion tube 12 mm in cutside diameter is mounted, with a small gap, on the bottom of the whole test section. In the dispersion tube, there are 580 orifices in one row, which are locat-

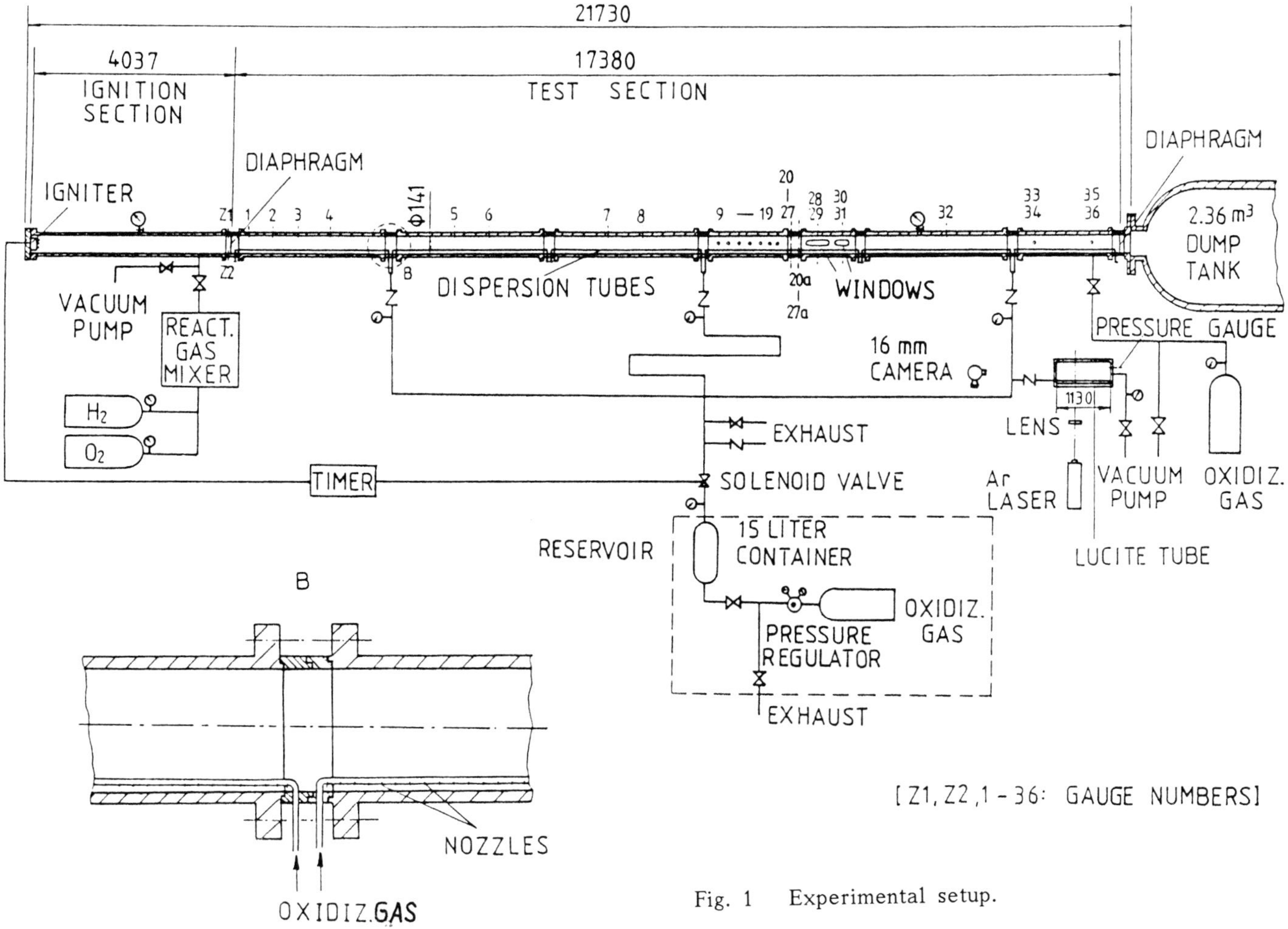

Fig. 1 Experimental setup.

ed opposite to the bottom. Dust is homogeneously fed into the bottom of the test section, in accordance with required concentrations. After evacuation the test section is filled with oxidizing gas up to certain pressure p_{01}, without disturbing the dust layer. Then, the dust dispersion over the cross section is achieved by an oxidizing gas flow, which comes from a reservoir through the orifices of the dispersion tube, until the pressure in the test section reaches the required initial pressure p_0. The dispersion time Δt, corresponding to the pressure rise from p_{01} to p_0, is set on a timer as initiation delay for the igniter.

Choosing the reservoir pressure to satisfy the Laval's pressure principle, the flow maintains the sound speed a_s at the orifices. Thus, the volume flux there reaches the maximum l_m and remains constant

$$l_m = \frac{V_t}{p^*} \frac{dp}{dt} = const \qquad (1)$$

in which V_t denotes the volume of the test section, dp/dt the dispersion pressure rate in the test section, and p^* =101.325 kPa. Here the influence of the particles upon the pressure is neglected and $p_0 - p_{01}$ assumed to be small. Hence, the orifice number n and its diameter $\varnothing$ can be obtained by using Eq. (1), i.e.,

$$n\, \varnothing^2 = \frac{4\, V_t}{\pi\, a_s p^*} \frac{dp}{dt} . \qquad (2)$$

From Eq. (1) follows

$$\frac{dp}{dt} = const \qquad (3)$$

whose value is determined by the experimental linear dispersion pressure-time profile, recorded by piezoresistive pressure transducers. Integration of Eq. (3) gives

$$p_0 = p_{01} + \Delta t \ (\ dp/dt \ ) \qquad (4)$$

Using this linear relation one can choose p_{01} and Δt at a given p_0.

The whole dispersion process is detected in an auxiliary lucite tube, which is connected to the test section at the last gas feed entrance (Fig. 1). This tube has the same dispersion arrangement, and therefore the same constant dispersion pressure rate as in the test section. During the dispersion process, the scattered laser light from the particles in different cross sections is recorded by a camera (Fig. 2). Figure 3 shows some photographs of a dispersion process. The dispersion time Δt should be taken greater than the duration in which the particles move from the bottom to the top in the cross section. In order to keep the suspension ahead of the combustion wave, the dispersion process finishes just after the wave propagates through the whole test section, which takes normally less than 25 ms. This causes a difference of less than 1.2% in the initial pressure p_0 between both ends of the test section. In summary, the above method provides a satisfactory homogeneous dust-gas suspension both, along the tube length and within the cross section. It supplies the desired and fairly accurate local dust concentration.

Two kinds of organic dusts were studied, i.e., corn starch and anthraquinone. The particle size distributions and some properties are presented in Table 1. Pictures of the particles obtained by electron microscope are shown in Fig. 4.

Signals were recorded by 32-channel transient recorders with a sampling rate of 500 ns. Pressure and wavefront velocity were measured by Kister 603 B transducers. The flame was detected by pin silicon photodiodes and ion gauges. A double response gauge was developed to measure the arrival of both the shock and reaction front (Fig. 5). It consists of a thin piezoelectric film of 2 mm diameter, and an ion gauge separated from the film by a 3 mm gap. The film responds to the arrival of the shock front, while the ion gauge reacts only on the arrival of the reaction front. This gauge is particularly suitable for the study of local ignition delay times and the wave front structure. The initial temperature is $T_o = 293 \pm 3\ K$.

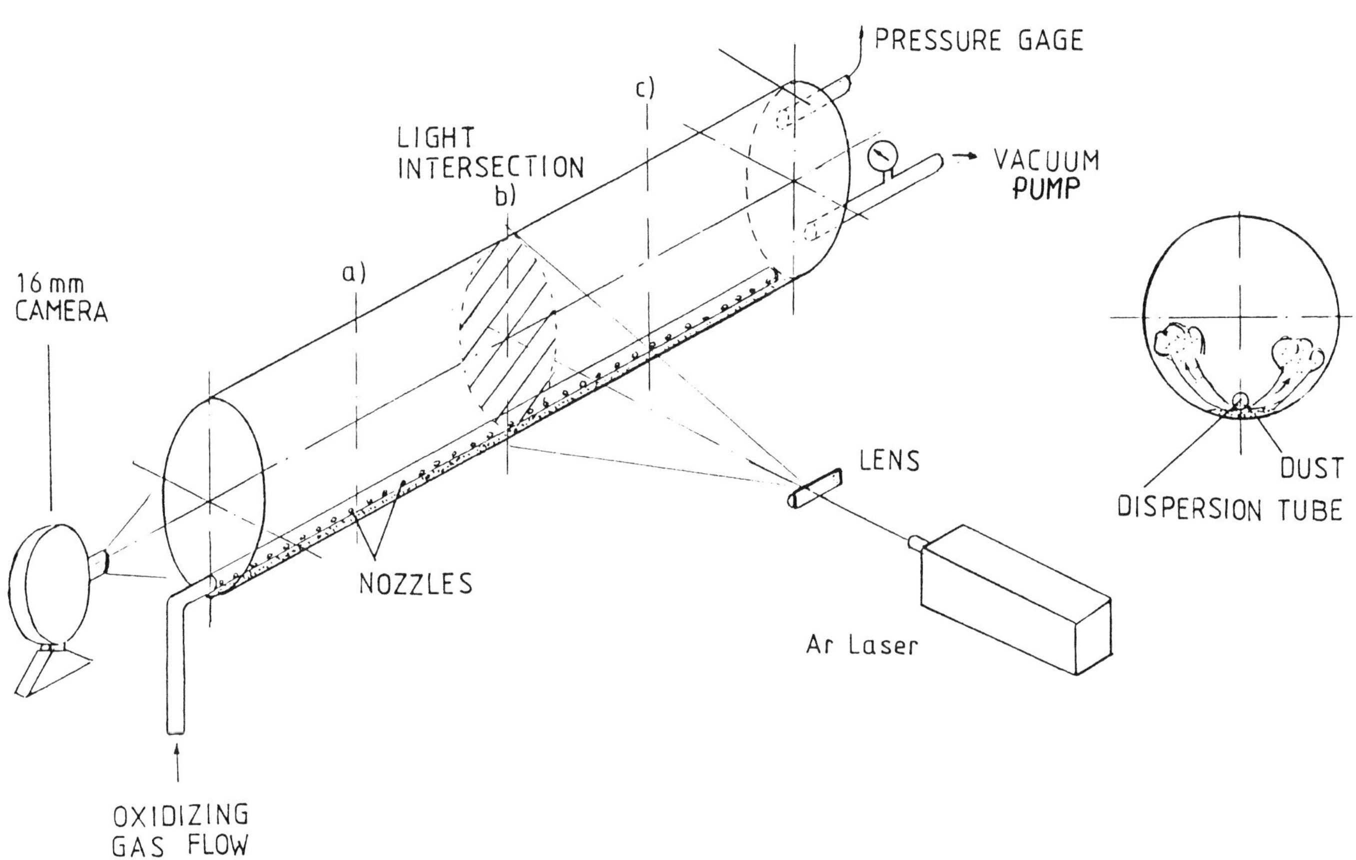

Fig. 2 Scheme of auxiliary apparatus for determining dispersion parameters.

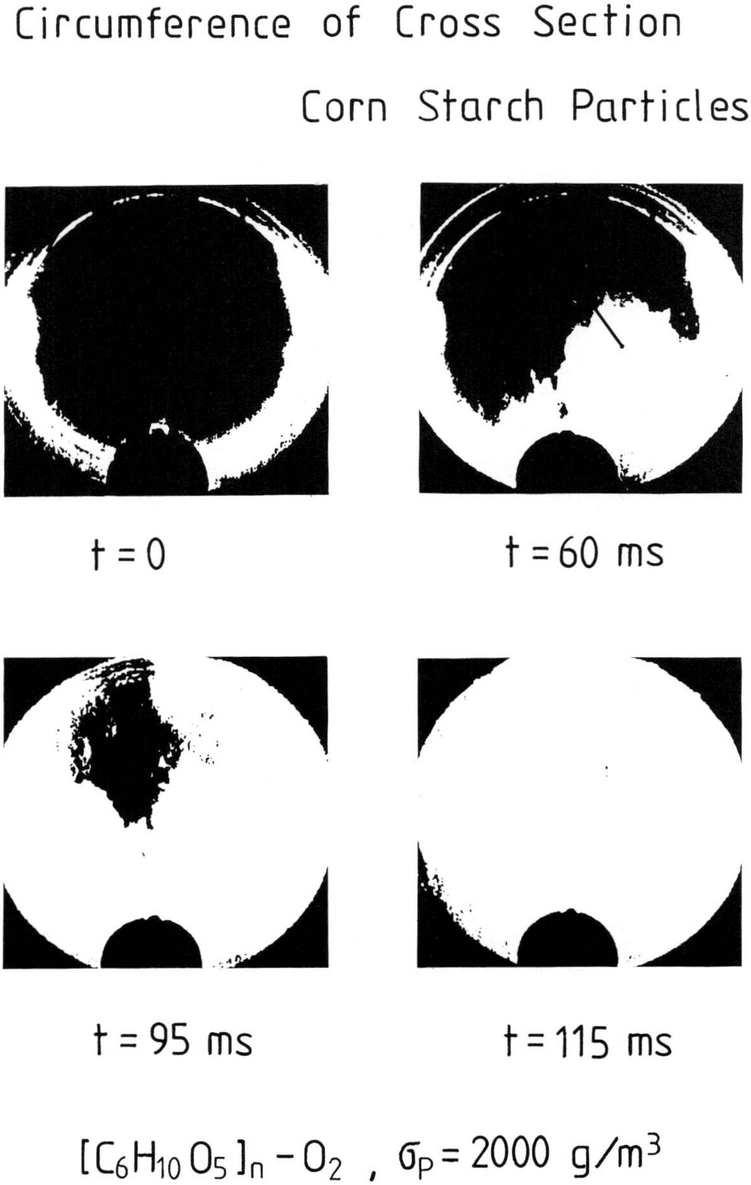

Fig. 3 Photographs of a dispersion process.

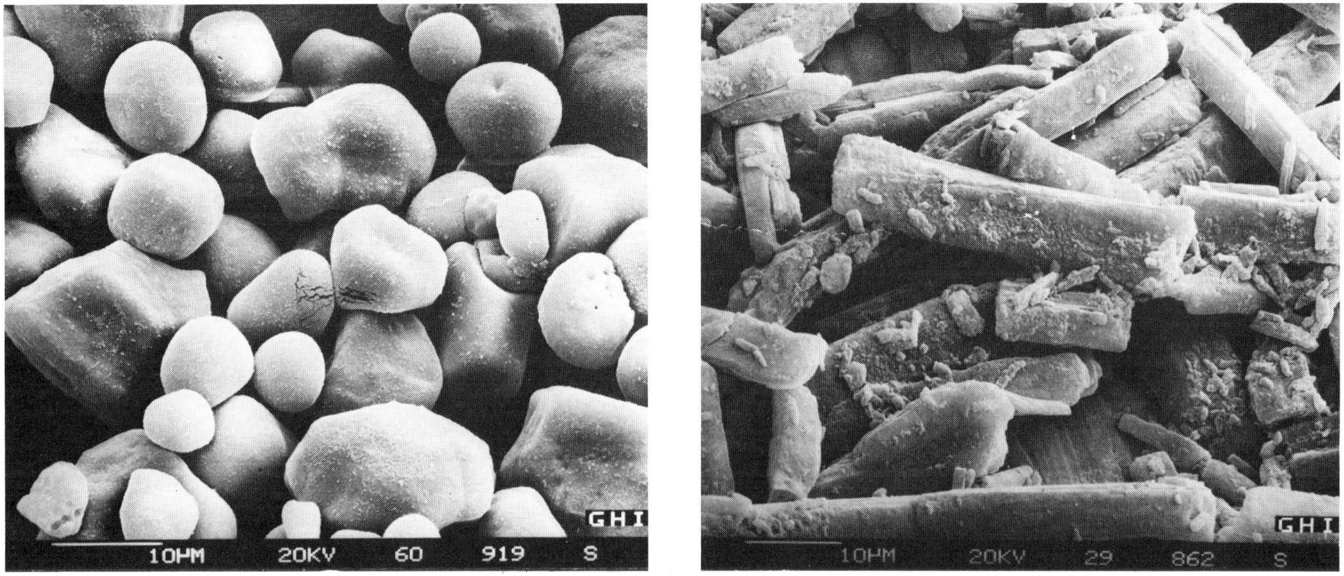

a) CORN STARCH

b) ANTHRAQUINONE

Fig. 4 Photographs of dust particles obtained by electron microscope.

To check the particle distribution and the propagation of the planar initiation wave, corn starch particles-nitrogen mixtures are used at an initiator pressure of p_i = 1 bar. Typical results are presented in Fig.5, 6 and 7. The agreement of the pressure profiles in the cross section implies a symmetrical homogeneous two-phase suspension. A nearly steady propagation region is achieved beginning from x/d = 70 to x/d = 123.4 (x: propagation distance; d: tube diameter). This region has a velocity of 475 m/s and an equilibrium overpressure of 1.8 bar. The two-phase relaxation time within this region is 0.99 ms. The small equilibrium pressure and the corresponding propagation velocity show that the planar initiation wave is not able to play the role of the leading shock wave within a detonation. The short relaxation time is attributed to the small particle size and density.

3. ESTIMATION OF INITIATION ENERGY

Many theoretical models predicting the critical initiation energy of detonation in homogeneous reactive mixtures were developed. Some of them are found to correlate well with the experimental critical charge data.[16] By estimating the effective critical initiation energy of detonations in heterogeneous mixtures in tubes, Tang et al.[17] applied a shock tube technique for a blast wave initiator, which works similarly as the planar wave initiation in the present paper. Their model is based on the assumption that the pressure becomes uniform in the initiation section before the rupture of the diaphragm between the initiation and the test section.

In practical cases a thin plastic diaphragm is often used. Therefore , the following 1 - D model is assumed for the planar wave initiation: As the blust wave emerges from the initiation section, the energy delivered to the mixture in the test section, is equal to the work done by the interface separating the combustion products of the initiation section, from the mixture originally in the test section, i. e.,

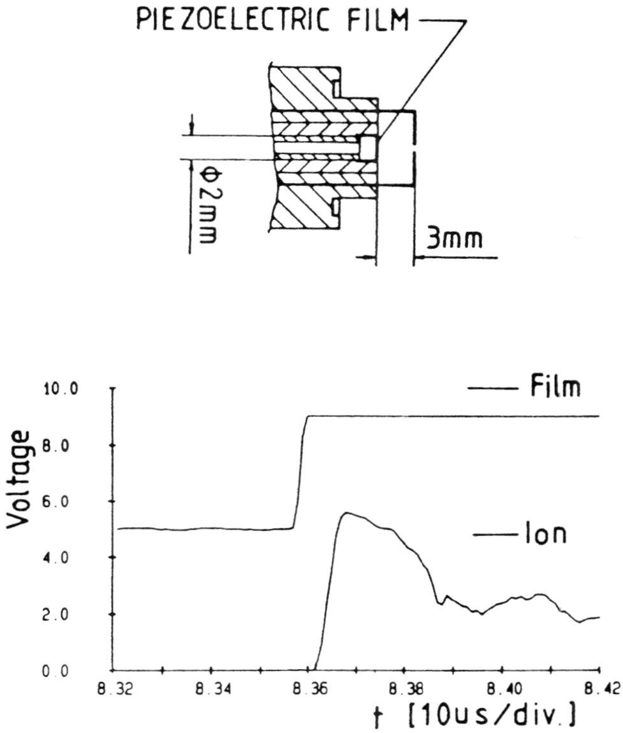

Fig. 5 Double response gauge and its signals, lower: arrival of the reaction front; upper: arrival of the shock front (signal saturated).

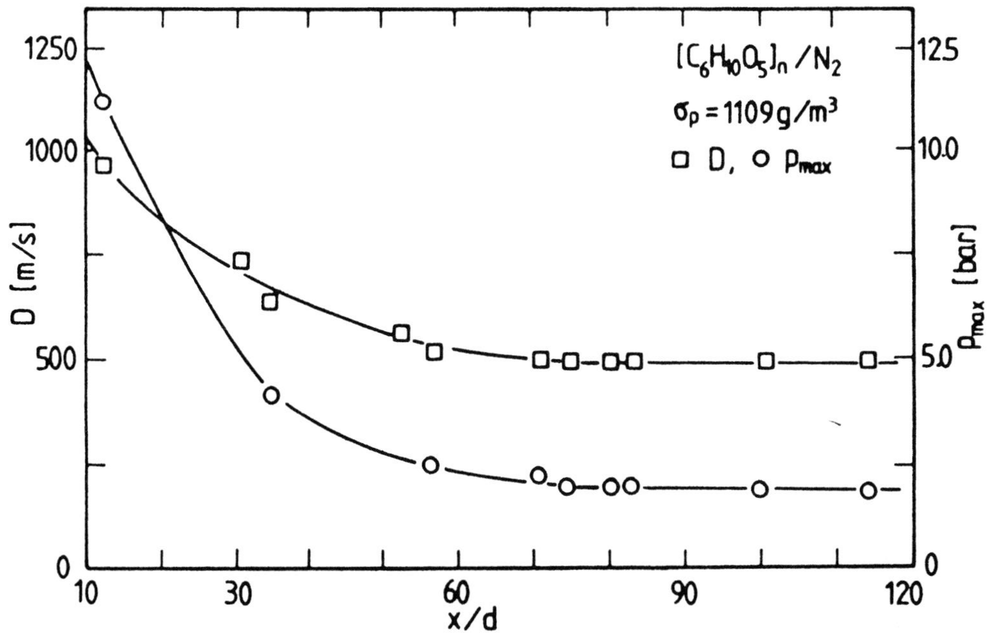

Fig. 6 Propagation velocity and overpressure of planar initiation wave in a inert corn starch dust-nitrogen mixture.

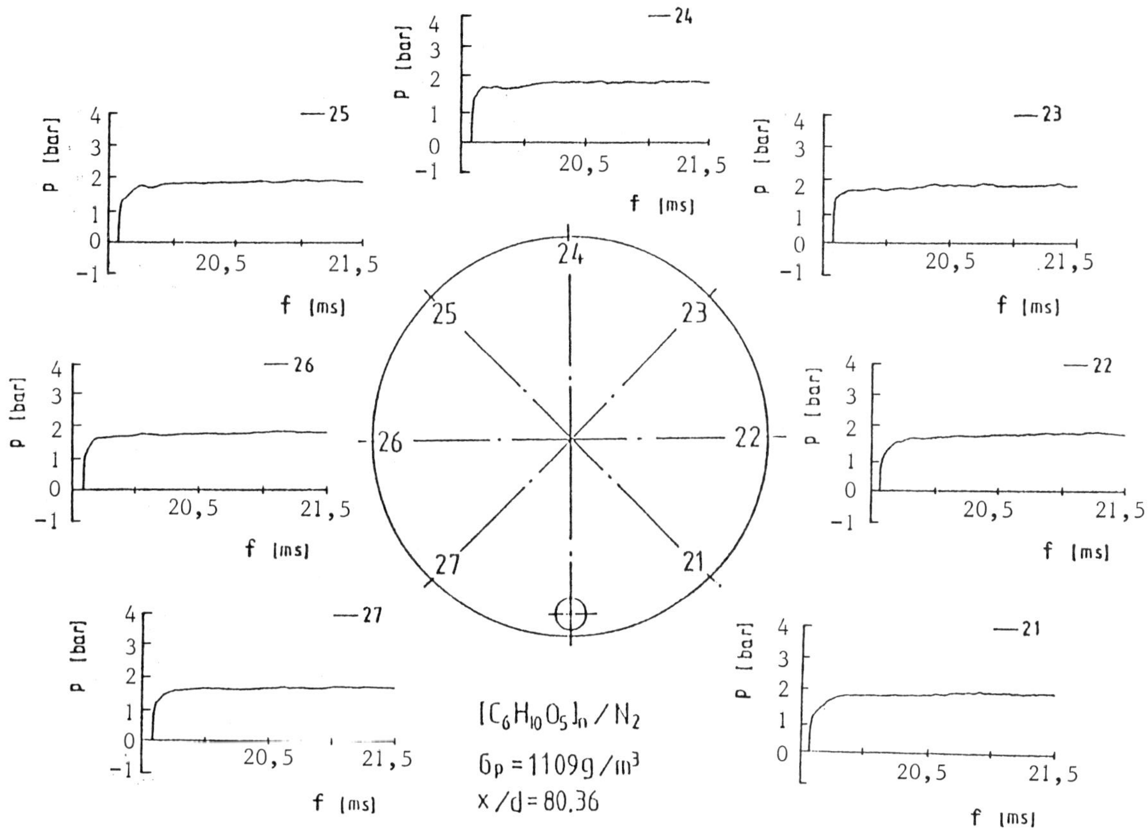

Fig. 7 Pressure profiles at x/d = 80.36 of planar initiation wave in a inert corn starch dust-nitrogen mixture.

$$E_{ec} = \int_0^{\tau_o} p\ v\ dt \tag{5}$$

where E_{ec} denotes the effective critical initiation energy, p and v the pressure and velocity at the interface, and τ_o a reaction time which corresponds to a stable reaction regime of the detonation in the heterogeneous mixture.

To get the values of the state variables at the interface, one-dimensional Euler equations, together with initial and boundary conditions, are solved by means of the random choice method.[18] The initial conditions in the initiation section can be derived by using Taylor's theory[19] as below

$$\frac{x^*}{L} = 1 - \frac{\gamma + 1}{2}\left(1 - \frac{a_{CJ}}{D_o}\right), \tag{6}$$

$$v = \left[D_o - a_{CJ} - \frac{2D_o}{\gamma + 1}\left(1 - \frac{x}{L}\right)\right] H(x - x^*) \tag{7}$$

for $0 < x \le L$,

$$\frac{p}{p_{CJ}} = \left[1 - \frac{\gamma-1}{2}\left(\frac{D_o}{a_{CJ}} - 1\right)\right]^{\frac{2\gamma}{\gamma-1}} H(x^* - x) +$$

$$+ \left[1 - \frac{\gamma-1}{\gamma+1}\frac{D_o}{a_{CJ}}\left(1 - \frac{x}{L}\right)\right]^{\frac{2\gamma}{\gamma-1}} H(x - x^*) \tag{8}$$

for $0 < x \le L$,

$$\frac{\rho}{\rho_{CJ}} = \left(\frac{p}{p_{CJ}}\right)^{\frac{1}{\gamma}} \tag{9}$$

where L denotes the length of the initiation section, D_o, a_{CJ}, p_{CJ}, ρ_{CJ}, γ, g, v, p and ρ are detonation velocity, Chapman-Jouguet value of sound speed, pressure, density, the ratio of specific heats, velocity, pressure and density of the combustion products in the initiation section. According to Eqs. (6) - (9) two of the main initiator parameters are the initiator pressure p_i and the length of the initiation section L.

Zel'dovich's criterion[20] establishes the connection between the critical initiation energy and the chemical induction time τ i.e.,

$$E_c \sim \tau^{\upsilon}$$

where υ = 1, 2, 3 for planar, cylindrical and spherical waves, respectively. This criterion is correct for normal gaseous detonations, since a large amount of the total reaction time is consumed within the induction reaction. But for the detonation in reactive heterogeneous mixtures the observed reaction time τ_o, which will be discussed in the next sections, is at least one order of magnitude greater than the induction time. Hence, if one assumes that the reaction time τ_o is measured from the arrival of the leading shock front to the point, where the gas velocity u relative to the shock front is equal to the local frozen sound speed a_f of the gas phase, then, for an ideal direct initiation of stable detonation in a heterogeneous mixture, the interface should be the gaseous frozen sonic plane with $u = a_f$, and the shock front should be located in a distance $x_o = D\tau_o$ from this plane, where D denotes the detonation velocity. Thus, the initiation wave must possess such a pressure profile that the transmitted wave in the heterogeneous mixture has a strength of $p > p^*$ with a duration not less than τ_o. At the interface the strength of the transmitted wave satisfies the condition

$$p = p^* \qquad \text{at} \qquad t = \tau_o \tag{10}$$

where p^* denotes the pressure at the gaseous frozen sonic plane of the detonation. For the spin detonation p^* can be taken as the periodic average pressure in the stable reaction zone. By using Eqs. (5) and (10) one can predict E_{ec}.

For simplicity both, the chemical reaction before onset of the detonation and the relaxation in the heterogeneous mixture are not considered at present. Therefore the estimated effective critical initiation energy is a conservative one. To justify the model, a comparison with an experiment is given in Fig. 8. It was made in a corn starch dust-air mixture, which had a dust concentration of $\sigma_p = 1100$ g/m^3 and an initial pressure of $p_o = 1$ bar, by means of the planar wave initiation with an initiator pressure of $p_i = 1$ bar. Figure 8 shows a good agreement except for the onset of the transmitted wave, because of the neglection of the relaxation in the heterogeneous mixture.

In Table 2 the results from the present model are correlated with experimental data of Tang et al., in 400 - μm decane drops-air mixtures.[17] Taking the critical initiator pressure p_{ic}, the length of the initiation section L and the detonation velocity D in Ref. 17, E_{ec} and τ_o are predicted. τ_o has an order of magnitude of 10^2 μs. The calculated source energy E_c in Ref. 17, which was released in the initiation section as a result of the constant volume combustion, is also presented in Table 2. It is noticeable that the ratio of the predicted E_{ec} to the source energy of the critical initiation energy E_c is approximately equal to $(\gamma_s - 1)/\gamma_s$, i.e.,

$$E_c \approx \frac{\gamma_s}{\gamma_s - 1} E_{ec} \tag{11}$$

where γ_s denotes the isentropic exponent of the burned mixture. The average effective initiation power is given by

$$P_{ec} \approx \frac{E_{ec}}{\tau_o} \tag{12}$$

The large estimate of $E_{ec}/E_c \approx 50\%$ in Ref. 21 from the model of Tang et al., is mainly due to the applied duration in which work is done at the interface. They used the duration, which was required by the supposed rarefaction wave to traverse the initiation section, reflect from the end wall, and intersect the shock front.

For the case of the free jet initiation an underexpanded jet is expected,[22] and the familiar jet pattern is shown in Fig. 9 with a single Mach disk. A spherical shock wave initially forms near the small sonic orifice and decays rapidly due to spatial expansion. Then the free jet

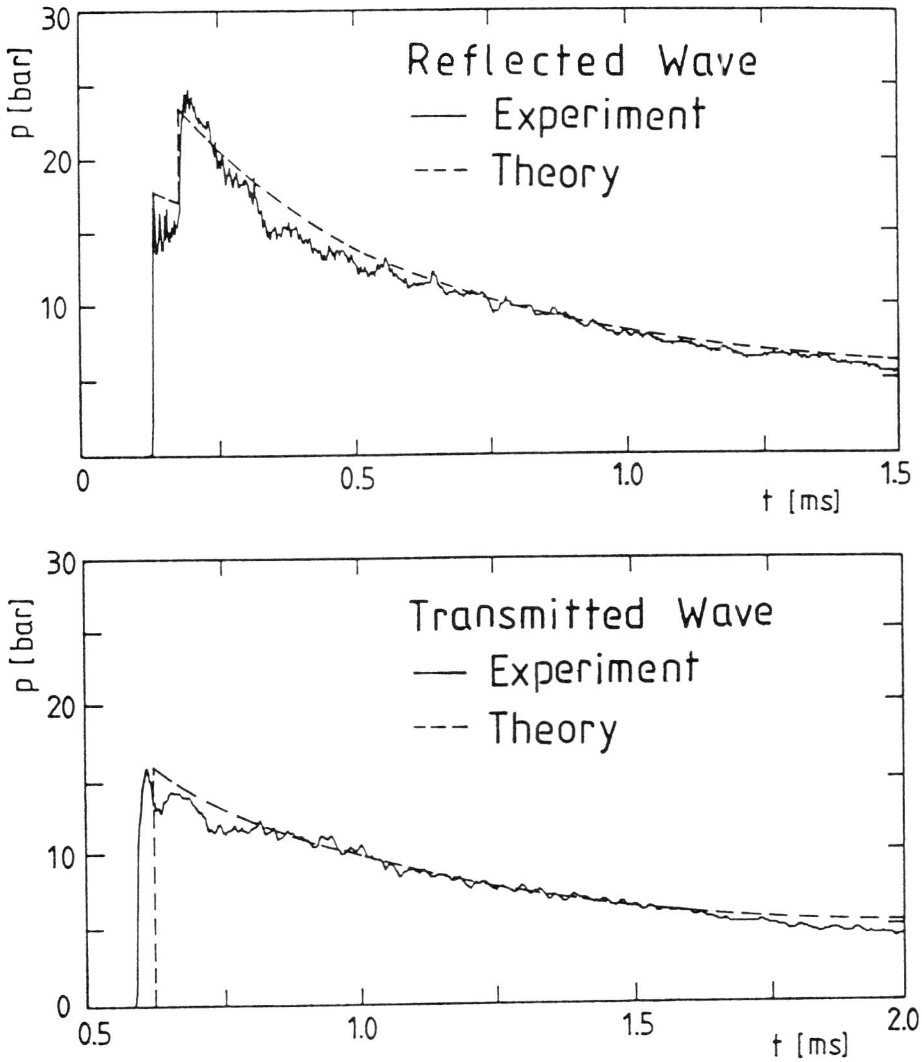

Fig. 8 Pressure histories from experiment and theory in a corn starch dust-air mixture: upper trace at x = − 0.037 m, lower trace at x = 0.478 m. x = 0 is the initial position of the interface.

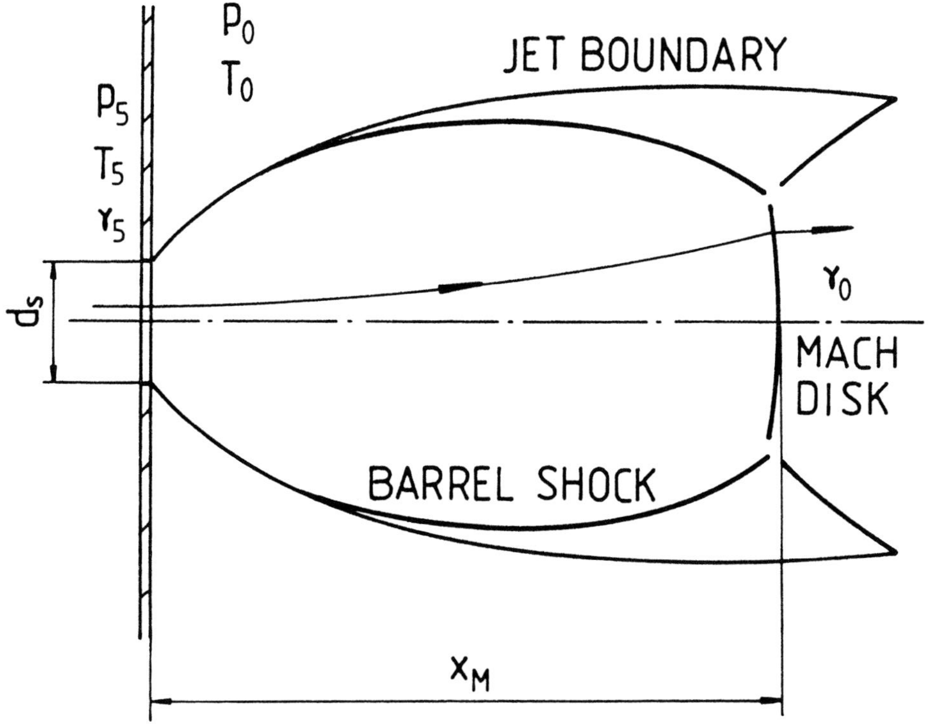

Fig. 9 Structure of underexpanded jet.

expands from the sonic orifice into the test section. The initiation takes place at first most probably in the shear layers where the unburned heterogeneous mixture is intensely mixed with the burned products of the initiation medium. Assuming that the flow at the circular orifice is uniform and sonic, the source energy E_o through the orifice is given by

$$E_o = \frac{\pi}{4} d_s^2 \int_o^t \rho_s \left(e_s + \frac{a_s^2}{2} \right) a_s \, dt \qquad (13)$$

where d_s, ρ_s, e_s and a_s denote the diameter, density, specific internal energy and sound speed at the orifice respectively. By using the energy conservation equation and the isentropic relations for an ideal gas one derives

$$E_o = \frac{\pi}{4} \alpha \, d_s^2 \int_o^t p_5 \, T_5^{\frac{1}{2}} \, d\,t \qquad (13^*)$$

with

$$\alpha = R^{\frac{1}{2}} \gamma^{\frac{1}{2}} \frac{\gamma^2 - \gamma + 2}{\gamma^2 - 1} \left(\frac{\gamma+1}{2} \right)^{-\frac{\gamma+1}{2(\gamma-1)}} \qquad (14)$$

where p_5 and T_5 are pressure and temperature of the reflected state at the end wall of the initiation section, R and γ the specific gas constant and the ratio of the specific heats of the combustion products of the initiation medium. To get the reflected state, the one-dimensional Euler equations, together with initial and boundary conditions including Eqs.(6)-(9), are solved similarly as in the case of the planar wave initiation.

A simple estimate for the jet length x_M, the distance from the sonic orifice to the Mach disk, is derived in a similar way as in Ref. 23. The method is based on the assumption that the expansion from the sonic orifice propagates as far as the recompression takes place over a shock front to the initial pressure p_o in the test section. As result one gets

$$\frac{x_M}{d_s} = \beta \left(\frac{p_s}{p_o} \right)^{\frac{1}{2}} \tag{15}$$

with

$$\beta = \left(\frac{1}{\gamma_o + 1} \right)^{\frac{1}{2}} \left(\frac{\gamma^2}{\gamma^2 - 1} \right)^{\frac{1}{4}} \left(\frac{2}{\gamma+1} \right)^{\frac{1}{2(\gamma-1)}} \tag{16}$$

where γ_o is the ratio of the specific heats for the mixture ahead of the Mach disk. For $\gamma = \gamma_o = 1.2$ one obtains $\beta = 0.71$.

In the following sections the source initiation energy E_o is estimated from Eqs. (5), (11) and (13) for the planar wave and the free jet initiation respectively, where integral (5) is to be multiplied by $\pi d^2/4$ (d: tube diameter) and γ_s is taken as 1.18. The upper limit of the integrals in Eqs. (5) and (13) is taken as 1 ms, a typical order of magnitude of initiation duration obtained from the pressure profiles at the beginning of the test section.

4. STABLE DETONATION

Corn starch particles-oxygen mixtures are first used to investigate the development of detonation. The initial pressure is p_o = 1 bar. Figure 10 represents a transition from deflagration to detonation for an initial particle concentration of σ_p = 500 g/m^3, with free jet initiation of p_i = 2 bar or E_o = 10.38 kJ. A detonation with pulsating combustion forms at x/d = 78.51. This detonation remains stable from x/d = 78.51 to the end of the test section (Fig. 11). Its velocity has an average value of D = 1860 m/s with a deviation of less than 2%. Figure 11 gives also the propagation velocity for the same initial particle concentration, but with planar wave initiation (p_i = 1 bar or E_o = 124 kJ). Its detonation velocity reaches the same value as that with free jet initiation and remains constant from x/d = 57.16 to the end. These results show clearly that a stable self-sustained detonation can be

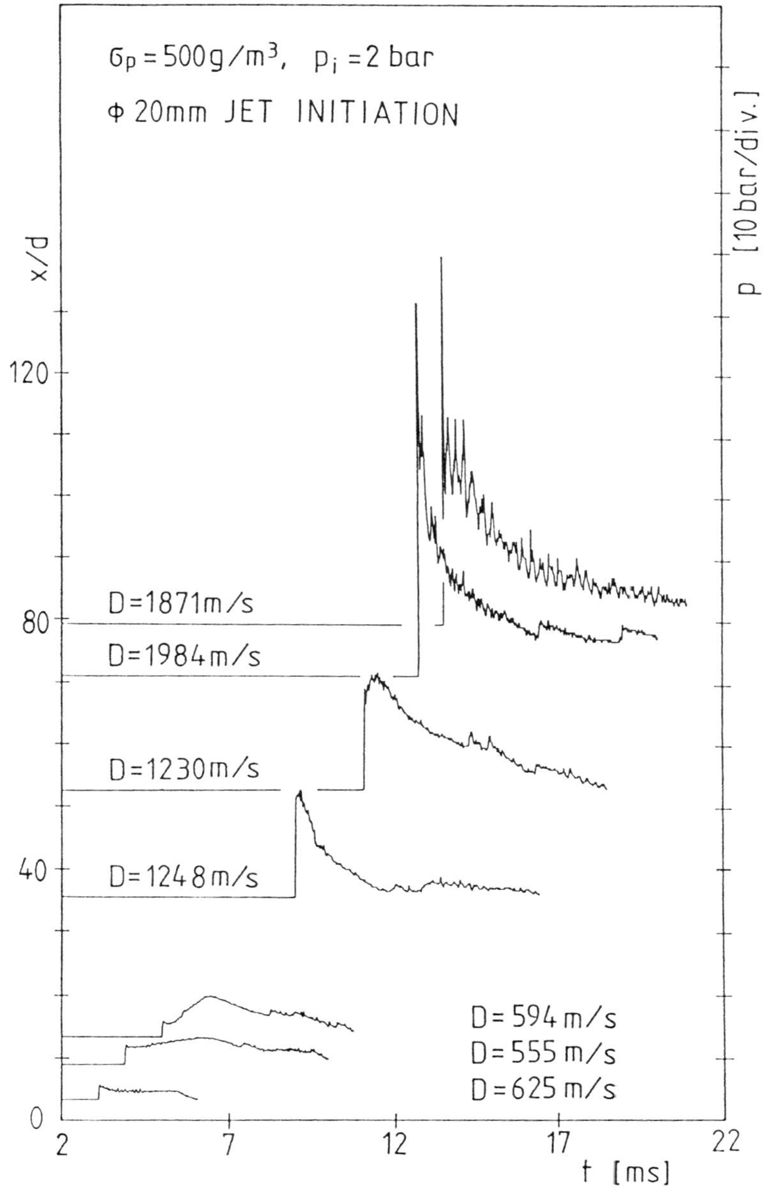

Fig. 10Transition to detonation in a corn starch dust-oxygen mixture of φ = 0.45 and p_0 = 1 bar with free jet initiation of p_i = 2 bar.

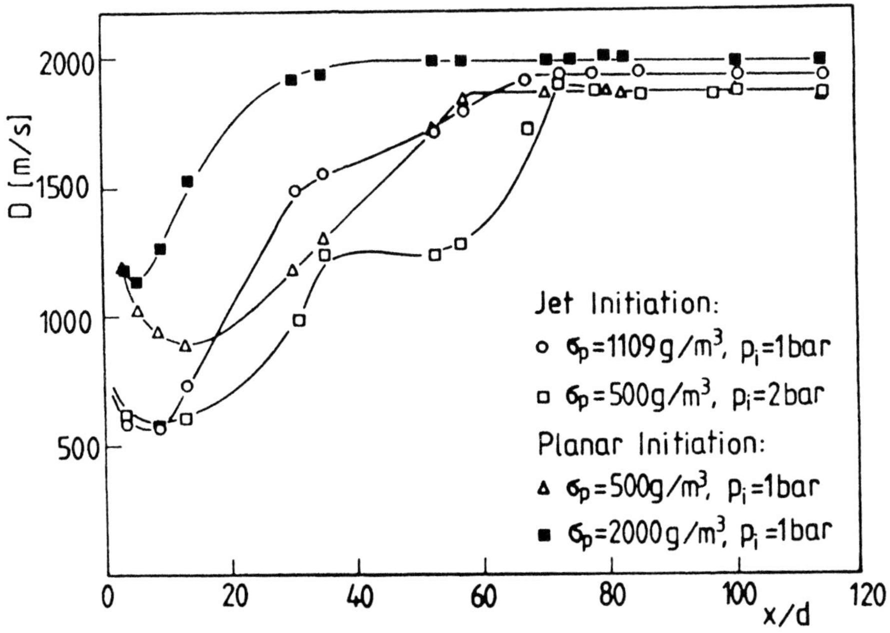

Fig. 11 Wavefront velocities versus propagation distance in corn starch dust-oxygen mixtures.

achieved in such a two-phase system and that its physical parameters at the wavefront are independent from source and magnitude of the initiation energy.

The same conclusion can also be drawn from a wide initial particle concentration range. Developments of the wavefront velocity for several initial particle concentrations are also shown in Fig. 11. The pressure records, for example in Figs. 10 and 12, show a stable region of pulsating combustion behind the wavefront. This region exists, depending on particle concentrations, for a duration of $\tau_o = 0.8 - 1.2$ ms in the experiments (Table 3). Thereafter the pressure decays obviously. If one considers that the pressure decay is due to the expansion from behind, the position $x_o = D\,\tau_o$ can be explained as that point, where the gas velocity u relative to the shock front is equal to the local frozen sound speed a_f of the gas phase. Thus, the flow beyond the gaseous frozen sonic locus cannot influence the subsonic region $0 < x < x_o$ in wave front fixed coordinates. Here the reaction time τ_o is of about two or three orders of magnitude greater than that of the C - J detonation in gaseous mixture. The stable detonation velocity D and the pressure peak p_s in the stable combustion region are given in Table 3. No upper detonation limit is observed up to an equivalence ratio $\varphi = 8.11$.

5. SPIN STRUCTURE NEAR THE TUBE WALL

The pulsating behavior of the stable detonation is associated with its three-dimensional structure. To study the detonation structure, a ring having gauges arranged on two circumferences 166 mm apart is used in the test section. In each circumference there are eight gauges spaced at 45 degree intervals. The double response gauges (Fig. 5), together with the pressure transducers yield detailed information of the periphery of the waves.

Typical pressure records on the periphery are shown in Fig. 12 which exhibits obviously a single spin detonation. The axial propagation of detonation velocity D and the spin track angle α are determined by the arrival times and the positions of the spin head between the two circum-

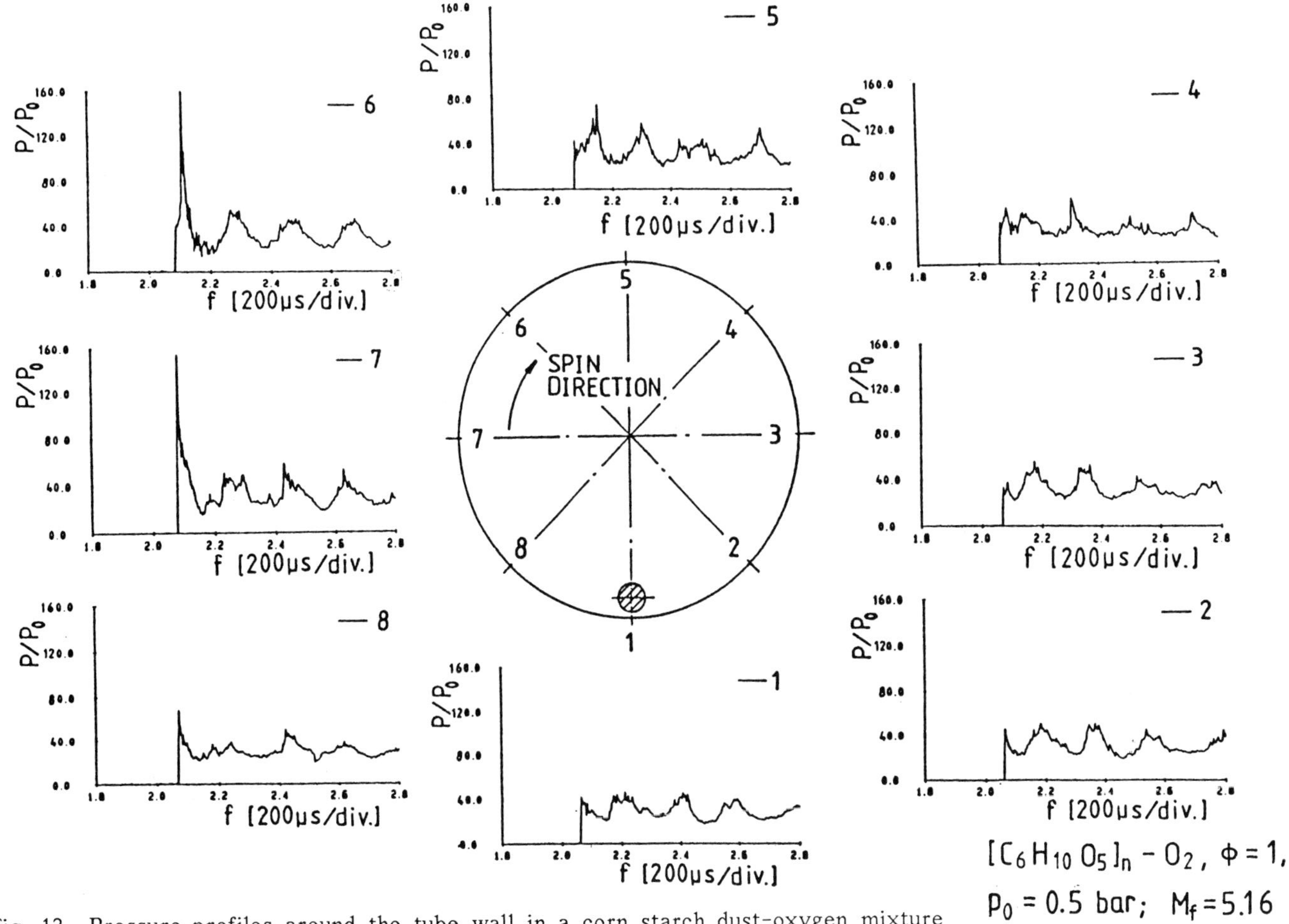

Fig. 12 Pressure profiles around the tube wall in a corn starch dust-oxygen mixture of $\varphi = 1$, $p_0 = 0.5$ bar. $M_f = 5.16$.

ferences. D is identical with the experimental detonation velocity. The instantaneous wave structure around the tube wall can be constructed from D, α, the arrival times and the gauge positions for each experiment. Figure 13 is derived from such data of four experiments with equivalence ratio $\varphi = 1$ and initial pressure $p_o = 0.5$ bar.

The basic framework of the wave structure in Fig. 13 consists of the leading shock wave AEFGA, the transverse shock wave BC, the reaction front AEHCB and the tail wave CI. There exist two triple points A and B. The front segment AE at the spin head A is oblique so that the normal velocity of the mass flow into it corresponds to an overdriven detonation with a value of 1.41 D, while the segment GA at A has a normal velocity of 0.85 D. Thus the normal flow velocity intc the shock front AEFGA reduces monotonically from 1.41 D to 0.85 D. The pressure distribution along the shock front AEFGA is shown in Fig. 14 in which the triple point B is assumed to coincide with A. The transverse wave BC and the segment AE near A, is accompanied by a reaction front which yields the strongest signal of the double response gauge, whereas the reaction behind the section HC, near C is very weak. This difference the reaction fronts is due to the different amount of unburned mixture influx. The rearward extension of transverse wave BC decays rapidly to a harmonic wave, which can easily be seen from the pressure records in Fig. 12.

The length of induction zone, behind the shock segment FGA is even bigger than the axial distance between the fronts FGA and EHC, because of the angle of incidence of mixture influx to the reaction front. On account of the two-phase relaxations, the induction zone length $x_i(P)$ behind a point P of the leading shock front should be defined as

$$x_i(P) = \int_0^{\tau_i} u_{px} \, d t \tag{17}$$

Here, the integral is taken along the particle path which begins at P immediately behind the shock front and ends at the point where the particle temperature reaches $T_p = T_i$, T_i denotes the ignition temperature

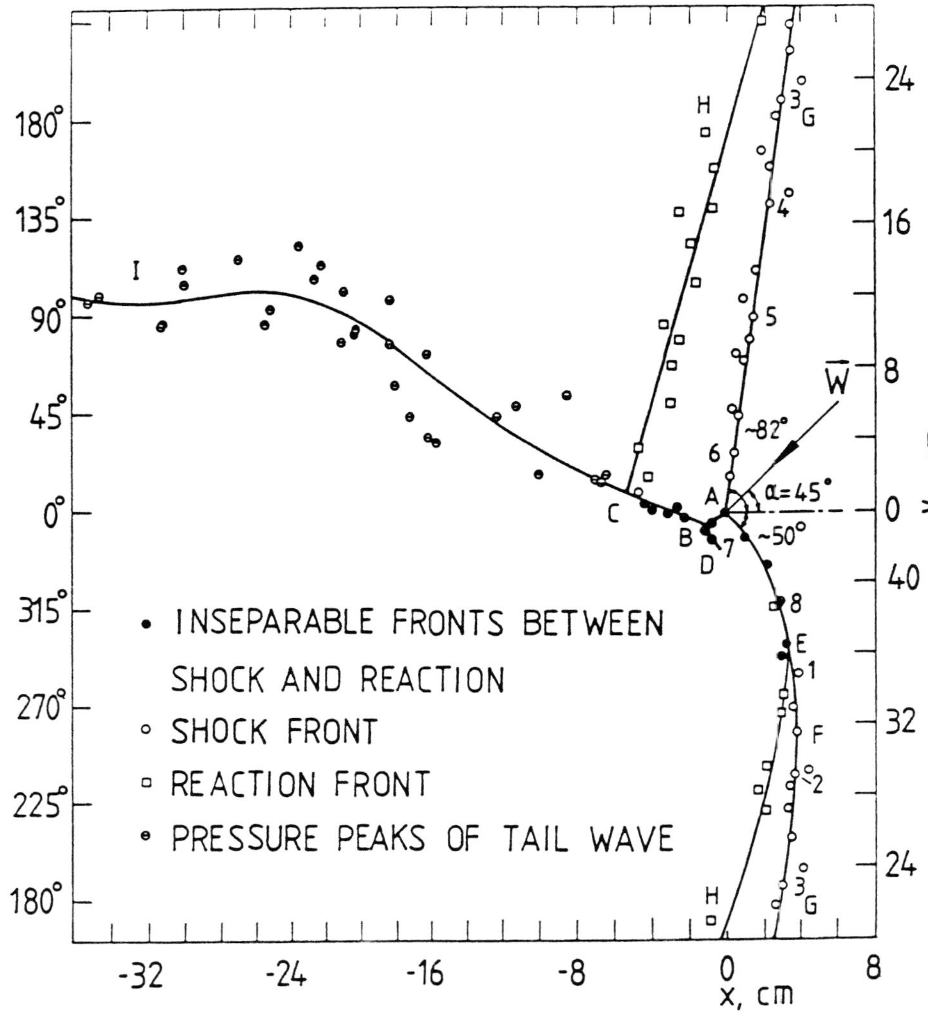

Fig. 13Instantaneous wave structure around the tube wall in a corn starch dust-oxygen mixture of $\varphi = 1$, $p_0 = 0.5$ bar. x: axial coordinate; y: circumferential coordinate. Points with numbers correspond to signals in Fig. 12.

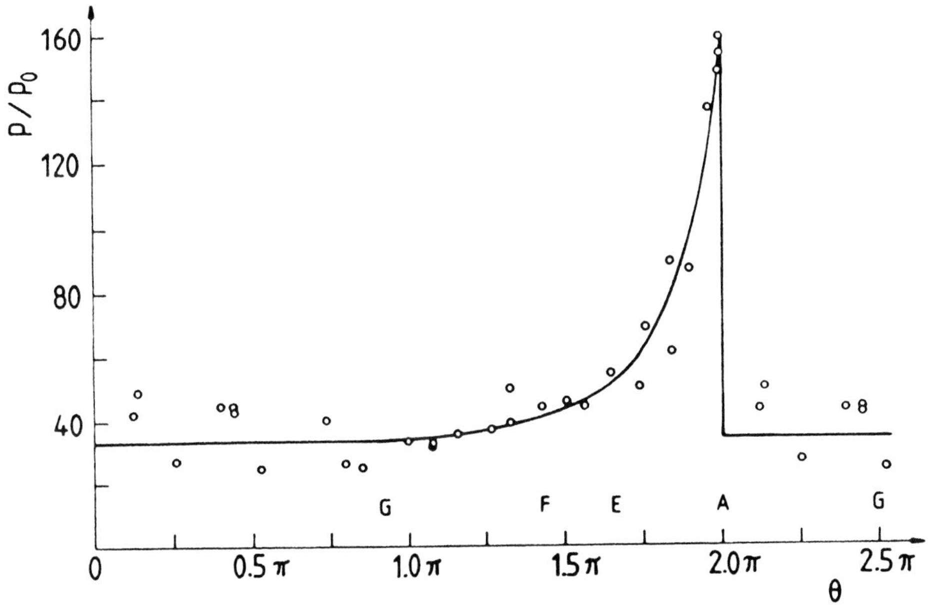

Fig. 14 Pressure distribution along shock front AEFGA in Fig. 13.

of the particle. τ_i is the induction time corresponding to the point $T_p = T_i . u_{px}$ describes the axial component of the particle velocity in wavefront fixed coordinates.

Newton's equation of motion for a spherical particle in y - direction (Fig. 13)

$$\frac{1}{6} \pi d_p^3 \rho_p \frac{du_{py}}{d t} = \frac{C_D}{4} \pi d_p^2 \frac{1}{2} \rho |u_y - u_{py}| (u_y - u_{py}) \tag{18}$$

is used for the foremost point F. Here d_p denotes the particle diameter, ρ_p the particle material density, and C_D the particle drag coefficient. u_y and u_{py} describe the circumferential components of the gas and particle velocity in wavefront fixed coordinates, respectively. At F the shock front is normal to the tube generatrix, thus, the normal velocity of the shock front is equal to D. Using the initial conditions

$$(u_y)_{t=0} = (u_{py})_{t=0} = V_o$$

where V_o denotes the transverse wave velocity on the periphery, one obtains from Eq. (18)

$$\left(\frac{d\ u_{py}}{d\ t} \right)_{t=0} = 0.$$

Thus, on the particle trace beginning at F, both u_y and u_{py} possess the equilibrium value and

$$u_{py} = (u_{py})_{t=0} = V_o \tag{19}$$

holds. Accordingly, from Eqs. (17) and (19) one obtains

$$x_i(F) = V_o \int_0^{\tau_i} \frac{d\ t}{tan\ \beta}, \qquad tan\ \beta = \frac{u_{py}}{u_{px}} \tag{20}$$

In absence of a detailed calculation only an estimate is made here. An approximation of Eq. (20) yields

$$x_i(F) = \frac{V_o}{(tan\ \beta)_m}\ \tau_i(F) \tag{21}$$

with

$$(tan\ \beta)_m = \frac{2\ V_o}{(u_{px})_{t=\tau_i} + (u_{px})_{t=0}}\ .$$

If $(u_{px})_{t=\tau_i} \approx (u_{px})_{t=0}/2$ according to Ref. 24, then $(tan\ \beta)_m \approx 4/3$ and correspondingly from Fig. 13 one gets $x_i(F) \simeq 20$mm. Thus, a value for $x_i(F)$ between 18 mm $[(u_{px})_{t=\tau_i} = (u_{px})_{t=0}]$ and 22 mm $[(u_{px})_{t=\tau_i} = (u_{px})_{t=0}/7]$ appears to be reasonable. From Eq. (21) the induction time τ_i lies therefore between 13 μs and 16 μs for point F, where the frozen Mach number amounts to $M_f = 5.16$, and the gas temperature is $T = 1792$ K immediately behind the shock front.

The single spin structure of the two-phase system here is strikingly similar to that of marginal gaseous detonations[25-26]. This supports the view that spin detonation appears to be the normal mode of propagation, whenever an induction zone of sufficient length exists. In a marginal gaseous detonation, a weaker mixture near the detonation limits, or a lower initial pressure, causes the induction zone length to increase. For instance, Schott[26] observed a single spin detonation in a mixture containing 5.0% acetylene, 7.5% oxygen and 87.5% argon at initial pressures near 18 Torr in a tube 4-in. in diameter. He gave an estimate of $x_i(F) = 5 - 10$ mm. In the reactive particles-oxidizing gas system studed here, firstly, the momentum and temperature relaxations and predominant immediately behind the shock front. There the gas velocity and temperature jump at once to the new state corresponding to the shock strength, while the particle velocity and temperature increase continuously from the frozen states, i.e., the initial state. This causes the induction zone length to increase. Secondly, the heterogeneous effects like pyrolysis, surface combustion etc. also cause the induction zone length to increase. Normally corn starch particles have a high percentage of volatiles. With the small particles used in these experiments, gasification behind the shock front may thus be sufficiently rapid to allow substantial reaction to occur in the gas phase. But the effect on the increase of the induction zone length due to the existence of particles is obviously not negligible. From this point of view the spin behavior of the detonation may also be expected in mixtures including particles with little percentage of volatiles like aluminium particles, etc. Recently Borisov et al.[27] reported their observation on spin detonations in aluminium dust-air mixtures. In the case of flake aluminium dust the obtained pressure-time profile is similar to that at x/d = 78.51 in Fig. 10.

Unlike the gaseous case, the single spin detonation in the present study appears in a wide concentration range with normal initial pressures. Two-head spin detonation was observed in the region of φ = 2.71 - 4.51 at p_o = 1 bar, and an example is shown in Fig. 15. The average experimental data of the spin structure at the tube wall are given in Table 3. Although the detonation of the two-phase system studied here shows spinning behavior, the pressure in the stable combustion region remains quite high. In Table 4, one can see that the reaction time τ_o is of an order of 10^2 - 10^3 μs, which corresponds to a stable reaction zone length of x_o = 4-5.5 λ (λ: cell width), whereas the induction time at the point F has an order of magnitude of 10 μs. Here x_o is obtained by counting the number of pressure pulses in the stable reaction zone.

The detailed spinning or cellular structure, discussed in this section, supports strongly the existence of a stable self-sustained detonation in dust fuel-gas oxidizer mixtures. As in the case of homogeneous gaseous mixtures, the transverse wave plays the dominant role in stable propagation of detonation in this heterogeneous system.

6. INTERIOR SPIN STRUCTURE

The interior structure of the spin detonation has been unclear until now even for gaseous mixtures. In order to try to reveal some details of the interior structure of the single spin detonation, an end-

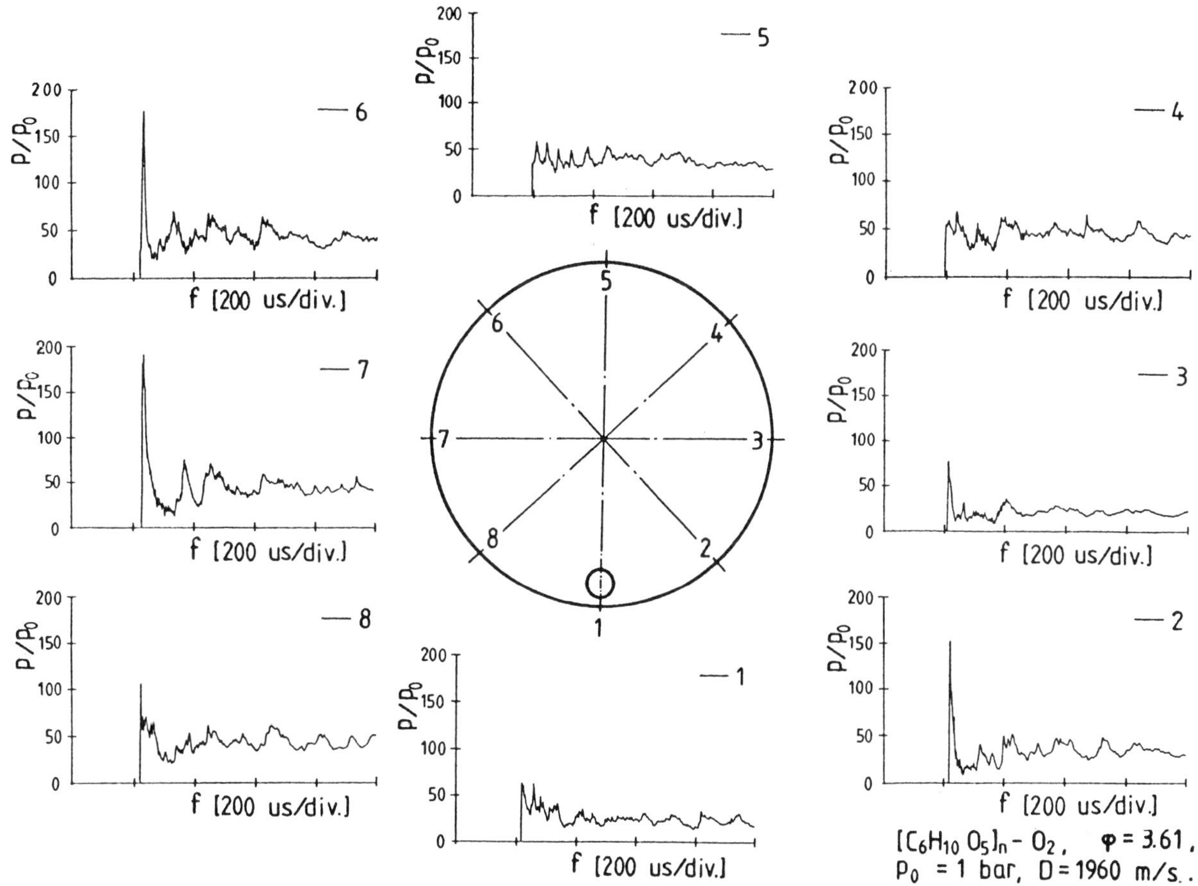

Fig. 15 Pressure profiles around the tube wall in a corn starch dust-oxygen mixture of φ = 3.61 and p_0 = 1 bar.

plate having 17 pressure gauges is used: eight on a circle of radius r = 61.5 mm; further eight on another circle of radius r = 30.5 mm and one at the center (Fig. 16). Experiments were carried out again with $\varphi = 1$ and $p_o = 0.5$ bar.

Typical pressure records from one experiment are presented in Fig. 16. The instantaneous structure of the leading shock front derived from two experiments for r = 0, 30.5 mm and 61.5 mm, respectively, are shown in Fig. 17. These results show the following facts:

1. Single spin structure exists also in the interior region of the tube cross section.
2. The pressure profiles from both, the inner and outer circle on a radial line differ with respect to their onset but are equal in phase. This means that the spins on both circles have a same constant angular velocity Ω (Some values of Ω are shown in Table 3).
3. Approaching the center, the pressure difference along the leading shock front and its curvature become smaller, the spin head moves forwards.
4. The transverse wave and the tail wave become weaker while approaching the center. At the center the tail wave disappears.

The incident flow field in wavefront fixed cylinder coordinates can be derived by using the Navier-Stokes equations and the equation of mass conservation

$$u_{ro}\frac{du_{ro}}{dr} - \frac{(u_{\vartheta o}+r\Omega)^2}{r} = -\frac{1}{\rho_{mo}}\frac{dp_o}{dr} + \frac{4}{3}\nu\left(\frac{d^2u_{ro}}{dr^2} + \frac{1}{r}\frac{du_{ro}}{dr} - \frac{u_{ro}}{r^2}\right), \quad (22)$$

$$u_{ro}\frac{du_{\vartheta o}}{dr} + \frac{1}{r}u_{ro}u_{\vartheta o} + 2\Omega u_{ro} = \nu\left(\frac{d^2u_{\vartheta o}}{dr^2} + \frac{1}{r}\frac{du_{\vartheta o}}{dr} - \frac{u_{\vartheta o}}{r^2}\right), \quad (23)$$

$$u_{ro}\frac{du_{xo}}{dr} = \nu\left(\frac{d^2u_{xo}}{dr^2} + \frac{1}{r}\frac{du_{xo}}{dr}\right), \quad (24)$$

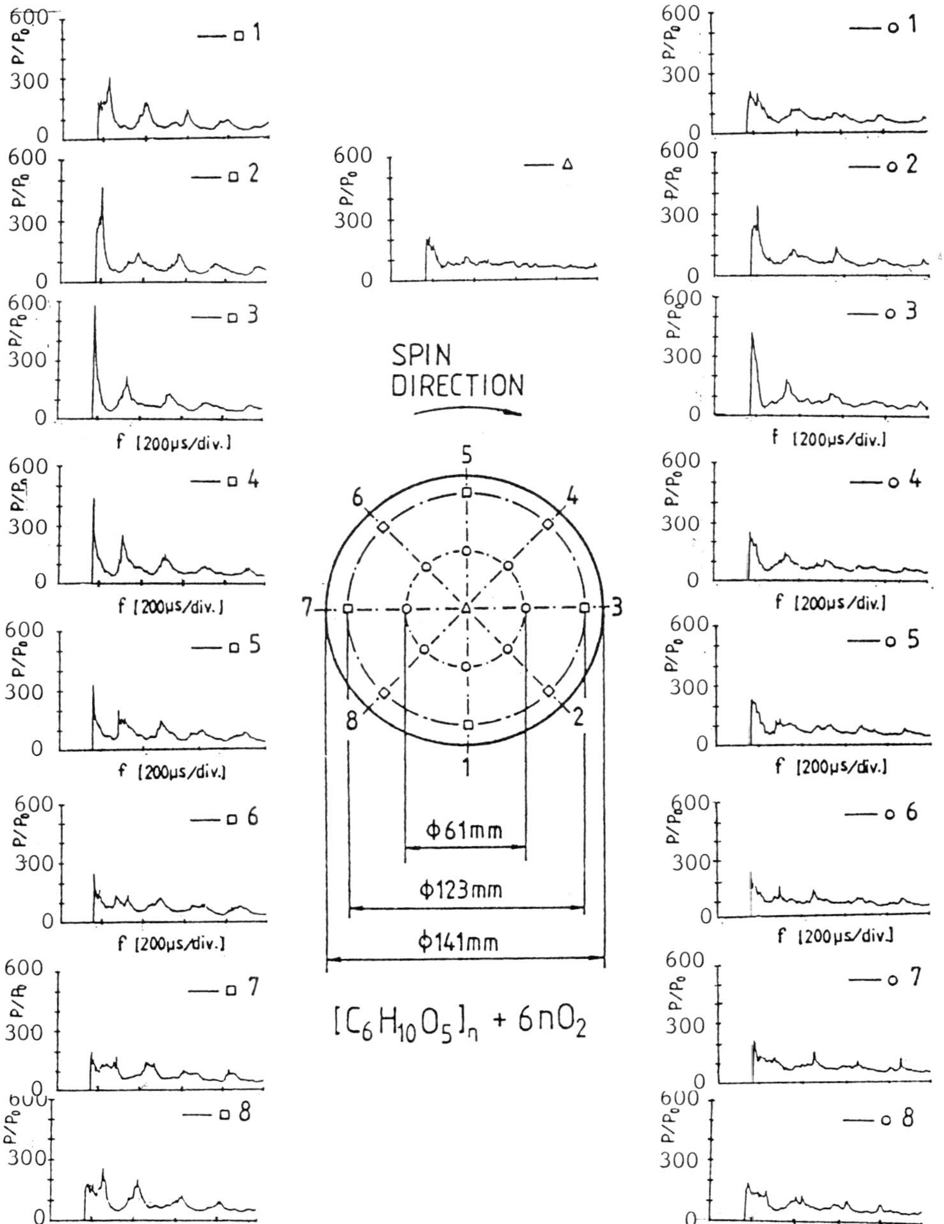

Fig. 16 Pressure profiles at the center and around inner circles in a corn starch dust-oxygen mixture of $\varphi = 1$, $p_O = 0.5$ bar. Δ : r = 0; ○ : r = 30.5 mm; □ : r = 61.5 mm.

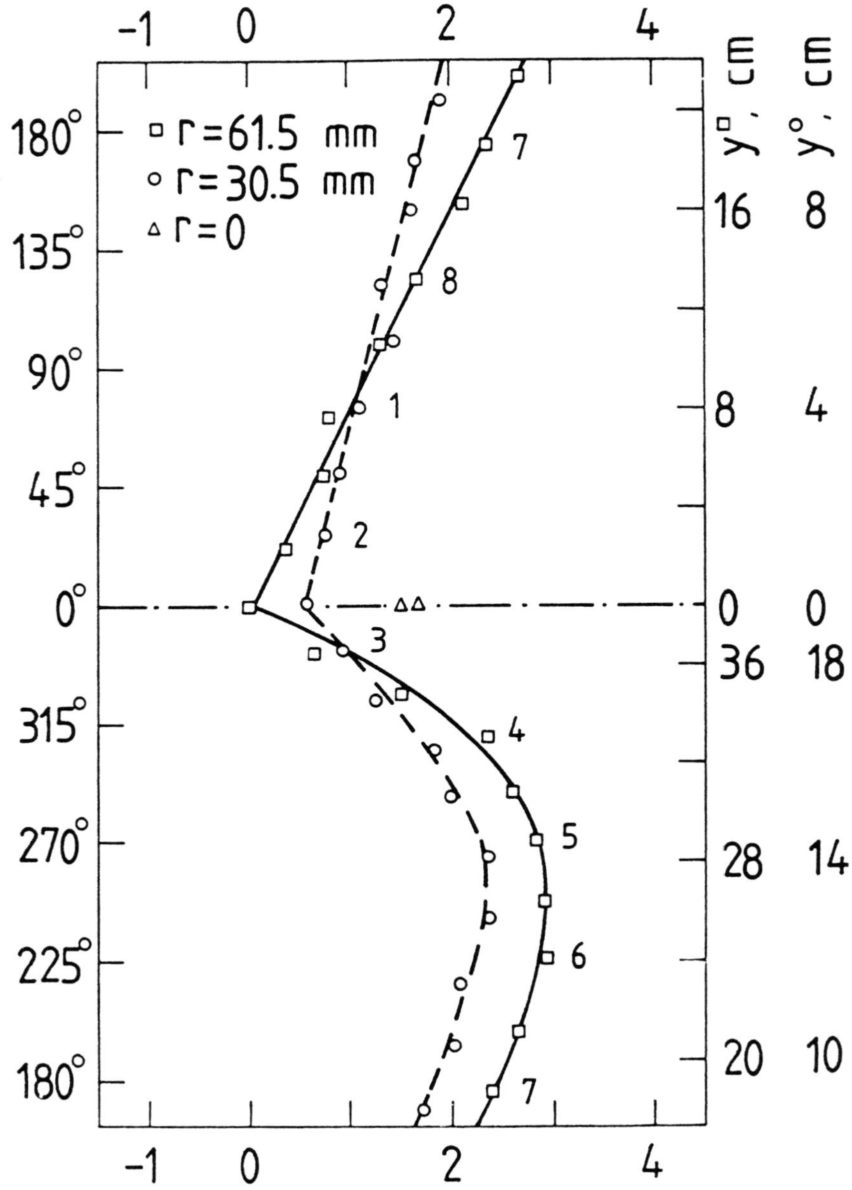

Fig. 17 Instantaneous structure of leading shock fronts for r = 0, 30.5 mm and 61.5 mm, in a corn starch dust-oxygen mixture of φ = 1, p_0 = 0.5 bar.

$$\frac{d}{dr}(r\, \rho_{mo}\, u_{ro}) = 0. \tag{25}$$

Here the system's origin is located on the tube axis, r, ϑ and x denote the radial, circumferential and axial coordinate respectively, u_{ro}, $u_{\vartheta o}$ and u_{xo} are the velocity components in r, ϑ and x direction respectively. ρ_{mo} denotes the mixture density and is defined as $\rho_{mo} = \sigma_p + \rho_o$ where σ_p describes the particle concentration and ρ_o the gas density ahead of the leading shock front. Here one has assumed that in wavefront fixed coordinates the incident flow is steady and variables are independent of ϑ and x. Using the boundary conditions at the tube wall and the conditions that u_{ro}, $u_{\vartheta o}$ and u_{xo} are finite at $r = 0$, the solutions of Eqs. (23)-(25) give the incident flow field

$$u_{ro} = 0, \quad u_{\vartheta o} = - r\,\Omega \quad \text{and} \quad u_{xo} = - D \tag{26}$$

Thus, the incident flow enters the leading shock front under the spin track angle

$$\alpha = tan^{-1}\, r\,\Omega\, /\, D \tag{27}$$

The angle α has a maximum at the tube wall and equals zero at the center. Equation (22), together with the solution (26), gives the uniform initial pressure ahead of the wavefront.

7. TRANSITION TO DETONATION

The transition to detonation in organic dust fuel-gas oxidizer mixtures can be shown in Figs. 10 and 18. Figure 10 represents a transition process in the corn starch dust-oxygen mixture of $\varphi = 0.45$ and $p_o = 1$ bar, with free jet initiation of $p_i = 2$ bar or $E_o = 10.38$ kJ. The overdriven detonation at $x/d = 71.42$ has a propagation velocity of $D = 1984$ m/s. At $x/d = 78.51$ it decays rapidly to a single spin detonation with $D = 1871$ m/s, which propagates stably to the end of the test section. The

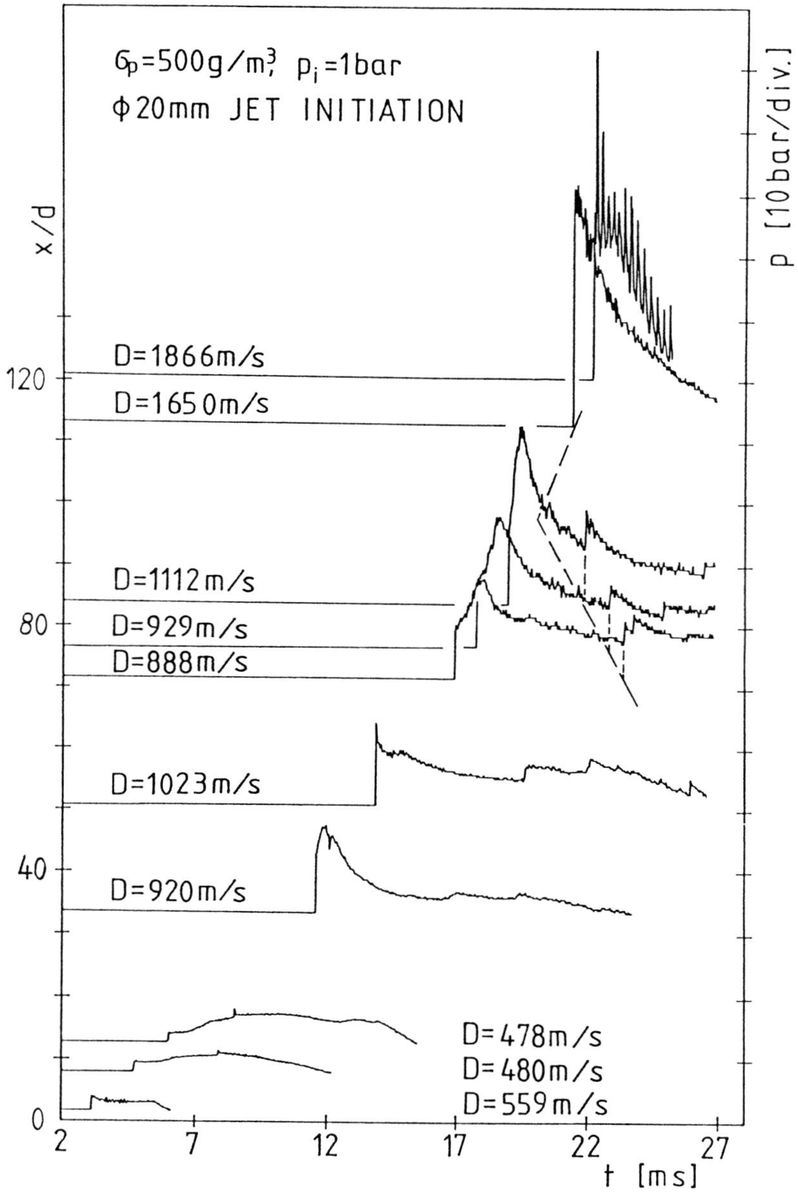

Fig. 18 Multistage transition to detonation in a corn starch dust-oxygen mixture of φ = 0.45 and p_O = 1 bar with free jet initiation of p_i = 1 bar.

run shown in Fig. 18 has the same conditions as in Fig. 10 except for p_i = 1 bar or E_o = 5.14 kJ. From these results the sequence of a complete detonation evolution can be physically distinguished in the following states:

1. initial particle ignition;
2. pressure wave amplification by coherent energy release;
3. unsteady reaction shock;
4. spin detonation.

The same transition evolution was also observed in a wide range of dust concentrations.

By decreasing the initiation energy the monotonic acceleration in stage 2 turns into a multistage one (Fig. 19). In Fig. 19 the transition distance and the Oppenheim velocity change from x_d/d = 57.16, v_{opp} = 1080 m/s at E_o = 124 kJ (planar wave initiation with p_i = 1 bar) to x_d/d ≈ 120, v_{opp} = 850 m/s at E_o = 5.14 kJ (jet initiation with p_i = 1 bar). This shows that the smaller the initiation energy, the larger the transition distance becomes for a certain region of initiation energy. The monotonic transition here is due to the influence of the initiation which amplifies the reaction acceleration. In consequence the energy release is higher than the losses during the propagation.

It is remarkable that in the multistage transition in Fig. 18 a shock front formed at x/d = 52.91, decays during subsequent propagation and is not synchronously amplified by chemical energy release. The shock wave at x/d = 52.91 has a frozen Mach number M_f = 2.93. Immediately behind the front the gas Reynolds number Re = $4.8 \cdot 10^6$ (characteristic length: tube diameter) and the gas temperature amounts to T = 760 K, which is of the same magnitude as the particle ignition temperature under atmospheric pressure (T_i = 720 K). Thus, one could postulate that the transition in this stage is rather due to turbulent mixing between the hot products and the unburned mixture than shock heating. During the decay of the leading shock wave a secondary explosion occurs at a certain distance behind the front (seen on the trace of x/d = 73.19). It excites the secondary acceleration in stage 2. In spite of the temporary decay of the leading shock wave, the pressure wave amplification is

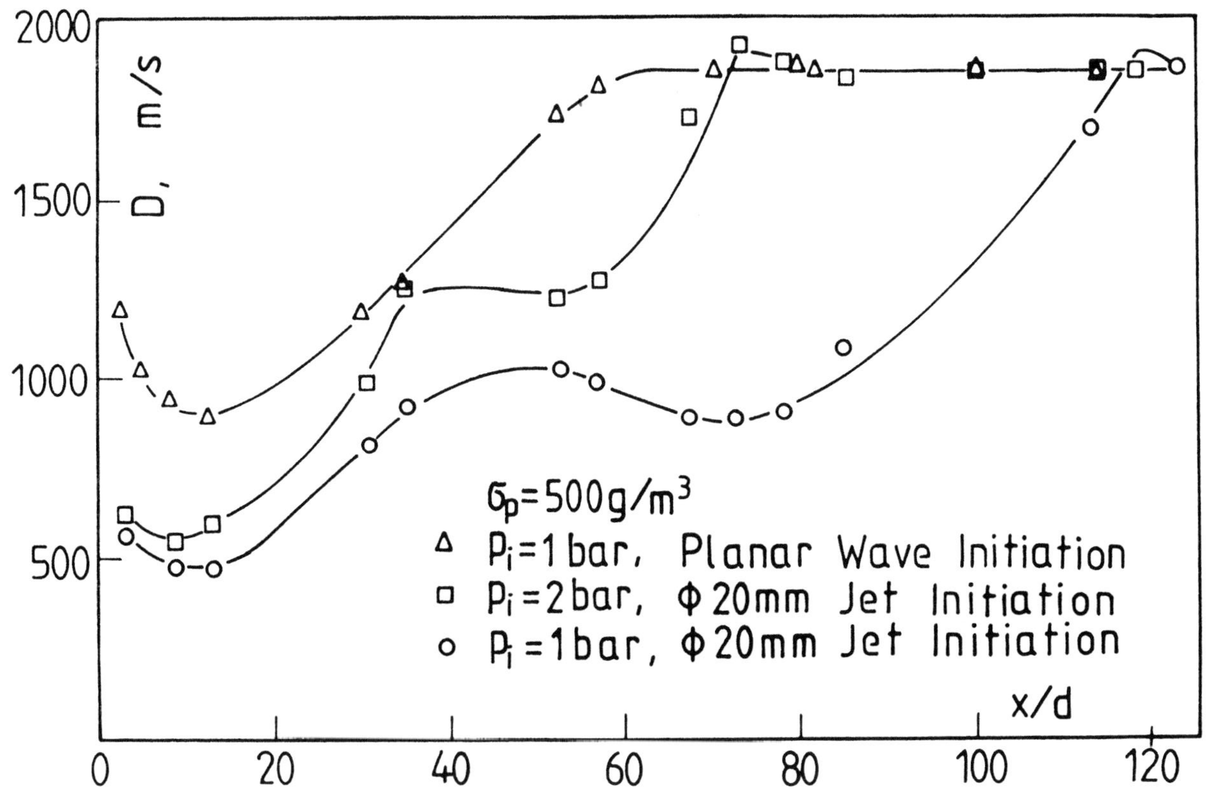

Fig. 19 Wavefront velocities versus propagation distance with different initiation energy, in a corn starch dust-oxygen mixture of $\varphi = 0.45$ and $p_0 = 1$ bar.

always synchronized with chemical energy release. For transition problems in dust cases, an initiation energy of about 10 kJ can be considered as weak initiation. Thus, the multistage acceleration process can be regarded as a kind of spontaneous confined deflagration-detonation-transition in organic dust mixtures.

Existence of the unsteady reaction shock as a transition stage, in which an exothermic reaction zone is closed coupled with the leading shock wave, is easily recognized, for example, at x/d = 114.64 in Fig. 18. The formation of the reaction shock in Fig. 18 occurs at the instant when the secondary compression waves overtake the wave front. This event produces a reflected wave which runs rearwards. During acceleration of the reaction shock, the induction time decays to the local value of the spin detonation. If one defines the spatial interval from the formation of the reaction shock to the appearance of the stable spin detonation as the propagation distance x_{rs} of the reaction shock, it has a value of $x_{rs}/d \approx 23$ in Fig. 18. It is remarkable that the shock wave in this stage, unlike in stage 2, is always amplified synchronously by chemical energy release, either smoothly (mostly in the first part of the stage like in Fig. 18), or violently as in Fig. 10. In coal dust-air mixtures Gardner et al.[7] obtain interesting pressure-time profiles at different locations along a tube, which has a diameter of 0.6 m and a ratio of length to diameter of L/d = 70. These profiles are shown in Fig. 20 and are similar to those in Fig. 10. The pressure profile at the tube end (D = 2850 m/s and p = 81.5 bar) observed by them, corresponds to that at x/d = 71.42 in Fig. 10, which is still overdriven or within the stage of the reaction shock.

A characteristic feature of the reaction shock is that within this stage, the symmetrical smooth wave front structure turns into a three dimensional one. This fact is clearly shown in the series of Figs. 21, 22 and 12, which give the pressure profiles on circumferences in the corn starch-oxygen mixture with φ = 1 and p_o = 0.5 bar. Figure 21 shows the beginning of the reaction shock, having a frozen Mach number of M_f = 3.66. All profiles start at the same time and are smooth, only one local explosion can be recognized on profile number 6. Spin appears in Fig. 22

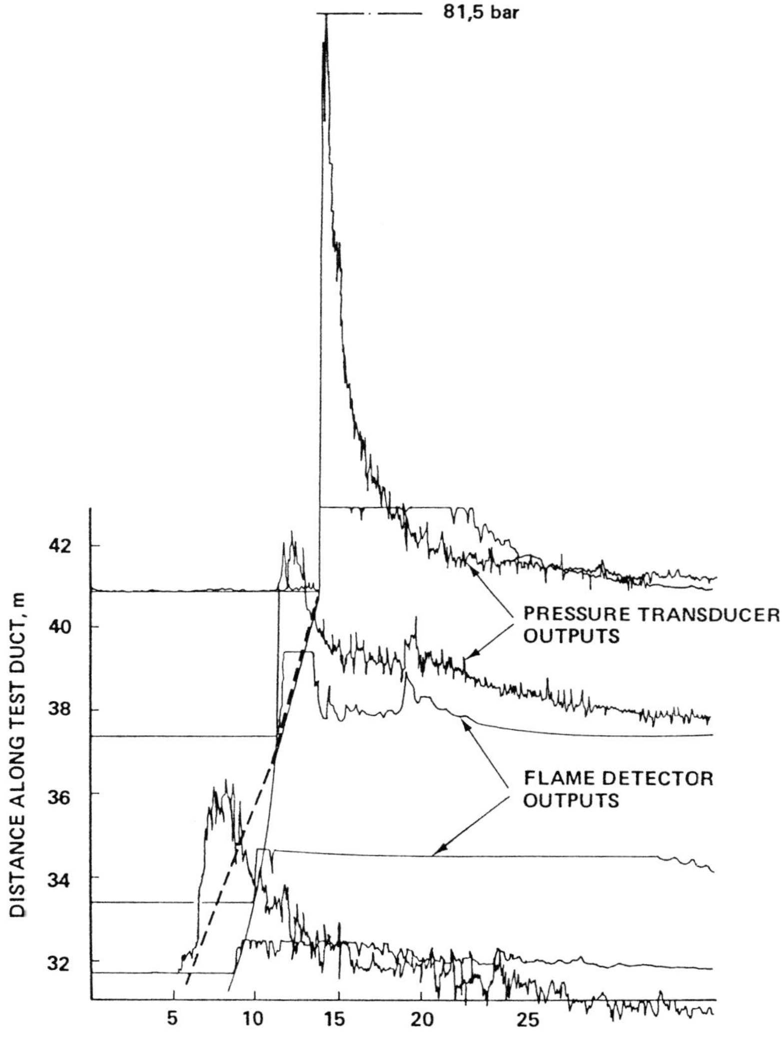

Fig. 20 Transition to detonation in a coal dust-air mixture. Courtesy of B.R. Gardner[7].

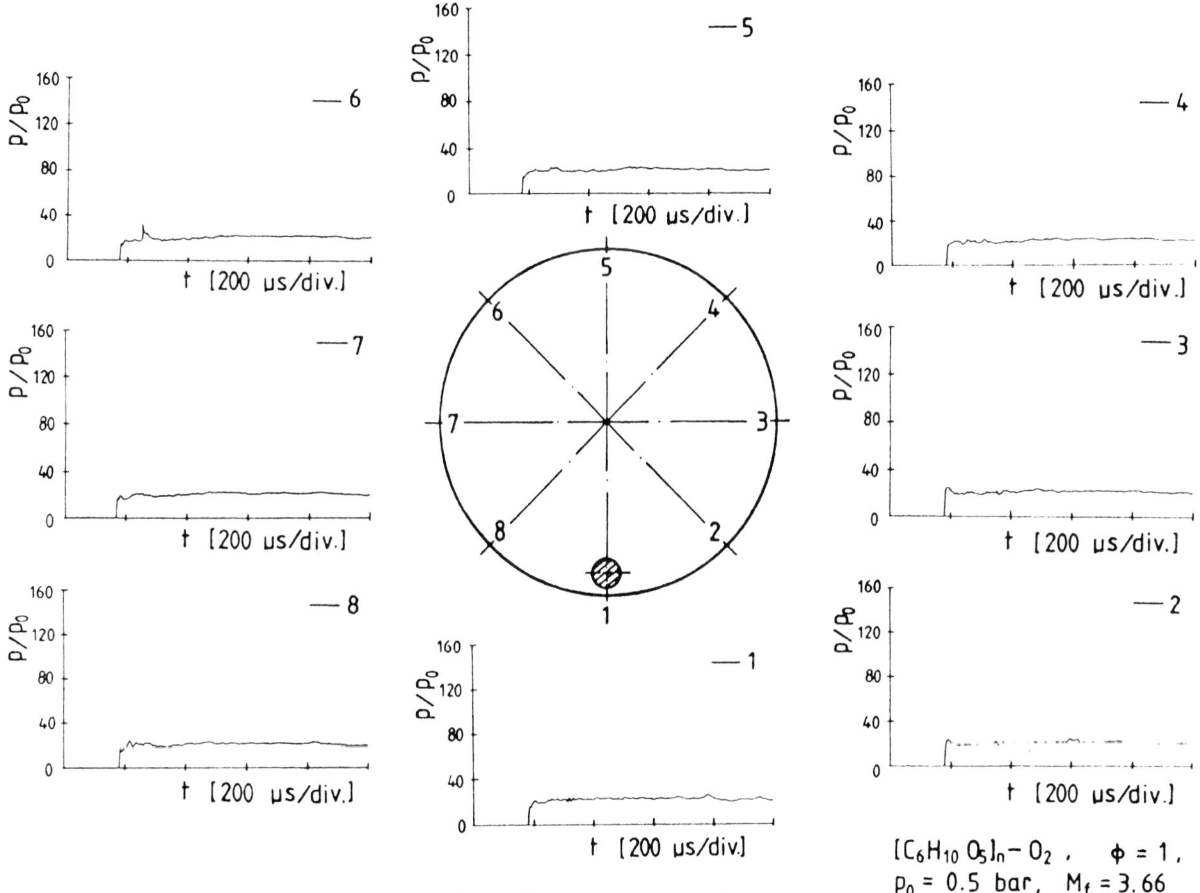

Fig. 21 Pressure profiles around the tube wall in a corn starch dust-oxygen mixture of $\varphi = 1$ and $p_0 = 0.5$ bar. $M_f = 3.66$.

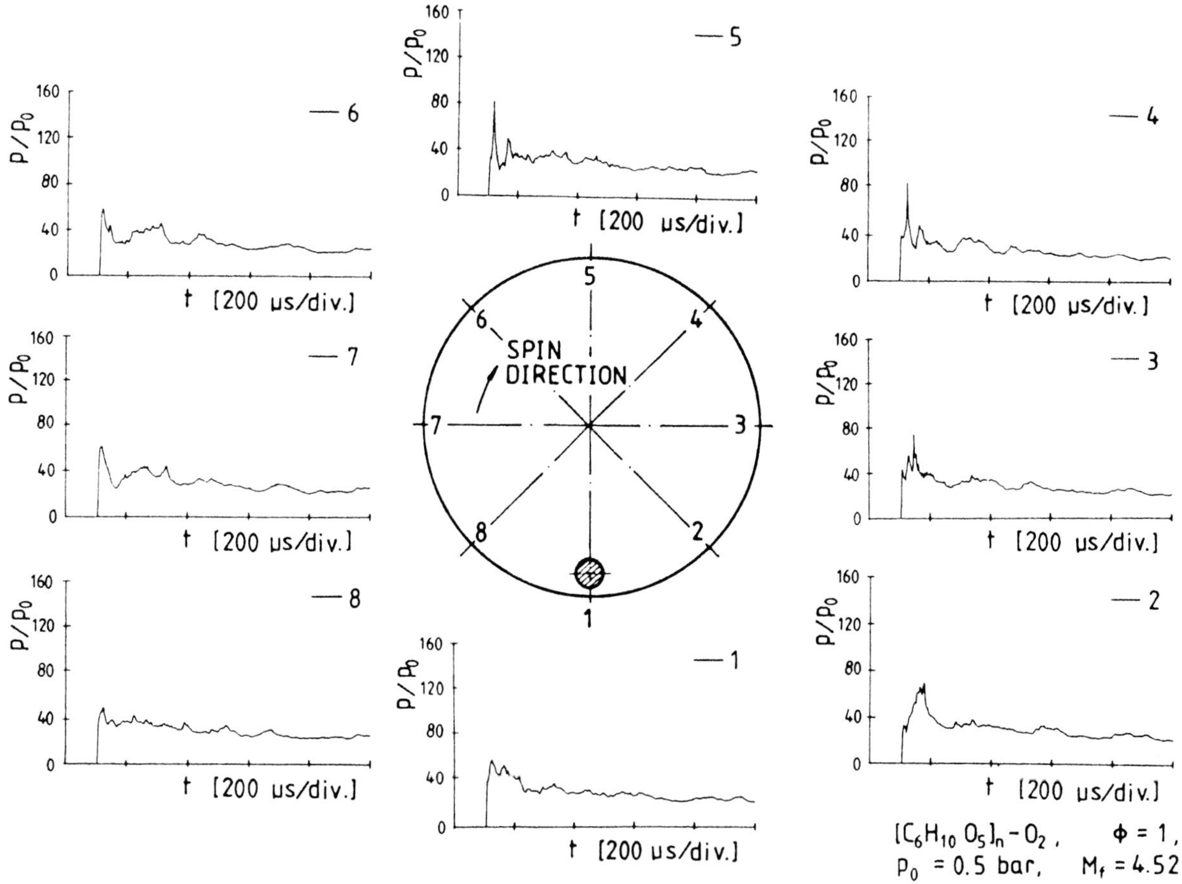

Fig. 22 Pressure profiles around the tube wall in a corn starch dust-oxygen mixture of $\varphi = 1$ and $p_o = 0.5$ bar. $M_f = 4.52$.

with M_f = 4.52, and is obviously in an immature state due to the lack of a stable reaction zone behind the shock front. At M_f = 5.16 a stable self-sustained detonation forms with a periodic tail wave behind the shock front (Fig. 12). The curvature of the shock front, as shown in Fig. 23, increases distinctly within a distance which corresponds to the growth of M_f = 4.52 to 5.16. The propagation distance of the reaction shock has a value of $x_{rs}/d \approx 20$ for this mixture. Thus, it is demonstrated that the formation of spin detonation takes place during the transition stage of the reaction shock.

From the above fact one kind of mechanism of spin detonation formation can be proposed. A local auto-explosion occurs firstly within the induction zone of the reaction shock. Since the induction zone is now quite long, a spherical blast wave originating from this local auto-explosion center is able to propagate further, being amplified by chemical energy release and possibly enhanced and deformed by other local explosions. Thus, Mach reflections take place due to the interaction of this spherical or deformed wave with the tube wall and the leading shock front. A transverse shock wave is subsequently developed, coupled with chemical energy release and amplified to a sufficient strength ($M_f \geq 5$ in the above example), thus resulting in spin detonation. The tail wave appears from the rearward extension of the spherical or deformed wave shaped by multiple wall interactions. Because of the violent reaction immediately behind the triple point configuration the flow there expands very strongly, thus causing forward curvature of the leading shock front. The local auto-explosion centers lie probably in the intense shear layer near the tube wall, as demonstrated by Urtiew and Oppenheim[28] and discussed by Lee et al.[29]

8. DETONABILITY IN DUST-AIR MIXTURES

From the practical point of view, it is important to explore the detonability of organic dust-air mixtures, which has not be solved up to date. One of the essential parameters of detonability is the cell size λ, or the minimum tube diameter d_{min} which is regarded as a necessary

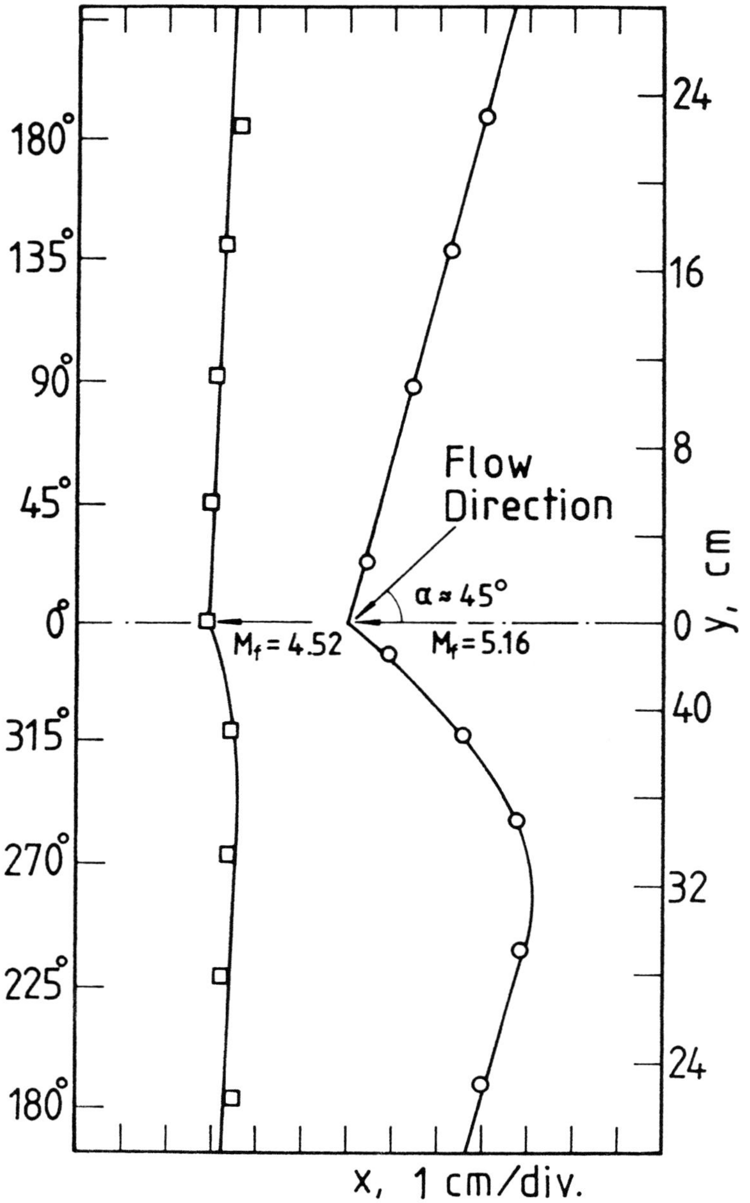

Fig. 23 Structure of leading shock fronts in a corn starch dust-oxygen mixture of $\varphi = 1$ and $p_o = 0.5$ bar for $M_f = 4.52$ and $M_f = 5.16$.

boundary condition for the transition to a stable self-sustained detonation. The following study is based on this point of view, since dust detonation also exhibits a cellular structure.

The equivalence ratio of $\varphi = 2$ is chosen for corn starch dust-air mixtures because the maximum detonation velocity lies closely at $\varphi = 2$. Figure 24 gives experimental results for this mixture with an initial pressure of $p_o = 1$ bar, under different initiation energy. With jet initiation of $p_i = 6$ bar or $E_o = 31.8$ kJ, deflagration propagation along the tube is quite mild. From $x/d \approx 80$ a rapid transition causes the propagation velocity to jump to a new level of $D \approx 1100$ m/s, which is maintained up to the end of the test section. The pressure in this propagation regime, illustrated in the pressure profile of number 5, displays a smooth structure. This phenomenon is similar to that in rough tube tests,[30] called "choked" regime due to $d_1/\lambda < 1$ (d_1 : orifice diameter of orifice rings spaced in the tube), where the deflagration velocity is found to be close to the sound speed of the hot products. By using the planar wave initiation with $p_i = 4$ bar or $E_o = 624$ kJ, the combustion shows a forced spin structure at the first stage as in the pressure profile of number 3, then decays quickly to the "choked" regime with the smooth structure of the wave front (pressure profile of number 4). Although the spin structure can be forced to exist in the whole test section by using very strong initiation, for example, planar wave initiation of $p_i = 6$ bar or $E_o = 964$ kJ in Fig. 24, the propagation velocity decays continuously. One can expect that up to a certain extended distance where the strong influence of initiation disappears, the spin structure will degenerate into the smooth structure, just as shown in the case of $p_i = 4$ bar. Thus, the forced spin induced by strong initial conditions is unstable due to $d < \lambda/\pi$.

According to the above results and the observation of single spin detonation in corn starch dust-oxygen mixtures of $\varphi = 1$ and $p_o = 1$ bar, one comes to the conclusion that in a tube of $d \leq 141$ mm a stable detonation is not able to be reached at normal atmospheric pressure. But the stable detonation can be expected by increasing the initial pressure p_o on the basis of the rule

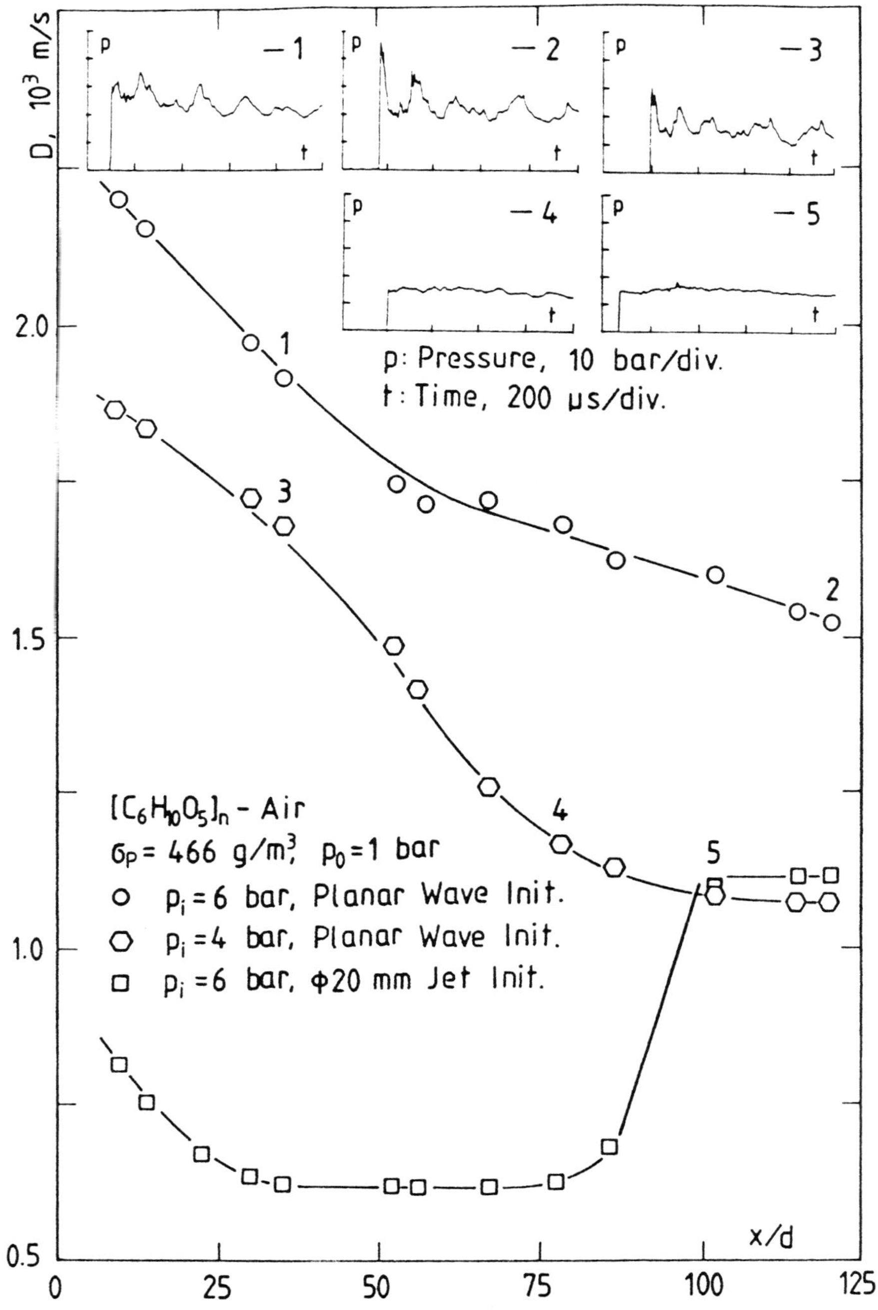

Fig. 24 Combustion propagation in a corn starch dust-air mixture of φ = 2 and p_o = 1 bar with different initiation energy.

$$d_{min} = \lambda / \pi \tag{28}$$

and the relation[31]

$$\lambda \sim p_o^{-\omega} \tag{29}$$

for gaseous mixtures where $\omega \approx 1$ under certain conditions, ω could be smaller in dust mixtures because the cell size is not as strongly dependent on p_o as in gaseous mixtures (see, Table 3).

Figure 25 shows the result of transition to detonation in a corn starch dust-air mixture of p_o = 2.5 bar and φ = 2, with planar wave initiation of p_i = 2.5 bar or E_o = 323 kJ. The transition process is the same as discussed in the dust-oxygen mixtures. Within the propagation of the reaction shock an abrupt onset of overdriven detonation accompanied by a retonation wave can be recognized. The transition distance has a value of $x_d/d \geq 120$, and the Oppenheim velocity is $v_{opp} \approx 795$ m/s. For p_o = 2 bar in the same mixture only the transition up to the reaction shock with D = 1400 m/s was observed within the tube length. From these results and the relations (28) and (29), an estimate of $\lambda \approx 0.88\text{-}1.10$ m and $d_{min} \approx 0.28\text{-}0.35$ m for the stable detonation at normal atmospheric pressure is derived.

The result of direct initiation of detonation in the same mixture of p_o = 2.5 bar and φ = 2 is shown in Fig. 26 with a critical initiator pressure of 6 bar or an effective critical initiation energy of E_{ec} = 7.92 MJ/m^2. In contrast to the instability of the forced spin in Fig. 24, the spin detonation here propagates stably. Concerning the critical energy for direct initiation of unconfined detonations, the surface energy model of Lee et al.[32]

$$E_c = \frac{2197}{16} \pi \rho_o v_{CJ}^2 I \lambda^3 \tag{30}$$

was found to correlate well with the experimental data in gaseous fuel-air mixtures.[17] Here ρ_o denotes the initial density of the mixture, v_{CJ} Chapman-Jouguet detonation velocity, and I the blast wave energy

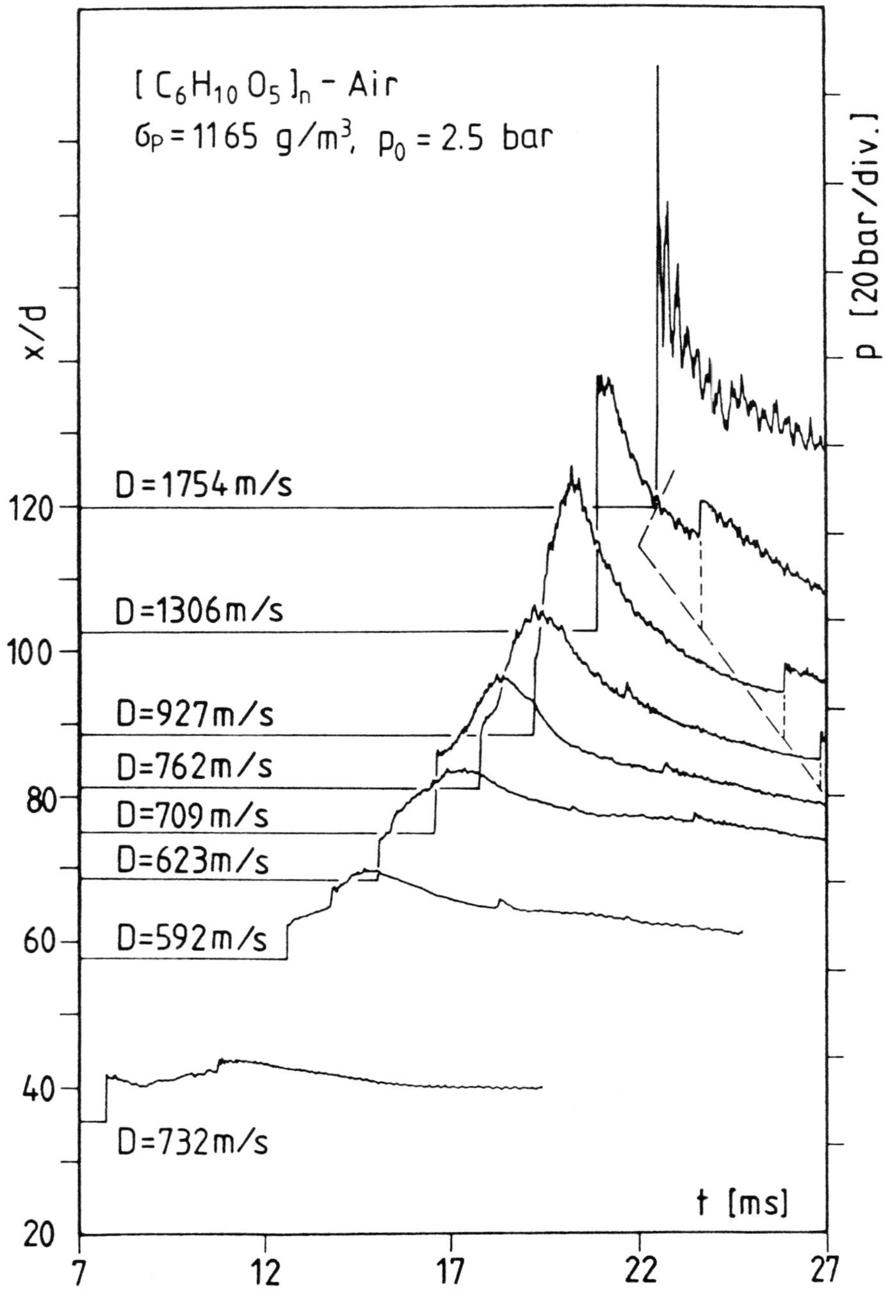

Fig. 25 Transition to detonation in a corn starch dust-air mixture of φ = 2 and p_o = 2.5 bar.

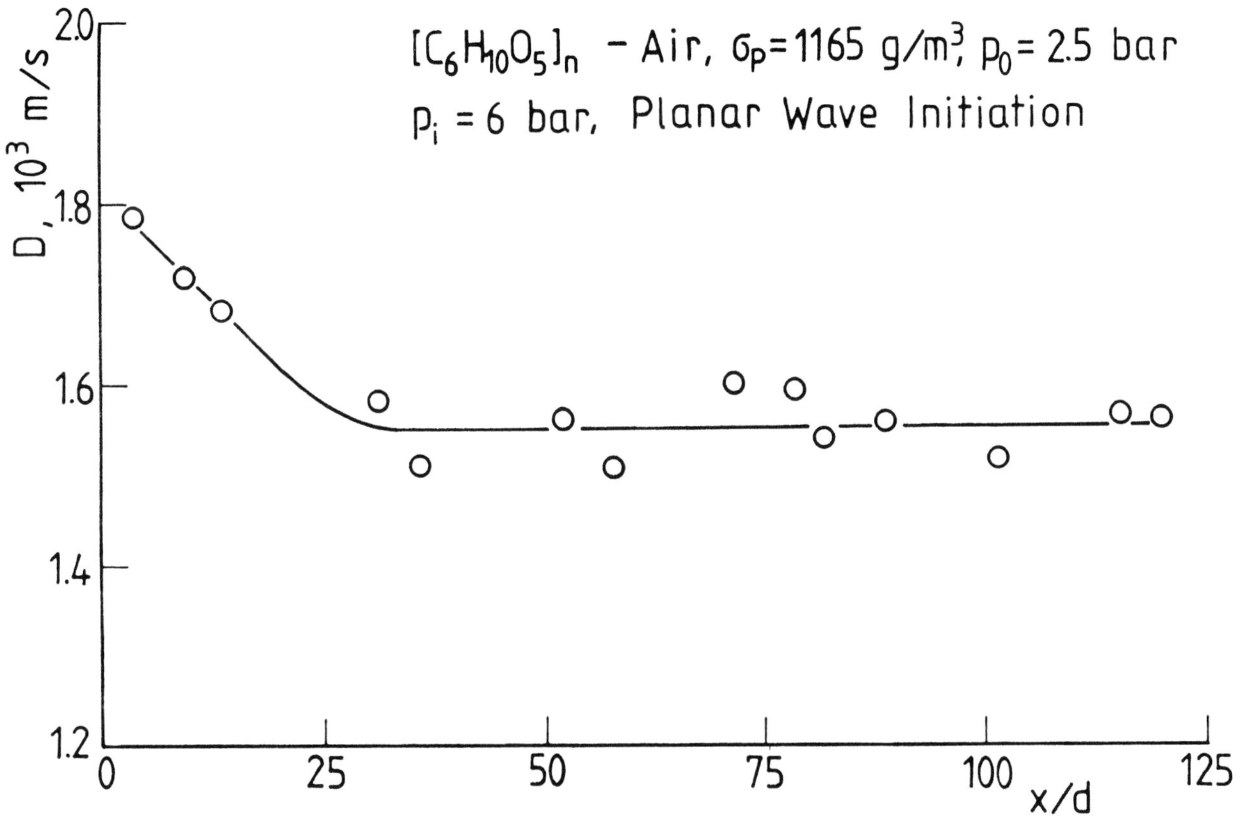

Fig. 26 Direct initiation of detonation in a corn starch dust-air mixture of $\varphi = 2$ and $p_o = 2.5$ bar.

integral. $I \approx 1$ holds for the ratio of specific heats equal to 1.2. By using Eq. (30) and the above estimate of λ, E_c = 1090 - 2130 MJ is derived for the unconfined detonation in corn starch dust-air mixture of φ = 2 and p_o = 1 bar. These values of E_c correspond to 225-499 kg of tetryl.

Transition to detonation in an anthraquinone dust-air mixture is given in Fig. 27 for p_o = 1.5 bar and φ = 3. The abrupt onset of detonation is obviously to recognize. The overdriven detonation caused by the onset decays continuously to the stable spin detonation.

9. CONCLUSIONS

The results of the experiments demonstrate that a dust fuel-gas oxidizer suspension can support a stable self-sustained detonation, and that the transverse wave is the key feature of the structure of the stable dust detonation. Single spin detonation exists both on the circumference and in the interior region, and possesses two constant parameters at the wave front: propagation velocity D and rotation angular velocity Ω.

The complete detonation evolution in some organic dust fuel-gas oxidizer mixtures consists of four stages, i.e., 1. initial particle ignition; 2. pressure wave amplification by coherent energy release; 3. unsteady reaction shock; and 4. spin detonation. By decreasing the initiation energy the monotonic transition turns into a multistage one, a kind of spontaneous confined deflagration-detonation-transition in such mixtures. During the acceleration of the reaction shock the symmetrical smooth wavefront structure is transformed into the three dimensional one. Thus, the spin detonation forms within this stage. The transition with an abrupt onset of detonation accompanied by a retonation wave was observed.

Under the initial conditions of p_o = 1 bar and T_o = 293 K, the minimum tube diameter for detonation in corn starch dust-oxygen mixtures of $\varphi \le 1$ is $d_{min} = \lambda/\pi$ = 0.14 m, the transition distance and the Oppenheim velocity are x_d/d = 68-120 and v_{opp} = 1150-850 m/s, respecti-

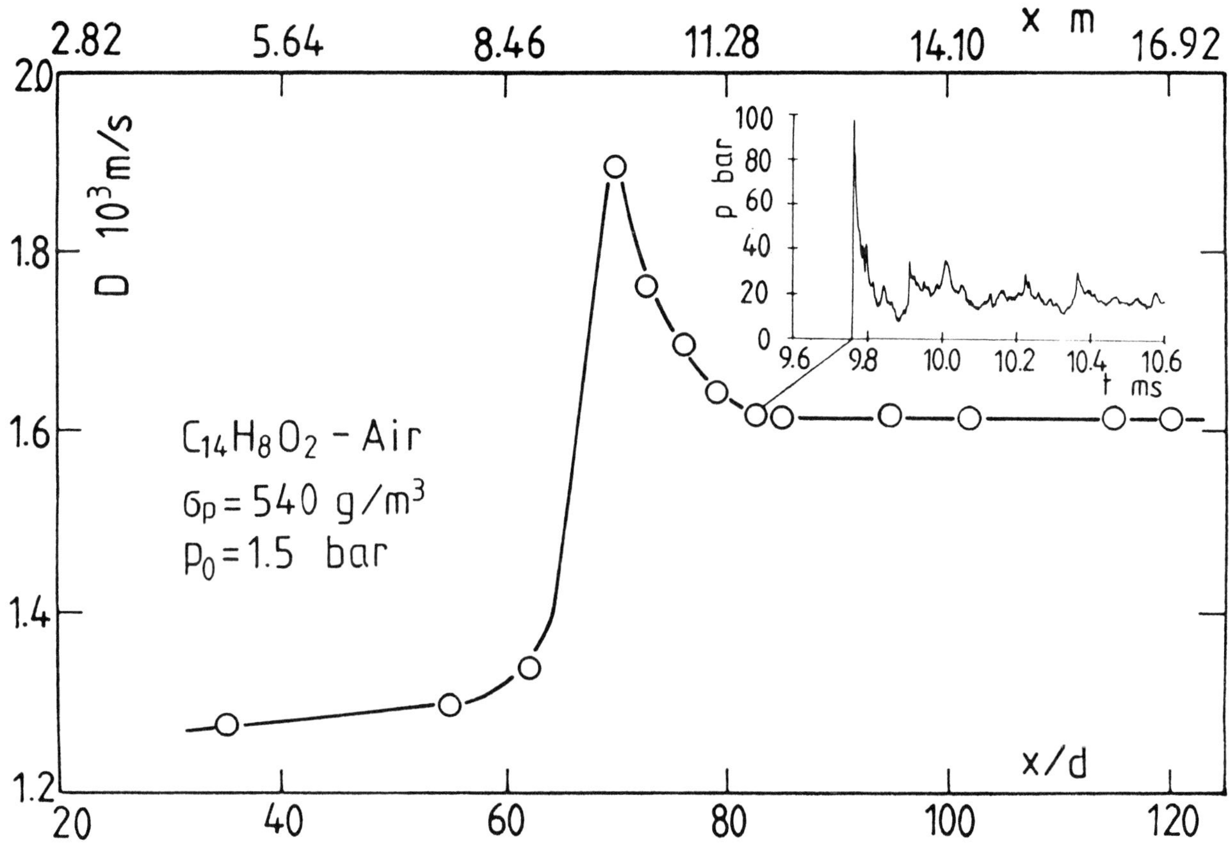

Fig. 27 Transition to detonation in an anthraquinone dust-air mixture of φ = 3 and p_o = 1.5 bar.

Table 1: Size Distribution and Heat of Combustion of Dusts

Corn starch Formula	Average Size μm	Heat of Combustion (P_o=1 atm, T_o=20°C)[33]
$[C_6H_{10}O_5]_n$	10	- 677 kcal/mol

Size Distribution (Particle Percentage)

< 5 μm	5 - 10 μm	10 -15 μm	15 - 20 μm	> 20 μm
18	50	29	2	1

Anthraquinone

Formula	Average Thickness × Length μm × μm	Heat of Combustion (P_o=1 atm, T_o=20°C)[33]
$C_{14}H_8O_2$	6 × 22	- 1554 kcal/mol

Size Distribution (Particle Percentage)

Thickness/Length	< 10 μm	10 - 20 μm	20 - 30 μm	> 30 μm
< 2 μm	22			
2 - 4 μm	15	15	4	
4 - 6 μm	3	9	3	1
6 - 8 μm	1	5	4	3
8 -10 μm		1	3	4
> 10 μm			2	5

Table 2: Critical Initiation Energy for Detonation in 400 - μm Decane Drops-Air Mixtures

φ		1.18	1.70	2.49	3.27
p_{ic}	bar	2.39	2.74	3.42	4.11
D	m/s	1543	1557	1569	1566
p^*	bar	14.3	15.0	15.9	16.5
E_{ec}	Mj/m^2	1.30	1.61	2.11	2.70
τ_o	μs	428	484	576	657
P_{ec}	MW/m^2	3037	3326	3663	4109
E_c (Ref. 17)	Mj/m^2	9.23	10.65	13.54	16.47
E_{ec}/E_c	%	14.1	15.1	15.6	16.4

Table 3: Properties of the spin structure at the tube wall (T_o = 293 ± 2 K)

φ	σ_p g/m^3	p_o bar	n	α	Ω 10^4 rad/s	D m/s	V_o m/s	$\frac{p_{AE}}{p_o}$	$\frac{p_{BD}}{p_o}$	$\frac{p_{AG}}{p_o}$	$\frac{p_{BC}}{p_o}$	$\frac{p_S}{p_o}$
0.64	357	0.5	1	45°	2.492	1757	1757	57	90	37	106	37
1.0	554	0.5	1	45°	2.555	1801	1801	78	143	39	150	47
0.27	300	1.0				1766			97			36
0.45	500	1.0	1	42°	2.376	1860	1675	83	118	37	108	48
1.0	1109	1.0	1			1945		77	145	48	147	55
1.80	2000	1.0				1988					154	65
2.71	3000	1.0	2			1965					214	67
3.61	4000	1.0	2			1961					239	66
4.51	5000	1.0	2			1911					283	65
6.31	7000	1.0	1			1846					238	51
8.11	9000	1.0	1			1795					206	43

Φ - equivalence ratio

σ_p - initial particle concentration

p_o - initial pressure

n - spin head number

α - spin track angle

Ω - spin angular velocity

D - propagation velocity

V_o - transverse wave velocity

p_{AE}, p_{BD}, p_{AG}, p_{BC} - pressure of wave fronts AE, BD, AG, BC near point A

p_S - pressure peak of the stable combustion region

Table 4: Reaction Time and Length of Stable Reaction Zone for Single Spin Detonations in Corn Starch Dust-Oxygen Mixtures (T_o = 293 ± 2 K)

φ	σ_p	p_o	D	τ_o	x_o
	g/m^3	bar	m/s	μs	λ
0.64	357	0.5	1757	875	4
1.0	554	0.5	1801	1100	5.5
0.27	300	1.0	1766	766	4
0.45	500	1.0	1860	863	4
1.0	1109	1.0	1945	966	5
6.31	7000	1.0	1846	1250	5.5

vely for φ varying from 1 to 0.45 by means of weak initiation (free jet, p_i = 1 bar). For corn starch dust-air mixtures under p_o = 1 bar and T_o = 293 K the estimation gives d_{min} = 0.28-0.35 m and $x_d/d \geq 120$.

ACKNOWLEDGEMENTS

This work was supported by the Stiftung Volkswagenwerk. The authors are indebted to A. van de Ven, P. Greilich and the technicians of the workshop of the Shock Wave Laboratory for their help during the experiments, and special thanks go to Bettina Odenthal for getting the paper typed on time.

REFERENCES

1. Strauss, W.A., AIAA J. 6, 1753 (1968).
2. Nettleton, M.A., and Stirling, R., Combust. Flame 21, 307 (1973).
3. Bartknecht, W., Explosionen. Springer, Berlin Heidelberg, FRG, 1978.
4. Kauffman, C.W., Wolanski, P., Arisoy, A., Adams, P.R., Maker, B.N., and Nicholls, J.A., Progress in Astronautics and Aeronautics, Vol. 94, AIAA, New York, 1984, pp. 221-240.
5. Tulis, A.J., and Selman, J.R., in Proceedings of the 19th Symposium (international) on Combustion, The Combustion Institute, Pittsburgh, 1982, pp. 655-663.
6. Peraldi, O., and Veyssiere, B., Progress in Astronautics and Aeronautics, Vol. 106, AIAA, New York, 1986, pp. 490-504.
7. Gardner, B.R., Winter, R.J., and Moore, M.J., in Proceedings of the 21wt Symposium (International) on Combustion, Combustion Institute, Pittsburgh, 1986, pp. 335-343.
8. Fangrat, J., Glinka, W., Wolanski, P., and Wolinski, M., archivum combustionis 7, 321 (1987).
9. Tulis, A.J., and Selman, J.R., Progress in Astronautics and Aeronautics, Vol. 94, AIAA, New York, 1984, pp., 277-292.
10. Veyssiere, B., Desbordes, D., and Lee, J.H., archivum combustionis 7, 185 (1987).
11. Lee, J.H., in Proceedings of the „16th International Symposium on Shock Tubes and Waves (H.Gronig, Ed.), VCH Weinheim, FRG, 1988, pp. 21-38. "
12. Zhang, F., and Gronig, H., in Proceedings of the 17th International Symposium on Shock Tubes and Waves, Bethlehem, PA, 1989. AIP Conference Proceedings (in Press).
13. Zhang, F., and Gronig, H., 12th ICDERS, Ann Arbor, MI, 1989. Progress in Astronautics and Aeronautics, AIAA, Washington, DC (in press).
14. Borisov, A.A., Gelfand, B.E., Timofeev, E.I., Tsyganov, S.A.,

and Khomic, S.V., Progress in Aeronautics and Astronautics, Vol.94, AIAA, New York, 1984, pp.332-339.
15. Sichel, M., Baek, S.W., Kauffman, C.W., Baker, B. and Nicholls, J.A., AIAA J. 23, 1374 (1985).
16. Benedick, W.B., Guirao, C.M., Knystautas, R., and Lee, J.H., Progress in Astronautics and Aeronautics, Vol. 106, AIAA, New York, 1986, pp., 181-202.
17. Tang, M.J., Nicholls, J.A., Lin, Z.C., Sichel, M., and Kauffman, C.W., Progress in Astronautics and Aeronautics, Vol. 106, AIAA, New York, 1986, pp., 474-489.
18. Zhang, K.Y., and Gottlieb, J.J., UTIAS Report No. 304, Institute for Aerospace Studies, University of Toronto, Downsview, Ontario, Canada (1986).
19. Taylor, G., Proc. Roy. Soc. A 200, 235 (1950).
20. Zel'dovich, Y.B., Kogarko, S.M., and Simonov, N.N., Sov. Phys. Tech. Phys. 1, 1689 (1956).
21. Tang, M.J., and Peng, J.H., Progress in Astronautics and Aeronautics, Vol. 114, AIAA, Washington, DC, 1988, pp., 201-208.
22. Thompson, P.A., Compressible-Fluid Dynamics. McGraw-Hill Inc., 1972, pp. 283-462.
23. Beylich, A.E.,Z.Flugwiss. Weltraumforsch. 3, 48 (1979).
24. Fan, B.C.,and Sichel, M., in Proceedings of the 22nd Symposium (International) on Combustion, The Combustion Institute, Pittsburg, 1988.
25. Voitsekhovskii, B.V., Mitrofanov, V.V., and Topchian,M.E., Zh. Prikl. Mekh. Tekn. Fiz., No. 3, 27 (1962).
26. Schott, G.L., Phys. Fluids 8, 850 (1965).
27. Borisov, A.A., Khasainov, B.A., Veyssiere, B., Saneev, E.L., Fomin, I.., Khomic, S.V., in Proceedings of the 23rd Symposium (International) on Combustion, The Combustion Institute, Pittsburg, 1990.
28. Urtiew, P.A., and Oppenheim, A.K., Proc. Roy. Soc. A 295, 13 (1966).
29. Lee, J.H., and Moen, I.O., Prog. Energy Combust. Sci. 6, 359 (1980).
30. Lee, J.H., Knystautas, R., and Chan, C.K., in Proceedings of the 20th Symposium (International) on Combustion, The Combustion Institute, Pittsburg, 1984, pp. 1663-1672.
31. Shchelkin, K.I., and Troshin, Y.K., Gasdynamics of Combustion. Mono Book Corp., Baltimore, Md., 1965.
32. Lee, J.H., Ann. Rev. Fluid Mech. 16, 311 (1984).
33. Weast, R.C., Handbook of Chemistry and Physics. 53rd, the Chemical Rubber Co., 1972, pp. D 230 - D 235.

On the Detonation of Aluminum Suspensions in Air and in Oxygen

Borisov A.A., Khasainov B.A., Saneev E.L., Fomin I.B., Khomik S.V.
Institute of Chemical Physics,
USSR Academy of Sciences, Moscow
and
Veyssiere B.
E.N.S.M.A., University in Poitiers, France

Abstract. The process of initiation and propagation in shock tube of stationary detonations in suspensions of the Al-particles of various size and shape has been investigated experimentally. The Al-suspensions of mean size which is more than 10 μm didn't ignited neither in air medium nor in the oxygen one. Al-air-suspensions with scales-like particles of thickness about 1 μm and diametrical size 10-15 μm have the same detonation ability as gaseous fuel-air mixture. Spherical Al-particles of diameter 1 μm had some more high detonation ability, i.e. detonation ability of Al-suspensions increases essentially with the particle size decreasing. It was no success to obtain stationary spherical detonation of Al-particles of diameter 3.5 μm dispersed in the oxygen-medium due to the finite size of the cloud.

The model for calculation of structure and parameters of non-ideal detonation wave of Al-particles suspensions in gaseous oxidizer is proposed. Satisfactory accordance of calculated data to the experimental ones is obtained. The effect of particle-size distribution and space orientation of scales-like particles on the detonation propagation is discussed. The results of the work show high detonation ability of the Al-particle suspensions of size approximately of 1 μm not only in shock

A. A. Borissov (ed.), Dynamic Structure of Detonation in Gaseous and Dispersed Media, 215–253.

tubes but also in space.

INTRODUCTION

The study of the detonation mechanism of two-phase media is necessary for the solving of the problem of their explosion-safety. From this point of view the detonation of Al-particle suspensions in gaseous oxidizer is interesting because aluminum is very power-intensive metal and can react not only with oxygen, but also with water vapor, carbonic acid gas and others.

First the detonation ability of the Al-particle suspension was studied by Strauss [1]. The experiments were carried out in the oxygen medium in the glass tubes with the length of 2.7 m and inner diameter from 20 to 55 mm. The scaly-particles of average diameter about 40 μm and the spherical ones of average diameter 5 μm were used. Initiation of detonation was carried out by pyrotechnical primers, detonators and also by the exploding of the Ag-wires, however, values of the initiation energy were not reported. The upper concentration detonability limit was not determined, since the used technique of the suspension generation didn't allow one to create dispersals with Al-concentration more than 64 %. The velocity of detonation propagation reached a stationary value at the distance ≈1.6 m from the initiation point for poor mixtures and ≈0.5 m for rich ones.

The detonation velocities independently on the tube diameter are in the range 1400-1600 m/s. In the case of spherical particles the detonation velocity was some higher and insignificantly decreased with the Al-concentration increasing (see fig. 7). The measured pressure of detonation, in spite of significant scattering of the detonation velocities, turned out to be close to 31 atm. Both the velocity and the pressure of detonations were smaller than corresponding thermodynamic values, calculated in [1] assuming, that initial mass of Al-powder contains 2 % of Al-oxide (ideal velocity of detonation decreases from 1750 to 1550 m/s with particle concentration growth within the range of 45-65 % and corresponding Chapman-Jouguet pressure increases from 33 to 43 atm.).

This effect was connected with incomplete combustion of aluminum. It was noted also, that detonation propagates in the spin regime, and evaluated size of the reaction zone is 3 cm (20 μs). Destruction of glass tubes behind the wave front took place at the distance of several centimeters. The control experiments, made with the use of more strong (due to thick metal casings) glass tubes (with 4.75 mm visualization slot) showed, that the destruction of tubes didn't effect to experimental results.

Detonation of Al-suspensions was investigated by Tulis and co-authors [2, 3] using shock tube of the 5.5 m length with inner diameter 152 mm. In all the experiments the detonation initiation was carried out by exploding of the blast material (RDX or tetril) by mass about 3 g. Suspensions with scaly Al-particles of thickness about 1 μm had maximal detonation ability. The most measured velocity of detonation in these suspensions was 1650 m/s, and thermodynamic velocity of Al air-suspensions detonation according to [3] is equal to 1850 m/s at stoichiometric concentration 330 g/m^3 (see curves *T* in fig. 6). However, in approximately half of experiments, the velocity of detonation in these suspensions at the end of shock tube reached only 1.3-1.5 Km/s. The measured pressure of detonation was equal to ≈50 atm, that is almost two times, more than thermodynamic pressure.

On the base of analysis of the glow recordings, obtained with the help of optical detectors, it was concluded, that detonation propagated in the spin regime (the same as in [1]). The ignition delay of the scaly Al-particles didn't exceed 1 μs.

In [2, 3] the spherical Al-particles were also used that allowed to reveal the effect of the particle diameter on the detonation ability of Al-suspensions. So, in the case of spherical particles with average diameter of 5 μm the detonation velocity at the end of shock tube reached 1380 m/s in a certain experiments, and the particle ignition delays were not less, than 14 μs. Suspensions with particles by diameter 10-15 μm under that conditions didn't detonate at all.

In [2, 3], as well as in [1] it is noted a weak influence of the Al-concentration to the velocity of detonation. Moreover, it is pointed out at the absence of upper concentration detonability limit (UCL) (suspensions with the Al-fraction up to 1000 g/m^3 were studied). It was suggested the significant incomplete combustion of the Al-particles in the reaction zone due to presence of large particles in the Al-powder, which remained inert up to Chapman-Jouguet plane.

Therefore, one has few experimental data showing that the detonation ability of the Al-suspensions depends strongly on the particle size and that stationary detonation can propagate in the Al-particle suspensions in oxygen. As for the Al-particle air-suspensions, experimental data due to their contradictions don't allow to predict with confidence the possibility of stationary detonations propagation in Al-air-suspensions and moreover to make any conclusions on the wave structure (although the detonation-like regimes are observed). Determining explosion hazard of the Al-air-suspensions, it would be also interesting to compare their detonation ability with the one of gaseous for instance, fuel-air mixtures. Besides, for the understanding of the process mechanism and conclusion on a possibility of the stationary detonation propagation under various conditions it is necessary to analyze and compare with experimental data the theoretical model of the detonation wave in the two-phase mixtures of Al with gaseous oxidizer.

In the section 1 of this paper the results of experimental determination of the detonation initiation energy in the Al air-suspensions with particles of various size are presented, which make possible such comparison and also to estimate the results [2, 3] in a near fashion, since in these works the initiator's energy was less, than minimal energy of the stationary detonation ignition determined in this work. Also the results of experiments on ignition of the Al-suspension detonation in medium of oxygen in space and in shock tube are described.

From theoretical point of view, the mechanism of detonation of the two-phase Al-mixtures with gaseous oxidizer, has been studied insufficiently. The results of thermodynamic calculations of the parameters detonation of Al-suspensions in oxygen [1] and in air [3, 4] give, naturally, no information on the detonation wave structure and don't explain causes of significant difference between experimental velocity of detonation and its ideal value.

The accordance between calculated and measured in [1] velocity of detonation was obtained in [5]. However, the structure of the detonation wave was not studied in that work and formula obtained for the detonation velocity didn't allow to estimate the effect of particle diameter on the detonation velocity.

Mathematical model of the detonation processes in suspensions of metallic particles in oxidizing gaseous atmosphere on the basis of the modern idea of mechanism of ignition and burning of Al-type metallic particles was formulated in [6]. However, the attempts to calculate using this model, the velocity and structure of the detonation wave in the Al-particle suspensions were not undertaken. But in [7, 9] comparatively simple model had been formulated using empirical data on ignition and burning of particles, which had allowed to obtain a reasonable accordance with experimental data on structure of the two-front detonation in the Al-suspensions in the detonation-able gaseous media [1] and to explain all the tendencies, observed in experiments. That is why, it is expedient to use the model [7-9] as a basis for calculation of the detonation waves structure in the two-phase Al-mixtures with gaseous oxidizer. Section 2 of this paper is devoted to description of this model and obtained results of calculations.

1. EXPERIMENTAL STUDY OF THE Al-SUSPENSION DETONATION

1.1. Technique of investigations

Study of initiation and propagation of detonation in the Al air-suspensions was carried out, in the main, in horizontal shock tube 4.2 m long with, inner diameter 122 mm. There were windows at the last two meters of the side surface of the tube through which photo registration of the process was carried out. Pressure transducers were mounted at the distances 1.25 m and 3.25 m from open face of the tube, where initiator was located (plastic blast material on the base of RDX and cap-detonator). In cross-section of the second transducer there also was placed ionization pickup. Vertical shock tubes 2 m long with diameters 145 and 55 mm, provided with the slot windows and pressure transducers were also used. In this case initiation began at lower face of the tube.

In experiments Al-powder of the PAP-type particles of which were covered by the hydrocarbon film and had the shape of scale with thickness about 1 μm. Approximately 75 % of particles (by mass) had diametrical size from 1 to 15 μm, i.e. there was studied the powder of the same type as in [2, 3]. Besides, powders of UDA-type (particle diameter was up to 1 μm), ASD-1 (average size of particles was 11-17 μm) and PA-4 (about 33 μm) were used. Powders were not subjected to special processing before experiment.

Suspension of particles inside the tube was prepared with the help of "cases" with the Al-powder, situated every other 25 cm along the tube length. The tube face, which is more remote from initiator, was closed before experiment by polyethylene-film. Approximately 0.5 s before exploding of electric detonator the "cases" were blown by compressed air from a reservoir with initial pressure ≈ 10 atm.

Mass of aluminum, leaving the tube volume through its open face and precipitating on walls of shock tube during the time, which is approximately equal to the exploding delay, was determined in special series of idle experiments in horizontal shock tube. For this purpose, special cylindrical copper inset 1 m long with diameter, which is equal to the shock tube inner diameter was put into the tube before experiment. The inset was moved out of the tube quickly in various time intervals after the suspension preparing. Quantity of aluminum precipitating on walls of shock tube during the time intervals of 1-10 s, practically remained constant. The average concentration of particles in suspension was determined by ratio of powder mass, entering to the tube volume in fact (i.e. minus mass of aluminum, precipitating at walls) per tube volume. Error in the concentration determination was ≈10 %. In vertical shock tube there was almost no precipitation of spherical particles at walls nevertheless the UDA-suspension concentration was not homogeneous along the tube length. Precipitation of the PAP-particles went on significantly slower, than of the UDA. The estimations and visual observations showed, that mixing of clouds from neighboring "cases" is sufficiently good.

The mass of initiating blast material was varied for each concentration of Al-particles to determine the minimal energy of the detonation ignition. If at the last two meters of the tube length velocity of process and pressure at the wave front were constant, one considered, that stationary detonation was ignited in mixture. If the process velocity decreased along the tube length, one considered, that initiation of stationary detonation didn't take place and it needed to increase blast material mass. Effective energy of the blast material explosion was determined on the basis of measured velocities of shock waves in inert medium and formulae for the point explosion using the technique of [11].

The detonation ability of the Al-suspensions in oxygen was studied in the Laboratory of Power Engineering and Detonation of the France National Centre for Scientific Investigations, N 193, Poitiers, France. Experiments were carried out in vertical 4,5 m long shock tube 6.9 cm in diameter with the Al-particles, 13 μm in diameter. Moreover the experiments on initiation of spherical detonation of the Al-particle suspension, 3.5 μm in average diameter (powder of the Alkan firm) were carried out using the technique which was identical to described in [12] one. The two-phase Al-oxygen mixture was prepared in cylindrical reservoir, 0.77 m in diameter and 1 m in height (volume was 0.385 m^3), limited by the polyethylene film. Initiator (TNT-charge with mass 125 g was placed at the reservoir axis at distance 38 cm from the earth surface. Pressure transducers were mounted in upper part of reservoir at its axis at distances 35.5 and 45.5 cm from the charge (correspondingly K1 and K2).

1.2. Results for Al air-suspensions

The results of experiments on determination of minimal detonation ignition energy in horizontal tube 122 mm in diameter and 4.2 m long are shown in fig.1. The initiation energy, at which propagation of stationary detonation takes place in the tube are shown by black points, and the initiation energy, at which detonations isn't stationary - by the hollow ones. It is seen, that for the scaly Al-particles (curve 1) there exists the optimal concentration of particles $\sigma \approx 330$ g/m^3, at which the initiation energy is minimal and equals to ≈ 3.4 MJ/m^2 (corresponding mass of the blast material is 12 g). This energy is about the value of the initiation energy of stoichiometric propane-air mixture [11] and is reached at stoichiometric concentration of the Al-particles in air. In the case of the UDA-spherical particles the initiation energy (curve 2) was significantly less, than in the case of the scaly particles. In this

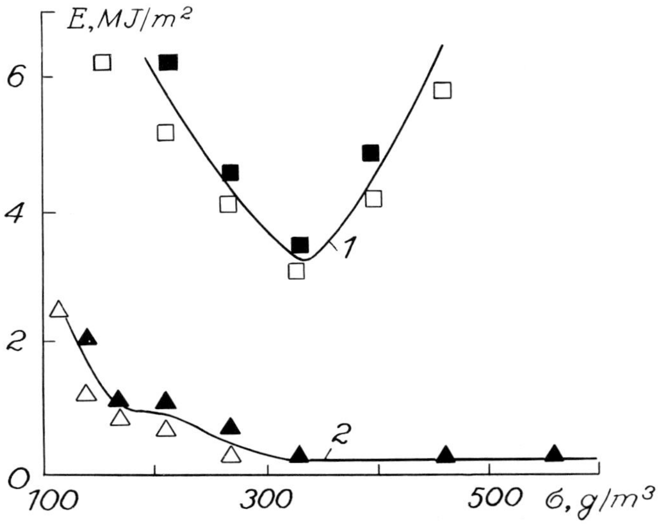

Fig. 1. Dependence of the initiation energy of air-suspension of Al-particles on the particles concentration. Hollow points correspond to attenuation of wave, the black ones - to stationary detonation. The powder type: 1 - PAP; 2 - UDA.

case the initiation energy of reach mixtures doesn't exceed 0.3 MJ/m^2 (charges with lesser effective energy are not used).

Photo-record of the detonation propagation process in the PAP air-suspension at $\sigma \approx 330$ g/m^3 is shown in fig. 2,a. It is seen, that the glow front oscillates with frequency ≈ 4.7 KHz, i.e. detonation propagates in the spin regime, as well as in the aluminum-oxygen mixtures [1]. The wave length of oscillating front Λ values of which are given in the

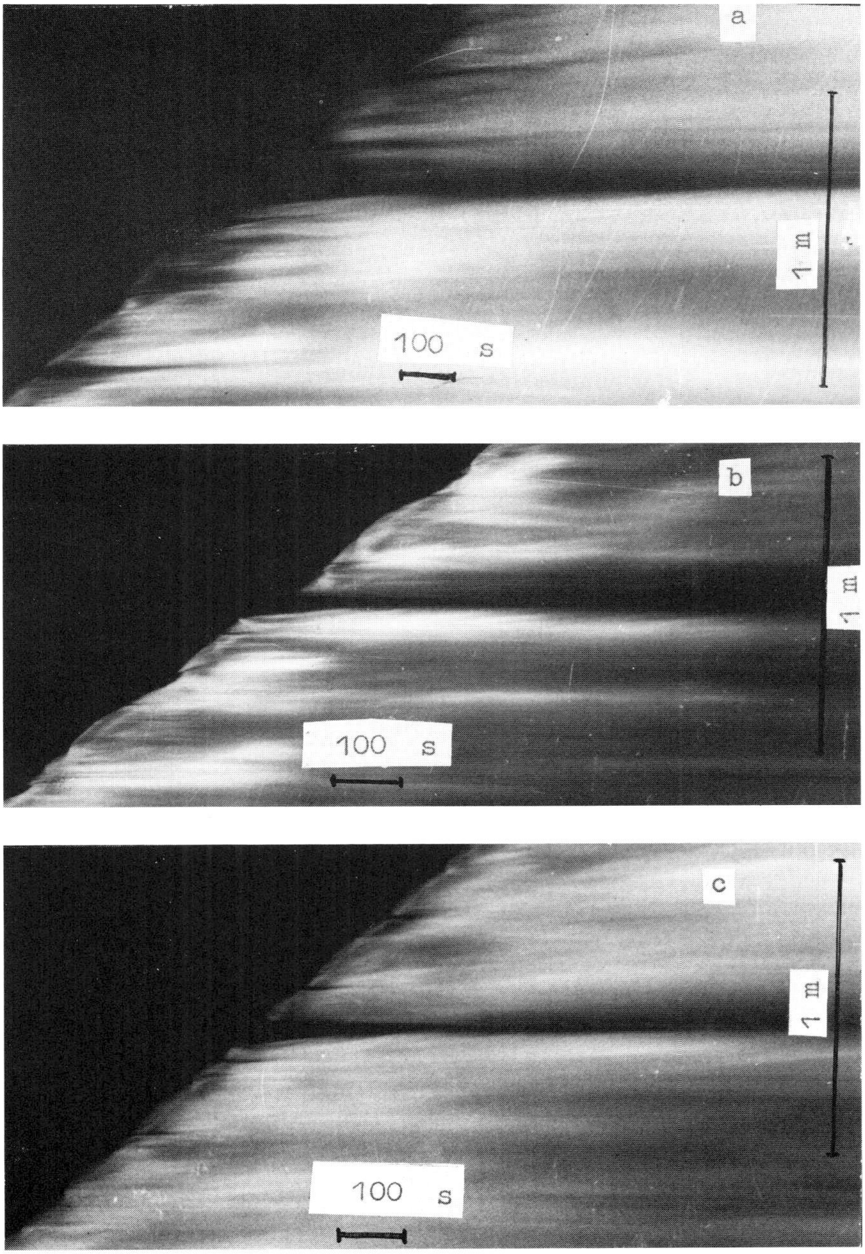

Fig. 2. Photo records of the detonation propagation process in aluminum air-suspension; a) PAP, σ=330 g/m^3, b) UDA , σ=170 g/m^3, c) UDA, σ=210 g/m^3. Energy of detonation initiation corresponds to fig. 1.

table 1, doesn't depend practically on concentration in the range 210-400 g/m^3 and approximately equals to $\approx$40 cm, i.e. about 3.3 of tube diameter. In the case of spherical particles the spin step decreases with the particle concentration growth (table 1, fig. 2,b and 2,c) and when $\sigma > 330\ g/m^3$ the oscillations of glow at the wave front are not observed. At the experimental profiles of pressure and conductivity (fig. 3), obtained with the photo record (fig.2,a) in the same experiment, there are also seen small jumps, following one after another in 200-300 μc, that corresponds approximately to period of the wave front oscillations τ (table 1).

Pressure at the wave front measured in various cross-sections of the tube (fig. 3,a and 3,b), doesn't change within the range of measurement error, that is an additive confirmation of the wave stationery. There is no sense to load for similarity of profiles in the vicinity of the wave front due to pulsating character of wave propagation and insufficient time speed of transducers ($\approx$ 30 μs). Due to the same reason the records of the mixture conductivity (fig. 3,c), which is $\approx 4\cdot10^{-5}$ Ohm m/mm^2 already at the wave front, don't allow to determine sufficiently reliably the Al-particle ignition delay behind the wave front. Values of the detonation velocity, measured on the basis of photo records and pressure, determined by extrapolations of the pressure profile on the wave 10 front are also given in the table 1 (and below, in fig.1).

Similarly to [1-3] the detonation propagation velocity in the case of the scaly particles doesn't depend on concentration of particles but in our experiments it is higher, than in [2, 3] and equals to 1800 ± 50 m/s, that is less, than thermodynamic velocity of detonation by 50-100 m/s. Measured pressure at the wave front is equal to 34 ± 3 atm, that is significantly less than in [2, 3]. In the case of spherical UDA-particles, 1 μm in diameter velocity of detonation and pressure increase noticeably when the concentration grows from 140 to 750 g/m^3 (respectively, by 400 m/s and 8 atm), thereby the most quick growth takes place at the small concentrations near lower concentration detonability limit (see below fig. 6). Detonation velocity of rich UDA-suspensions coincides with the one of the scaly aluminum, but the pressure at the wave

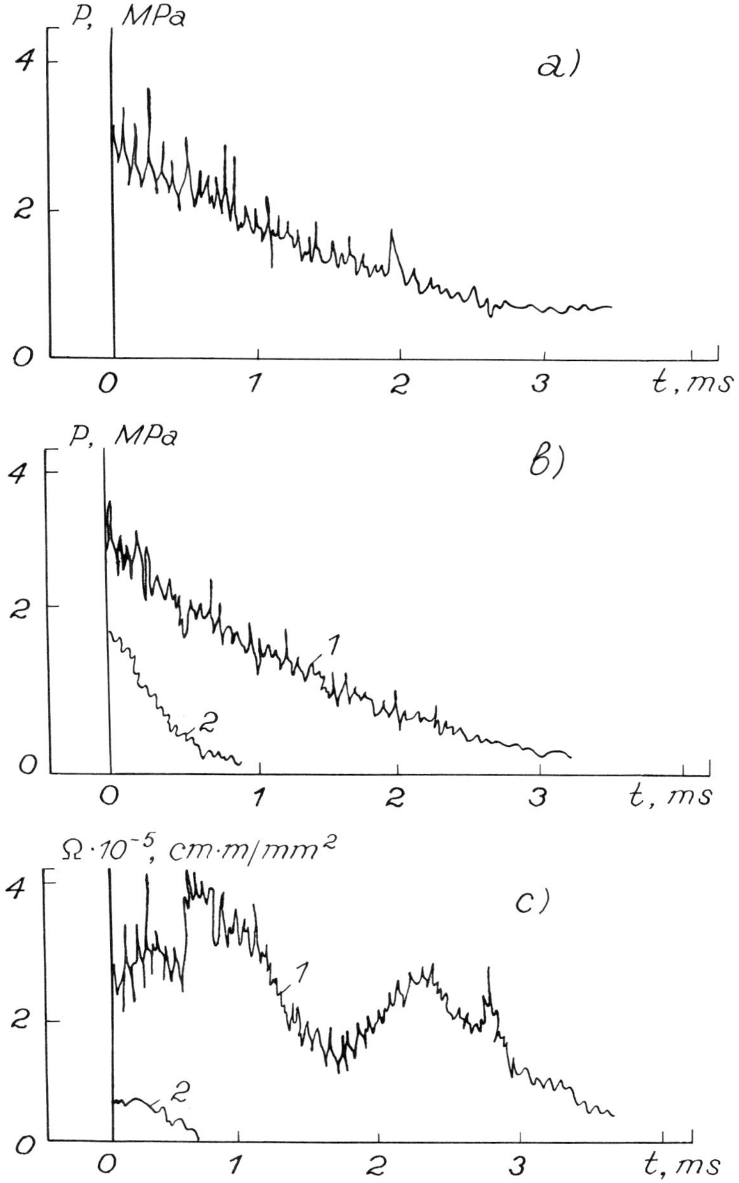

Fig. 3. Oscillograms of pressure at distance 1.25 (a) and 3.25 m (b) from the point of initiation and profile of conductivity (c) at distance 3.25 m from the initiation location in the case of the scaly particles for the same conditions as in the fig. 2a. Analogous profiles for stoichiometric propane-air mixture are denoted by number 2.

front is significantly lower, than for the PAP and equals to 26 ± 3 atm.

Studies of detonation of the PAP Al-suspensions in vertical tubes with inner diameters 55 and 145 mm and in more long one, 122 mm in diameter results in the same values of initiative energy (per unit of the tube cross-section area) and detonation parameters. There is no success to obtain detonation of the Al air-suspensions of the ASD-1 and PA-4 types in the used tubes.

1.3. Results for aluminum-oxygen mixtures

Experiments carried out in shock tube have shown, that suspensions of spherical Al-particles with average diameter of 13 μm don't detonate in the oxygen medium.

The Al-suspension concentration in the experiments on excitation of spherical detonation is chosen so, that pressure of detonation was as close as possible to maximal thermodynamic pressure of Chapman - Jouguet calculated by the program QUATUOR [13], and to take into account a non-uniformity of the particle size distribution, that can lead to only particle combustion of aluminum up to the Chapman-Jouguet point. Experimental values of velocity and pressure at the shock wave front and the calculated ones using [13] are given in the table 2. Pressure profiles for σ=4140 g/m^3 are shown in fig. 4. One can see that behind the detonation wave front secondary wave of compression is formed caused by burning of the Al-particles. One can conclude, comparing with the results of analogous experiments using the starch-particles, 25 μm in diameter [12], that used powder of aluminum is more able to detonate, but dimensions of the installation are insufficient to observe the transition of the wave to self-sustaining regime.

1.4. Discussion of experimental results

The analogy between photo records of the glow front propagation in the Al air-suspensions of the UDA- and PAP-types and photo records of the detonation propagation in gaseous fuel-air mixtures together with constancy of the wave propagation velocity and pressure at its front allow to confirm, that stationary detonation can propagate studied heterogeneous mixtures (in the main in spin regime in air mixtures).

Formerly, it was shown [14], that there is the relationship between step of the one-head spin Λ and tube diameter d

$$\Lambda/d = \left(\pi/1.841\right) \cdot \left(D/C\right)$$

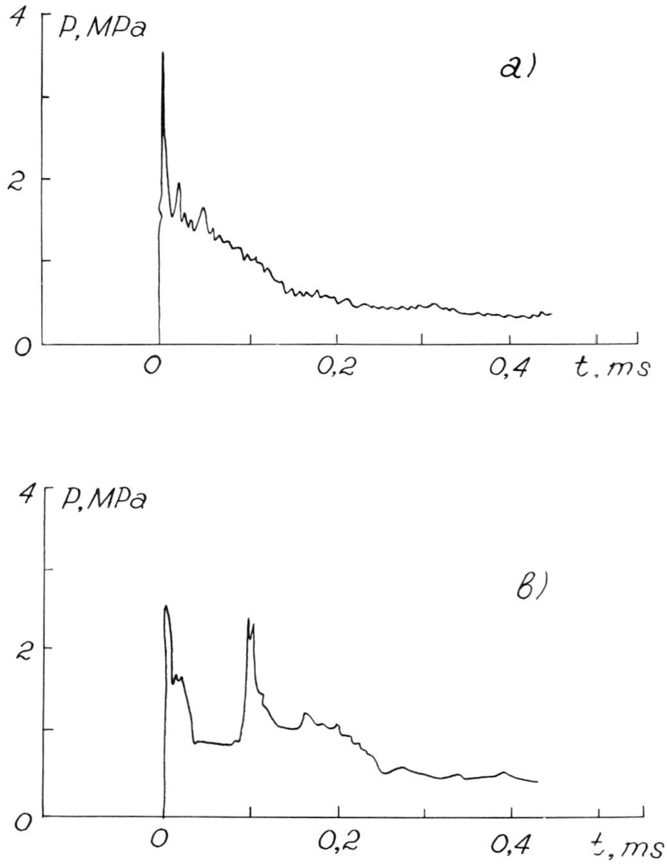

Fig. 4. Profiles of pressure at initiation of spherical detonation in the Al-particle suspension with size of 3.5 μm in the oxygen medium at σ=4140 g/m^3. Pressure transducers are located at distance 30 cm (a) and 40 cm (b) from initiator.

where D is velocity of detonation, and C is sound velocity in gas immediately behind the shock wave front. If the shock wave velocity equals to detonation velocity for Al air-suspension C ≈850 m/s than Λ/d ≈3.5. This value is close to experimental data of the table 1 for the scaly particles of the PAP-type, that confirms the spin character of the detonation propagation. Decrease of the ratio Λ/d to 1 when UDA-concentration increases testifies, that in the UDA-suspensions not only the one-head spin is observed but also regimes with several heads. Disappearance of the spin structure at large concentration is connected with the junction of glow of single pulsations at their small scale, most probably.

Maximal pressure at the front of detonation wave (p_f =34 ± 3 atm) propagating through the PAP scaly Al-suspension exceeds calculated thermodynamic pressure at the Chapman-Jouguet point p_{CJ} (taking into account deficit of wave velocity relatively to the ideal one), but for suspension of the UDA-spherical particles p_f is significantly less at practically the same velocity of detonation. This result can be explained by the fact that in the case of PAP transducers record also the zone of chemical peak with higher pressure, than p_{CJ}, but in the case of UDA the reaction zone is less, and chemical peak is not recorded at all by pressure transducers (one should note that the scaly aluminum is covered by thin hydrocarbon film, so the induction zone increases in addition). We'll discuss the question about the reaction zone width below, describing the detonation wave structure. Pressure of detonation up to ≈ 50 atm measured in [2, 3], are connected, apparently, with non-stationary character of the wave propagation. Factually in [2, 3] at the initiation energy less, than kinetic energy of direct initiation of detonation determined in the present paper (3g of blast material instead of 12 g in the case of optimal PAP concentration) apparently regimes of the gallop-type were observed, when the wave amplitude at some sections can essentially exceed p_{CJ} at low average wave velocity. Besides that, transducers, used in [2, 3] but with lesser inertia allowed to register peak pressure at reflection of transverse wave from the tube wall.

The presence of condensed phase in products, partial burning out of aluminum not burned out to the Chapman-Jouguet point and shaken off from the walls of shock tube leads to significant decrease of the pressure fall velocity behind the wave front. Profiles of pressure of the suspension detonation (curve 1) and of the gas detonation wave of stoichiometric propane-air mixture (curve 2) at approximately identical velocities of detonation are compared in fig. 3,b.

To estimate the reaction zone (and chemical peak) length the experiments on measurements of electric conductivity of the detonation products were carried out. Profiles of conductivity at heterogeneous detonation of the scaly Al-suspensions (curve 1) and at gas detonation (curve 2) are compared in fig. 3,c. Note, that conductivity at heterogeneous detonation is approximately five times greater as compared with the case of gas detonation and falls down extremely slowly. In whole this testifies that ionization has a thermic nature. Therefore the time behavior of conductivity should correspond to the temperature profile to a certain degree. The conductivity increase up to maximum goes during 20-30 μs and delay of the growth of ionization current behind the wave front is approximately the same, i.e. the length of chemical peak turns out to be about several tens of μs and can be recorded by pressure transducers. In a series of experiments the ionization current after quick initial growth increases again smoothly but more slowly to higher values, this fact, apparently, indicates that particles (shaken off from the wall of shock tube or passes Chapman-Jouguet plane being not burned down) burn out behind the Chapman-Jouguet plane.

Detonation parameters of the Al-powders (UDA and PAP) are not changed practically by variation of the shock tube diameter from 55 to 145 mm and don't depend on the distance from initiator. This fact confirms that effect of losses on the tube walls is weak and the length of the reaction zone is sufficiently small.

In order to be convinced of the absence of initiator effect on the detonation parameters, idle experiments were carried out with the purpose of determination of the character of air shock wave attenuation in shock tube, not containing the Al-suspension, at operation one of initiating blast material. Velocity of shock wave in idle experiments towards the end of shock tube of 4.2 m long decreases to 800 m/s at the initiator mass 12 g. The wave amplitude doesn't exceed 6 atm, i.e. the initiator contribution is small as compared with the contribution of the process of the Al-particle burning in detonation wave.

Lower concentration limit (LCL) of initiation and propagation of the UDA air-suspension is between 110 and 140 g/m^3, and in the case of the PAP Al-suspensions it lies between 150 and 200 g/m^3 (see fig. 1). In rich PAP-mixtures the initiation energy of detonation goes up with the concentration growth and upper concentration limit (UCL) is about 460 g/m^3. Unlike the PAP, the UCL for the UDA-mixtures exceeds noticeably the most of studied Al-concentrations (750 g/m^3), since of the UCL would be close to 750 g/m^3, then at this concentration a detonation would propagate in the spin regime, how it always happens near the concentration limits, however it is no success to observe spin in rich mixtures.

Therefore, the carried out experimental investigation shows, that the detonation ability of the scaly Al-particle (with thickness 1 μm) air-suspension is close to the one for the propane-air mixture when concentration of the Al-particles is stoichiometric 330 g/m^3. More poor and rich mixtures are ignited more difficult. The UDA-powder suspensions with spherical particles of diameter about 1 μm have significantly higher detonation ability and more wider concentration limits, than suspensions of the scaly PAP-powder, which average diameter of particles is about 6 μm, and thickness is about 1 μm. Since the Al-powder suspensions with particles, diameter of which is more, than 10 μm, doesn't detonate neither in oxygen, nor in air, one can conclude that the less Al-particle size the more its detonation ability. This conclusion is in agreement with the conclusions in [2, 3], but contradicts to the data of [1], where application of the flaky powder with average diameter 40 μm instead of the granule Al-powder with the granular diameter 5 μm doesn't effect practically on the detonation characteristics, what's more, according to our data the Al-suspension with 13 μm - particles in oxygen turns out to be not able to detonate. This contradiction can be connected with the fact that in the flaky particles [1] one of the characteristic dimensions was, apparently, significantly less, than their mean diameter[1]. Then in the fact, the flaky particles would warmed up, indeed, more quickly, than spherical particles of the same effective diameter would. So, the detonation ability of the scaly particles can not correlate with their mass mean diameter.

The attempt to initiate a spherical detonation of the 3.5 μm Al-particles didn't result in stationary detonation due to limitedness of dimension of cloud used in experiments.

Detonation of aluminum air-suspensions studied by us propagates, in the main, in the spin regime, that corresponds to the conclusions of [1-3]. In [2, 3] the initiator energy was significantly lower, than minimal energy of the detonation initiation, that is why, the process observed in those works was not a stationary detonation.

Analogy between the laws of initiation and propagation of detonation in the Al-suspensions (with the Al-particle diameter about 1 μm) in air and oxygen media and laws of detonations in gaseous fuel-air mixtures allows to conclude, that such suspensions of aluminun can detonate not only in shock tubes, but also in space.

2. MODELING OF DETONATION OF Al-SUSPENSIONS

2.1. Description of a model

The model of stationary detonation of the Al-suspensions in gaseous oxidizer used in the paper is based on the model, formulated in [7] and defined more precisely in [9], which allowed to describe all the experimentally observed [10] laws of propagation of the two-front detonation in the Al-suspensions in capable of detonating various gaseous fuel mixtures (hydrogen, ethylene and acetylene) with oxygen and nitrogen in different ratios. In such mixtures [10] under certain conditions behind the leading detonation wave, connected with quick transformation of gaseous explosion substance the secondary detonation wave can propagate with the same velocity due to reaction of aluminum with products of the head detonation. Calculations [7-9] are in satisfactory agreement with

1 Unfortunately, thickness of scales in [1] is not reported.

experimental data on the concentration limits of the two-front detonation propagation and its structure.

The main physical grounds of this model, based on theory of the two-phase flows [15], are following. There is considered one-dimensional plane stationary flow of carrying gaseous phase, submitting to equation of the perfect gas state with metallic particles, volume fraction of which is negligible comparatively with the one of gas. Real polymodal size distribution of the powder particles is replaced by discrete N-modal distribution. It is taken for granted, that particles are incompressible, aren't crushed and don't agglomerate behind shock wave front, and the maximal radius of the particle is so small, that the volume of the particle is warmed up uniformly.

Delay of ignition of the each particle fraction behind the wave front $(\tau_{ign})_i$ is determined by their warming-up to ignition temperature T_{ign}. For calculation of the combustion time of fractions i=1, 2,...N at t $(\tau_{ign})_i$ empirical law of combustion is used in the form [16]

$$(\tau_b)_i = K\,(d_{pio})^n \big/ \Theta^{0.9}$$

where d_{pio} is initial dimension of the Al-particles, N is number of fractions, Θ is sum of volume fractions of O_2, H_2O and CO_2 in gas phase, K and n are empirical constants. The effect of slipping in velocity of particles on their combustion time [7] is also taken into account. Effective heat release of the Al-combustion is calculated proportionally to initial mass fractions of O_2, H_2O and CO_2 in gas phase and individual heat effects of the Al-reaction with O_2, H_2O and CO_2.

Dimensions of the Al liquid oxide drops, generating as a result of vapor-phase burning of aluminum, are assumed so small, that their temperature and velocity don't differ from corresponding parameters of gas phase, i.e. products of the Al-burning adapt instantaneously to the current state of gas phase. According to [16] it is supposed, that the Al-burning process becomes endothermal after the running to temperature of "boiling" of the Al oxide T_{ob} in gas phase. In fact gasification of Al_2O_3 after the running to this temperature leads to generation of gaseous AlO and atomic oxygen, since Al_2O_3 doesn't exist in gaseous form [17]. There are two different points of view relatively to the value T_{ob} in the literature, according to which T_{ob}=3350 K or 3800 K (see [16]). A function of boundary layer type is used for dependence of Q on gas temperature, then the jumping character of the reaction thermal effect Q changing from $Q_1>0$ to $Q_2<0$ at the oxide boiling temperature doesn't lead to extreme inconvenience of differential equations

$$Q = \begin{cases} Q_1 , & T \le T_{ob} \\ Q_2 + (Q_1-Q_2) \exp [-(T-T_{ob})/ \Delta T] , & T > T_{ob} \end{cases}$$

The value 10 K is used as a characteristic dimension of boundary layer ΔT , and decrease of ΔT doesn't change the computed results, but leads to noticeable increasing of the computer time expenditure.

It was assumed, when taking into account of momentum and energy losses out of the reaction zone on shock tube walls, that the tube walls have "technical" roughness, i.e. characteristic size of the sand roughness is 1/2300 of inner radius of shock tube [18]. (When modeling of experiments in glass pipes [1] they were considered smooth ones, but radiation losses which action turned out to be negligible were taken into account in addition).

Tabulated dependencies on temperature [19] are used for coefficients of viscosity and heat conductivity of gas phase (air) and heat capacity of four components of gas phase) O_2, H_2O, CO_2 and N_2. Dependence of heat capacity of gas phase on content of the liquid Al oxide in it, i.e. dependence of adiabatic exponent of gas phase γ is considered not only on its temperature, but also on composition. The last circumstance is an important addition of the model [9], since dependence $\gamma = \gamma(T)$ acts strongly on the detonation velocity and in the considered case gas phase temperature varies in more wide range , than in two-front detonation. Heat capacities of aluminum and liquid oxide of aluminum, reaction heat, melting and vaporization heats of aluminum are taken from [16, 19] and considered to be independent on temperature and pressure.

System of governing equations [7, 9] describing a stationary flow of the two-phase medium behind the shock wave front, has a special point of the saddle-type, in which according to theory of non-ideal detonation [20] must be valid two conditions: 1) Mach number is equal to 1 ($M = v_g/C = 1$, where v_g is the gas velocity relatively to the wave front, and C is local sound speed) ; 2) rate of heat release is equal to effective rate of heat losses, caused both by interaction of a flow with the shock tube wall, and by loss due to acceleration and warming-up of large particles, if $N > 1$, where N is number of fractions. Velocity of non-ideal detonation was numerically calculated using method of shooting in such a manner, that integral curve, starting at the wave front, would approach as close as possible to this special point. The detonation velocity is computed with error 0.25 m/s, and parameters of non-ideal detonation in the Chapman-Jouguet point are determined by extrapolation. Values of entry parameters of the model are given in [7-9], and taken partially from [16].

2.1'. Choice of the temperature of combustion initiation

The value 1350 K was used in [19] for effective temperature of the Al-particle ignition T_{ign} . This choice provided a reasonable accordance to known experimental data on structure of the two-front detonation, although computed delays of ignition were higher than the experimental ones [10]. Let us note, that this difference isn't connected with inhomogeneity of the particle warming-up, since characteristic time of the Al-particle warming-up $\tau_H = \rho c(d_{po}/2)^2/\lambda$ even in the case of the 10-μm particles doesn't exceed 0.35 μs, that is much more small than typical [10] delays of the Al-particle ignition in shock waves, which are equal to tens micro seconds (here ρ, c and λ are density, heat capacity and heat conductivity of particles).

The present paper results show, that effective temperature of the Al-particle ignition in shock waves must be close to the Al-melting temperature 933 K. Actually, the shock wave velocity action on computed ignition delay of the Al-particles of various dimension is shown in figure 5. The curves 1, 2 and 3 are obtained for mixtures $Al - O_2$ at

T_{ign} =950 K, the particle concentration σ=1500 g/m^3 and d_p = 1, 5 and 13 μm , respectively[2], where d_p is the particle diameter. It is seen, that the ignition delays grow strongly with the particle dimension growth and the shock wave velocity decreasing. In shock waves, velocity of which is less than ≈1400 m/s the particle ignition doesn't occur at all. The curve 2' is obtained under the same conditions, that the curve 2, but at T_{ign} =1350 K. It is seen, that increase of T_{ign} from 950 K to 1350 K

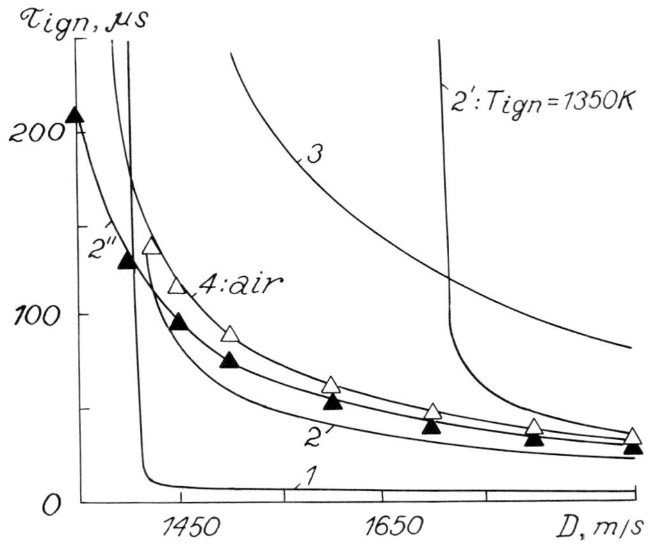

Fig. 5. Dependence of delay of the Al-particle ignition on velocity of shock wave. Curves 1, 2 and 3 are calculated for the aluminum - oxygen mixture at σ=1500 g/m^3, T_{ign} =950 K and for the particle sizes 1, 5 and 13 μm, respectively. Curve 2' is analogous to 2, but for T_{ign} =1350 K, curve 2" is analogous to 2, but at σ=300 g/m^3. Curve 4 corresponds to the Al-air mixture at σ=300 g/m^3, T_{ign} =950 K and the particle diameter of 5 μm.

2 Remind that the particle diameter 1 μm corresponds to UDA, the PAP, which is equal to 6 μm. The 13 μm - particles are used by us at study of the Al-suspension detonation in oxygen in shock tube.

leads to the growth of the threshold velocity of shock wave, able to initiate the Al-particles, from 1400 m/s to 1700 m/s. We remind in connection with this fact that the measured in[1] velocities of detonation of the aluminum-oxygen mixtures liein the range 1400-1500 m/s. That is why, when using incomputations T_{ign}=1350 K, the minimally possible velocity of detonation in stoichiometric mixtures of aluminum + oxygen would be higher, than 1700 m/s, that would be in contradiction with the data [1]. That is why, one assume here that T_{ign} = 950 K. From physical point of view, this means, that almost at once after melting behind the wave front the particles lose their integrity (as a result of difference of thermal coefficients of expansion of 5 μm-particles are used in [1], and besides, this diameter is also close to effective mean mass diameter of the scaly particles of aluminum and the oxide film, covering it [16]) and some part of aluminum contacts directly with oxygen, i.e. the beginning of ignition takes place.

Comparison of curves 2 and 2" (fig. 5) illustrates the effect of the particle concentration (σ=1500 and 300 g/m^3) on the ignition delay. The concentration decrease in "strong" shock waves leads to increase of the ignition delay and in the "weak" ones - vice versa, since the smaller concentration of particles the less energy is spent for their warming-up and acceleration and the less the threshold velocity of shock wave, able to provide the particle ignition. At last, curve 4 corresponds to the same conditions, that curve 2", but is obtained for air. Difference between these curves is not significant, i.e. noted tendencies can be applied equally also to the Al-air-suspensions.

2.2. Results of computations

First of all, we have to be convinced that presented model doesn't contradict to computed results of thermodynamic parameters of detonation. Curves, denoted by the letters Q and T in fig. 6, correspond to ideal velocities (a), pressures (b) and temperatures (c) of the aluminum air-suspension detonation, calculated with the help of program QUATUOR [13] and reported by Tulis [3]. Let us note, that ideal parameters of

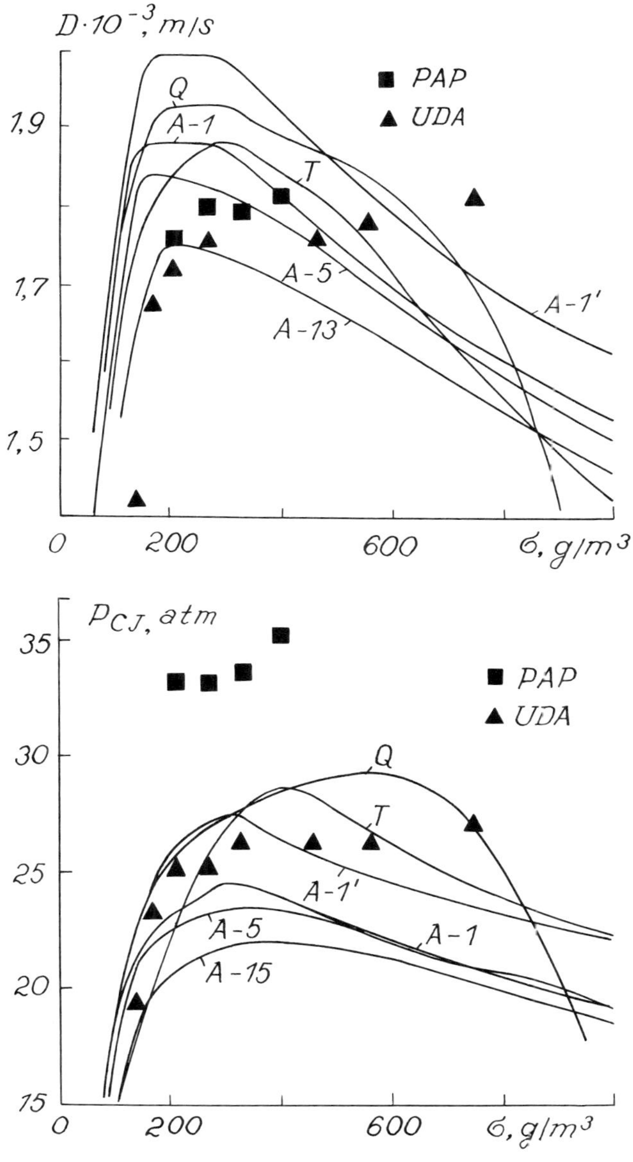

Fig. 6. Velocity (a), pressure (b) and temperature (c) of the Al-air-suspension detonation. Thermodynamic parameters are denoted by letters *Q* [13] and *T* [3]. Curves A-1, A-5 and A-13 are calculated in this paper for particles of diameter 1, 5 and 13 μm, respectively, at T_{ign}=950 K and T_{ob}=3350 K. Curves A-1' are analogous to A-1, but T_{ob}=3850 K. The tube diameter is 122 mm. The points correspond to data for PAP and UDA.

detonation of the aluminum-air mixtures, computed by different methods, on the whole correspond to each other satisfactorily, and difference (in the detonation velocity the difference reaches 100 m/s) is connected with the fact, that in [3] as distinct from [13] a possibility of generation of condensed products of reaction was taken into account.

Curves A-1 and A-1' are computed on the base of presented here model of non-ideal detonation for mono modal particles of aluminum of diameter 1 μm in shock tube with inner diameter of 122 mm and "technical" wall roughness of 25 μm at the Al-boiling temperature T_{ob}=3350 K and 3800 K, respectively. Comparison of curves A-1 and A-1' shows, that increase of T_{ob} leads to the growth of computed velocity of detonation approximately by 100 m/s, since at lesser T_{ob} the earlier endothermal stage connected with "boiling" of aluminum oxide begins. Nevertheless, both values of T_{ob} are in good accordance (especially, in the case of poor mixtures) with thermodynamic parameters of detonation, at computation of which the equilibrium of 10 various components of reaction is considered (in the case of so small particles velocity and thermal relaxations are finished long before the carrying out of the Chapman-Jouguet conditions, and that is why, curves A-1 and A-1' can be surely compared with curves, corresponding to the equilibrium thermodynamic computations). Pressures and temperatures in the Chapman-Jouguet plane calculated on the base of presented here model conform with thermodynamic parameters better at higher values of T_{ob}. But since the higher value of T_{ob} will lead to certain exceeding over ideal equilibrium velocity of detonation, we will use[3] T_{ob}=3350 K, although it is shown in [9] , that substitution of T_{ob}=3350 K by 3800 K at T_{ign}=1350 K in a several mixtures leads to the better accordance with experimental data on structure of the two-front detonation.

Now let's compare curves A-1 having intermediate location between curves Q and T in fig. 6a with the data for the UDA. On the whole, an accordance in the detonation velocities is quite satisfactory and besides, experimental points lie slightly on the right from calculated

3 Taking into consideration a simplicity of this model, one could see that attempts to restore the dependence of the Al-oxide boiling temperature on pressure with the help of the adjusting of the data calculated here using the experimental ones are not useful.

curve, that can be connected with the overrating of the particle concentration in suspension in experiments, caused by imperfection of the two-phase mixture generation technique. Measured pressures for the UDA (fig. 6b) are also close to curve A-1. We note, that completeness of combustion of aluminum is close to 100 % in the case of poor mixtures (i.e. losses on the tube walls in this case are insignificant and detonation is close to the ideal one) and decreases at the particle concentration growth to 40-50 % due to deficiency of oxygen.

For the aluminum-oxygen mixtures our model also gives satisfactory accordance with thermodynamic velocity and pressure of detonation, that confirms physical well-foundation of our model (in fig. 7 curve O-1 corresponds to the 1 μm diameter particles). Temperature in the Chapman-Jouguet plane T_{CJ} , determined by our model, agrees well with the thermodynamic one in poor mixtures, but in rich mixtures again $T_{CJ} = T_{ob}$ that is noticeably lower than thermodynamic temperature.

The Al-particle dimension increase (numerals near curves in fig. 6 and 7 correspond to the particle dimensions, μm) leads to decrease of velocity and pressure of detonation and to the narrowing of the concentration detonability limits[4]. It was no success to obtain detonation of the 40 μm diameter of particles even in oxygen. Comparison of curve O-5 with the Strauss's experimental data for spherical particles of diameter 5 μm (fig. 7) shows that computations give more higher velocities of detonation. This exceeding, remaining significant also taking into account content of 2 % Al-oxide in the powder [1], apparently is connected with the fact, that destruction of glass tubes, which takes place at distance of only several cm behind the wave front, can change the detonation velocity (see below).

2.3. On the detonation wave structure

Calculated structures of non-ideal detonation wave of the Al-particle air-suspension with mean diameter of 1 μm at σ=300 and 750 g/m^3 (corres-

4 Note that the particle ignition temperature increase leads to the detonation velocity growth at lower and upper concentration detonability limits and to corresponding narrowing of these limits.

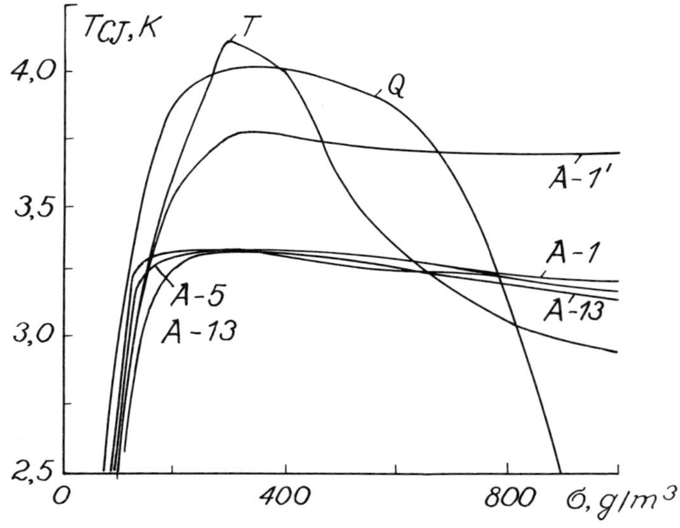

Fig. 6c.

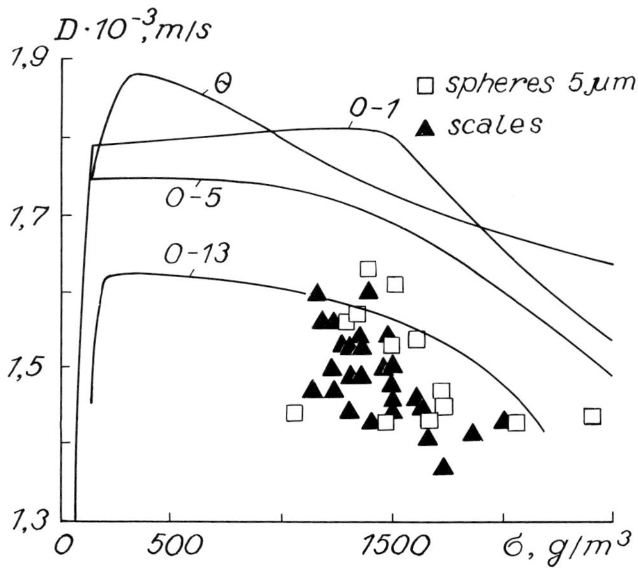

Fig. 7. Detonation velocity in the aluminum oxygen mixtures. *Q* - thermodynamic calculation by the program [13], curves O-1, O-5 and O-13 correspond to the particle diameters 1, 5 and 13 μm and the tube diameter 26 mm, the points - for data from [1].

ponding to D_{CJ}=1869.1 and 1622.6 m/s) including profiles of pressure, velocity and temperature of gas and particles, and also profiles of combustion completeness of aluminum f and volume fraction of oxidizer θ are shown in fig. 8 and 9. Immediately behind the shock wave front pressure grows up to moment of ignition and begins to fall only after the beginning of chemical reaction. Comparison of fig. 8 and 9 shows that the greater concentration of particles the greater the particle melting time (plateau is seen at T_p=933 K in fig. 9) and ignition delay and the slower relaxation processes. After the time of ignition delay due to high heat release of aluminum combustion a sharp increase of gas and particle temperature and synchronic sharp fall of pressure take place. But at once after approach of gas temperature to the Al-oxide boiling temperature the pressure decrease rate and, therefore, the heat release rate fall down sharply, since the pressure gradient modulus is proportional to effective heat release rate and inversely proportional to $(1 - M^2)$.

Such behavior of the heat release rate confirms a presence of two stages of heat release at the Al-particle burning in shock waves - the first is the stage of quick heat release at which chemical energy of the Al-burning releases in the main, and the second stage after attainment of the Al-oxide boiling temperature in products, the stage of slow heat release, at which heat of the Al-combustion is already small. This effect is typical for stoichiometric and rich mixtures, since temperature of products in very poor mixtures remains below than Al-oxide boiling temperature.

Narrow zone of abnormally high pressure in immediate proximity to the wave front isn't felt by used pressure transducers. If a gently sloping part of shown pressure profiles one attempts to extrapolate to the wave front, then in the first case, when the reaction zone is rather narrow (fig. 8), extrapolated pressure is only insignificantly more, than pressure in the Chapman-Jouguet plane p_{CJ}, and in the second case (fig. 9) p_f can be considerably higher, than p_{CJ}. Evidently, this is connected with the fact that in the case of the PAP-particles mean mass diameter of which is greater than of the UDA-ones pressure p_f is noticeably

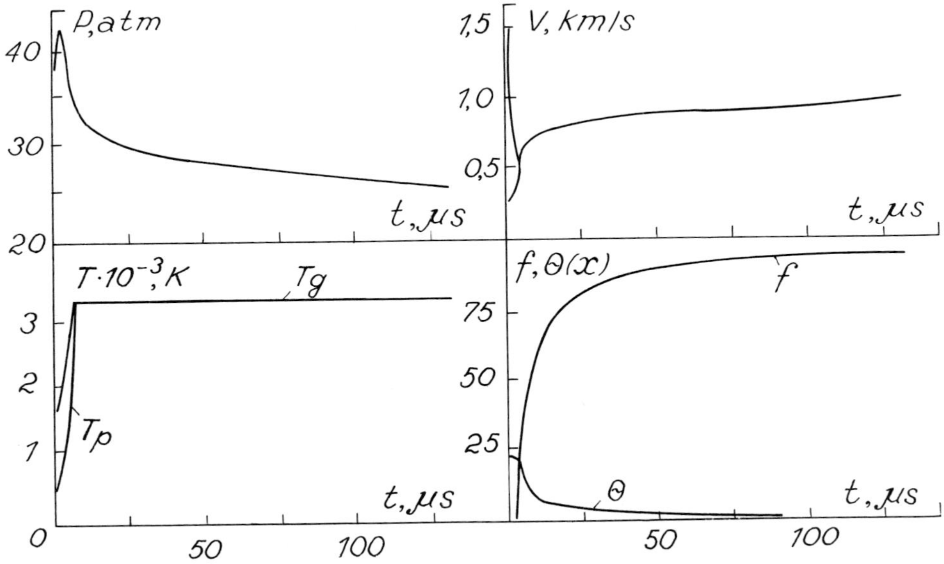

Fig. 8. Structure of the detonation wave in air-suspension of the Al-particles, σ=300 g/m^3, the particle diameter is 1 μm, D_{CJ} = 1869,1 m/s (f is completeness of combustion of aluminum, θ is volume fraction of oxygen)

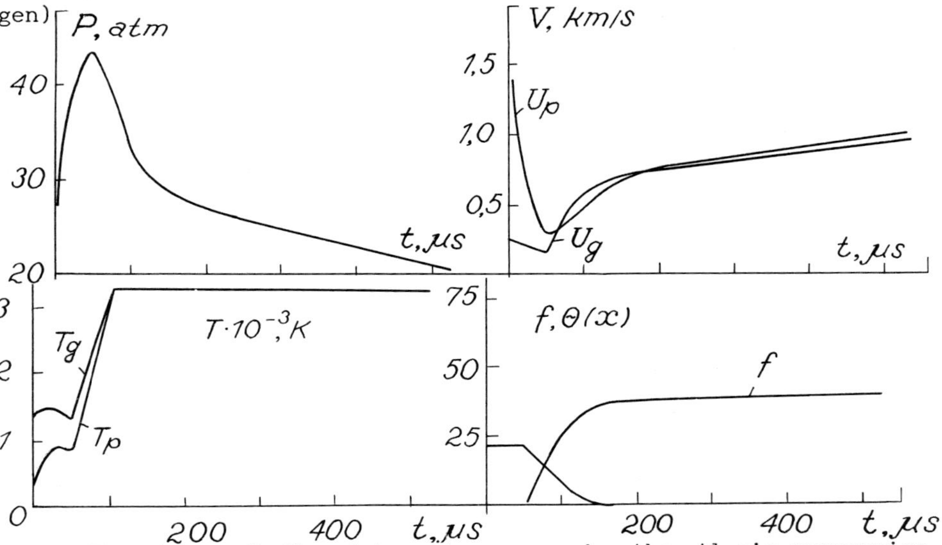

Fig. 9. Structure of the detonation wave in the Al-air-suspension. σ=750 g/m^3, the particle diameter is 1 μm, D_{CJ}=1622.6 m/s.

higher, than for the UDA (fig. 6b).

An important characteristic of the detonation, ability is not only the particle combustion ignition ability (fig. 5), but also the size of the heat release zone X_{CJ} between the shock wave front and the Chapman-Jouguet plane[5], since the higher X_{CJ}, the greater length of the detonation tube necessary for the obtaining of really stationary process. However, by force of the process of heat release is two-stage in the case of the Al-suspensions it should be noted that the detonation ability, in the main, is determined not by whole extent of the heat release zone, but by size of the quick heat release zone, which is connected with attainment of the Al-oxide boiling temperature in products and is several times less than whole length of the heat release zone X_{CJ}.

Calculated dependencies of X_{CJ} on concentration of particles for non-ideal detonation of the Al-suspensions in air (curves A-1 and A-5, tube diameter of 122 mm) and in oxygen (O-1, O-5 and O-13, tube diameter of 26 mm). are shown in fig. 10, and besides numerals near curves again correspond to the particle diameters. It is of interest, the dimension of the heat release zone is maximal at stoichiometric concentration. This is connected with the fact, that the reaction velocity in the zone of quick heat release at this concentration is the highest and the heat loss rate can be equal to the heat release rate in the second slow stage at larger distance from the front, than in the case of poor or rich mixtures. Due to the same reason X_{CJ} of non-ideal detonation grows at decrease of losses on the shock tube walls (see below), and what's more, in this case the increase is caused in the main by the growth of length of the slow heat release zone. In the case of curve A-5 the heat release zone length grows as the concentration increases also near the upper concentration detonability limit.

So, figure 10 shows, that heat release zone size grows strongly with

5 Heat release at non-ideal detonation, naturally, can take place also behind the Chapman-Jouguet plane in the zone of non-stationary rarefaction wave, and besides the temperature fall in the rarefaction wave (at significant under burning of aluminum before the Chapman-Jouguet plane and at presence of oxidizer) can again cause strong exothermicity of the Al-burning and the arising of secondary pressure shocks behind the Chapman-Jouguet plane. However, it should be used non-stationary model of detonation to analyze this situation.

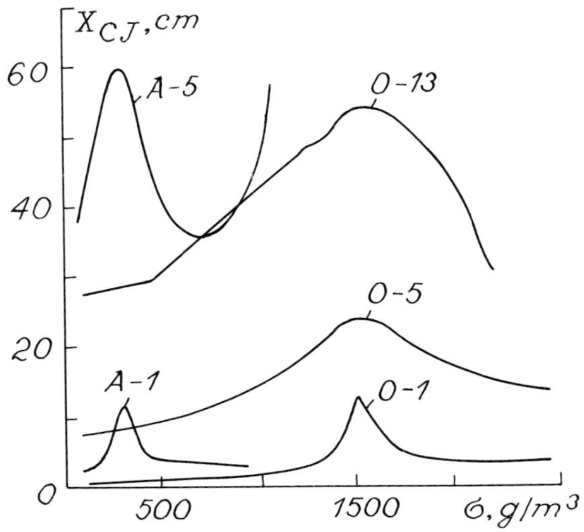

Fig. 10. Extent of the heat release zone at detonation of the Al-mixtures with air and oxygen versus the Al-particle concentration: curves A-1 and A-5 correspond to air and the particle diameters of 1 and 5 μm, curves O-1, O-5 and O-13 - to oxygen and radius of particles 1, 5 and 13 μm, respectively (T_{ign}=950 K and T_{ob}=3350 K).

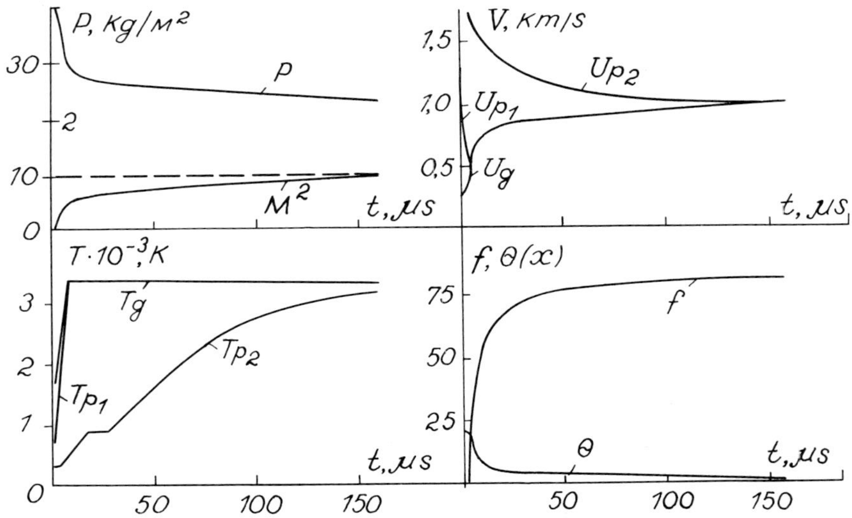

Fig. 11. Structure of the detonation wave in bidispersive air-suspension of the Al-particles (80/20 %, 1-10 μm), σ=300 g/m^3 (M^2 is profile of Mach number squared). Subscripts 1 and 2 correspond to particles of small and large fractions.

the particle diameter growth (especially, in air), that the Al-powder suspensions with dimension about 1 μm have approximately the same detonation ability both in air and in oxygen but already the 5 μm-particles detonate much worse in air, than in oxygen. At detonation of the 13 μm - particles in oxygen X_{CJ} is already about 0.5 m, but in air for the same particles X_{CJ} > 1 m (!), i.e. the detonation ability of this powder is very low even in the oxygen medium, that is found out experimentally.

In connection with exceeding of curve 0-5 in fig. 7 over corresponding experimental velocities of detonation [1] we note that in the case of the 0-5 curve X_{CJ} is 10-20 cm (depending on particle concentration), but destruction of glass tubes in [1] occurred at the distance of several cm behind the wave front. That is, even in check experiments [1] using of glass tubes, strengthened by metallic cover, destruction of vision slit could lead to the detonation velocity reduction due to side scatter of the reaction products.

At the 3.5 μm - diameter of the Al-particles and stoichiometric Al-concentration in oxygen X_{CJ} =41 cm at tube diameter of 122 mm, and after the 10-times increase of the tube diameter the dimension of the heat release zone grows (due to the length growth of slow heat release zone) up to 145 cm (here the detonation velocity increases insignificantly from 1767 to 1802 m/s). So high values of the heat release zone size for almost ideal detonation allow us to understand, why it is necessary to use the installations of significantly larger scale than in [12], when investigating of spherical detonations of such aluminum particles. It will be shown in the next section, however, that a presence of noticeable fraction of large particles in suspension can lead to a significant reduction of the reaction zone length at practically constant velocity of detonation.

It is interesting, that the proposed model allows to obtain also the estimation of detonation initiation energy by value of whole energy, accumulated by gas and particles in the zone between the shock wave front and the Chapman-Jouguet plane. For this aim the value

$$E_{\Sigma} = \int_{0}^{X_{CJ}} \left[\rho_g \, (e_g + U_g^2/2) + \sigma \, (e_p + U_p^2/2) \right] dX$$

is calculated, where U_p and U_g are velocity of particles and gas in laboratory coordinate frame, e_p and e_g are their internal energies. Dimension of the quantity E_{Σ} is J/m^2, that is, it coincides with dimension of energy of direct initiation of detonation. Dependencies of E_{Σ} on the particle concentration σ are analogous to dependencies X_{CJ} on σ shown in fig. 9 and by the value E_{Σ} is equal to 1.8 and 8.8 MJ/m^2 for stoichiometric concentration of aluminum in air at the particle dimension of 1 and 5 μm, respectively, i.e. these values are by several times greater, than experimental energy of initiation. Agree between calculated and experimental energy of detonation can be improved, if one suppose, that the detonation ability depends not on X_{CJ}, but on dimension of the quick heat release zone, which is by several times less than X_{CJ}.

2.4. On the effect of distribution and shape of the Al-particles

In connection with the assumption [2, 3] of under burning of the Al-particles dimension of which is significantly higher than mean distribution diameter of particles let us study a possible effect of such particles on structure of the non-ideal detonation wave. Note, that such particles can be generated as a result of agglomeration of the Al-particles in initial powder and also in a flow of hot gas behind the shock wave front.

So, let in the Al-particle suspension side by side with particles of main "small" fraction significantly larger particles present. Consider, for example, such two-fraction suspension in which d_{p1}=1 μm and d_{p2}=10 μm, and mass part of "large" fraction is 20 %. Structure of such wave for the Al-air-suspension at σ=300 g/m^3 is shown in fig. 11. The Chapman-Jouguet velocity in this case is 1847.4 m/s, that is only slightly lower, than in mono-dispersive suspension of the 1 μm - diameter particles. However, in this case computation was carried out in the absence of any loss on the tube walls. As it is seen from fig. 11, where together with pressure the profile of Mach number squared is shown, small particles are accelerated and ignited more quickly, than the large ones (τ_{ign}=2.7 and 27.5 μs, respectively). As a result, large particles "burn" under conditions, when small particles have completely burned, and the gas phase temperature is close to the Al-oxide boiling temperature T_{ob} . Here the current heat effect of the Al-burning becomes less and less and, at last, the reaction becomes endothermal, i.e. the burning of large particles at temperature, which is close to T_{ob} , can lead to change of sign of the heat release rate.

Large particles aren't ignited at all already at σ=600 g/m^3 before the Chapman-Jouguet plane. Note, that at σ=300 g/m^3 X_{CJ} is only 14 cm, E_{Σ}=1.8 MJ/m^2, and completeness of the Al-burning in the Chapman-Jouguet plane is f=0.82. At σ=1000 g/m^3 D_{CJ}=1563 m/s (that differs slightly from the detonation velocity of mono-dispersive suspension of particles again due to specificity of the Al-burning at high temperature), X_{CJ}=2 cm, E_{Σ}=0.3 MJ/m^2 and f=0.30). It is of interest, that in spite of significant incompleteness of the Al-burning (1-f) at detonation of rich mixtures the detonation velocity doesn't strongly differ from the corresponding thermodynamic value, that is connected again with the fact, that main heat release takes place in the zone of quick burning of aluminum. Approximately the same tendencies were obtained at content of main "small" fraction 90 and 95 % for the 5 and 10 μm - particles of "large" fraction, and besides, intensity of energy "losses" decreases with the decrease of large particle fraction and that is why, velocity of non-ideal detonation and the reaction zone size grow, but increase of D_{CJ} is much less, than the X_{CJ} and E_{Σ} - ones.

Therefore, availability of significant fraction of large particles in suspension can be an effective origin of the heat losses and even more powerful, than the shock tube walls, i.e. parameters and structure of non-ideal detonation in such suspensions don't depend practically, on losses on the tube walls. Analogous possibility of the non-ideal detonation propagation in the two-fraction suspensions of particles with different mass in the case of unitary fuel was demonstrated in [21]. Therefore, laws of the non-ideal detonation propagation in the Al-suspensions with large particles have to be the same not only in shock tubes but also in space.

It was stated above, that in the case of the scaly particles of aluminun the suspension detonation ability can depend in the main on the smallest characteristic dimension of the scaly particles, i.e. on their thickness. In the model described above only effective mean mass diameter of spherical particle is used. It is clear, that behavior of the scaly particles behind the shock wave front can be essentially differ from behavior of spherical particles.

Really, chaotic orientation of the scaly particles in space allows to conclude that the particles, which planes at the moment of the shock wave front fall on them are directed along gas flow, keep their orientation for a some time, but all the other scaly particles with non-zero angle of attack are "turned around" quickly by shock wave in the way, that their planes become parallel to the shock wave front. It is clear, that particles, sharp side of which is directed "to meet" gas flow, will accelerate much slower, than the scaly particles, which plane is parallel to the shock wave front, since at the same mass of these particles their projections differ by an order of magnitude, and besides, friction coefficient of longitudinal flow of a plate is 3.7 times smaller than of cross-flow of the plate [18]. This means, that friction force and acceleration of scales-particles, slipping along gas flow, is approximately 35-50 times smaller than the ones of "badly-flowed" scaly particles, oriented by the plane "to meet" a flow. As a result, the particle velocity, slipping along a flow, will grow (in laboratory coordinate frame) much slower, than the velocity of "badly-flowed" particles. This provides strong relative increase of coefficient of heat transfer from gas to "well-flowed" particles due to high velocity of their slip. What's more, heat transfer coefficient at sharp side is higher, than at the flowing of "badly-flowed" bodies, [18]. Further, heat transfer from gas to particles, slipping along a flow, is determined by stagnation temperature of gas on the forward edge of particles T_{gs}, which is approximately, two times more than gas temperature T_g fare from the particles. But the stagnation temperature of gas at the particles which plane is parallel to the shock wave front plane, falls down quickly to gas temperature T_g, since their velocity becomes equal to gas velocity very quickly. The sum of all these tendencies provides an earlier ignition of the particles slipping along the flow.

Therefore, in the first approximation one can accept, that suspensions with the scaly particles can be considered as bidispersive ones of the same mass with different friction coefficients, projections and heat transfer coefficients, differing from standard values for spherical particles, but in this case "small", i.e. "well-flowed" particles are only, small fraction of the total mass of particles. That is why, by analogy to above-considered case of the two-fraction suspension one can wait for, that the burning of particles, oriented by sharp edge "to meet" a flow, leads again to sharp increase of gas temperature up to the Al-oxide boiling temperature T_{ob}, and "large", i.e. "badly-flowed" particles will burn at temperature close to T_{ob}, when exothermicity of the Al-burning decreases sharply.

Indeed, calculations have confirmed, that time difference of ignition of various "fractions" of the scaly particles mounts to 1.5-2 times[6], in spite of acceleration of warming-up of "badly-flowed" particles due to quick gas temperature growth after ignition of "well-flowed" particles. Even when the mass fraction of particles slipping along the flow isn't very high the great value of combustion heat of aluminum provides growth of gas temperature up to 3000 K and because of this the temperature of products reaches the value T_{ob} even after combustion only little part of "badly-flowed" particles. As a result the thermal effect of aluminum combustion decreases sharply and heat release rate falls down providing Chapman-Jouguet condition for non-ideal detonation even in absence of losses on the shock tube walls. Changing the mass fraction η_1 of particles slipping along the flow in the range 10-50 % shows that η_1 practically doesn't effect on the dependence of detonation velocity on particle concentration in rich mixtures, but when decreasing η_1 the

6 Approximately the same difference takes place also between distances from the shock wave front to the points of ignition of both fractions.

lower concentration detonability limit (LCL) is shifted to right.

CONCLUSION

Process of initiation and propagation of detonation in air-suspensions of the Al-particles in the form of sphere with diameter 1 μm (UDA) and scales with thickness 1 μm and cross dimension 10-15 μm (PAP)) in shock tube have been experimentally studied. The Al-suspensions with particles more, than 10 μm by size doesn't detonate both in oxygen and in air. The spherical detonation of the Al-particles with diameter about 3.5 μm, dusted in the oxygen medium, isn't obtained due to limitation of the cloud dimension.

The detonation ability of the PAP-air-suspension and gaseous propane-air stoichiometric mixture are close. The detonation ability of the Al-suspensions increases with the particle dimension decrease, and the concentration detonability limits widen. In the PAP-case detonation propagates in the spin regime, its velocity depends weakly on the particle concentration and equals to 1800 ± 50 m/s. In the UDA-case the detonation velocity of poor mixtures grows with the particle concentration growth andbecomes 1800 ± 50 m/s in rich mixtures. In the certain range of concentration of the UDA spin detonation is also observed and with increase of particle concentration the number of "heads" at the wave front increases. At last, fluctuations of glow on the front become undistinguished at the Al-concentration more, than 330 g/m^3.

The analogy between laws of initiation and propagation of detonation in the Al-particle suspensions (about 1 μm in the air and in oxygen medium and in gas fuel-air mixtures allows to conclude that such suspensions of aluminum can detonate not only in shock tubes but also in space.

The model for calculation of structure and parameters of non-ideal detonation wave of the Al-particle suspension in gaseous oxidizer is proposed. Computed data are in satisfactory accordance with the experimental ones obtained at the detonation study of the Al-suspensions in air and oxygen.

The sign change of thermal effect of the Al-combustion after the

Table 1

Experimental values of velocity and pressure of detonation, of spin step Λ , Mach number and period of glow pulsations τ.

type	σ, g/m^3	D, m/s	p, atm	Λ, cm	Λ/d	M	τ, μs
PAP	210	1760	33.0	43	3.52	6.26	240
	270	1800	33.0	40	3.28	6.29	220
	330	1790	33.5	38	3.11	6.31	210
	400	1810	35.0	38	3.11	6.32	210
UDA	140	1420	19	40	3.28	4.89	280
	170	1670	23	33	2.70	5.78	200
	210	1720	25	13	1.07	5.98	80
	270	1750	25	15	1.23	6.11	90
	330	1800	26	11	0.90	6.31	60
	460	1760	26				
	560	1780	26				
	750	1810	27				

Table 2

Results of experiments on excitation of spherical detonation in suspension of Al-particles of diameter 3.5 μm in oxygen.

Concentration	Calculation by [13]		Experimental values		
g/m^3	D_{CJ} m/s	p_{CJ} atm	D_{K1-K2} m/s	p_{K1} atm	p_{K2} atm
2700	1615	52	1430	43	31
4150	1427	53	1220	39	30

"boiling" beginning of the Al-oxide leads to the two-stage heat release at the Al-particle burning past the shock wave front - full value, in the sense of release of chemical energy, combustion of aluminum occurs only at the first quick stage of heat release (and namely, this heat release provides, in the main, propagation of the detonation wave) determined by an attainment of the Al-oxide boiling temperature in products. That is why, the detonation velocity of stoichiometric and rich mixtures is not enough sensitive to completeness of the Al-combustion in the detonation wave. The time and space extent of a slow stage of heat release is significantly more, than of the first stage.

Structure of the detonation wave is strongly dependent on the particle dimension distribution, since large particles are ignited later than small ones and burn at temperature, which is close to the Al-oxide boiling temperature, when thermal release already is small and even negative. In spite of significant under burning of large particles up to the Chapman-Jouguet plane, velocity of non-ideal detonation in such suspensions doesn't differ strongly from corresponding thermodynamic value. In the presence of noticeable mass fraction of such particles in suspension laws of propagation of non-ideal detonation are the same both in shock tubes and in space. Analogous effects take place also in the case of the scaly particles, since their different orientation in a space leads to non-simultaneity of ignition of different particles (particles, slipping along a gas flow, are ignited earlier, than other scales).

REFERENCES

1. Strauss W.A. Investigation of the Detonation of Aluminum Powder-Oxygen Mixtures // AIAA Journal, 1968, 6, N 9, 1753-1756.
2. Tulis A.J., Selman J.R. Detonation Tube Studies of Aluminum Particles Dispersed in Air// Proceedings of the 19th Symposium (International) on Combustion - The Combustion Institute, 1982, pp. 655-668.
3. Tulis A.J., Fochtman E.G., Heberlein D.C. Experimental methods for Assesing Detonation/ Deflagration in Pyrotechnical Dusts // The 7th (International) Pyrotechnics Seminar, vol. 2, IIT Research Institute, 1980, pp. 859-877.
4. Vasilyev V.M., Vol'pert A.I., Klychnikov L.V., Petrov Yu.M., Salakatova L.S., Stesik L.N. Calculation of detonation parameters of the burning material-air mixtures // Phyzika Goreniya i Vzryva (Combustion, Explosion and Shock Waves), 1980, 16, N 3, p. 127. (in

Russian).
5. Medvedev A.E., Fedorov A.V., Fomin V.M. Mathematical modelling of the metal particle ignition in high-temperature flow past shock wave // Phyzika Goreniya i Vzryva (Combustion, Explosion and Shock Waves), 1982, 18, N 3, p. 5. (in Russian).
6. Afanasyeva E.A., Levin V.A. Al-particle ignition and burning past shock and detonation waves // Phyzika Goreniya i Vzryva (Combustion, Explosion and Shock Waves), 1987, 23, N 1, p. 1-14. (in Russian).
7. Khasainov B.A., Veyssiere B. Analysis of the Steady Double-Front Detonation Structure for a Detonable Gas Laden with Aluminum Particles // Archivum Combustionis, 1987, 7, N 3-4, pp. 333-352.
8. Khasainov B.A., Veyssiere B. Steady Plane Double-Front Detonations in Gaseous Detonable Mixtures Containing a Suspension of Aluminum particles // AIAA Progress in Astronautics and Aeronautics, 1988, 114, pp. 284-289.
9. Veyssiere B., Khasainov B.A. A Model for Steady, Plane Double-Front Detonations in Gaseous Detonable Mixtures with Aluminium Particles in Suspension // Combustion and Flame, in press.
10. Veyssiere B. Structure of the Detonations in Gaseous Mixtures Containing Aluminum Particles in Suspensions // Dynamics of Explosions, AIAA Progress in Astronautics and Aeronautics, 1986, 106, pp. 522-544.
11. Borisov A.A., Khomik S.V., Mikhalkin V.N., Saneev E.L. On the Critical Energy of Direct Detonation Initiation in Gaseous Mixtures // Presented at the 12th ICDERS, 1989.
12. Veyssiere B., Desbordes D., Lee J.H. Preliminary experiments for direct initiation of spherical detonations in two-phase mixtures // Archivum Combustionis , 1987, 7, N 1-2, pp. 185 - 196.
13. Heuze O., Bauer P., Presles H.N., Brochet C. The equation of State of Detonation Products and their Incorporation into the Quatuor Code // The 8th Symposium (International) on Detonations, Preprints, Albuquerque, New Mexico, 1985, p. 103-110.
14. Fey J.A. Mechanical Theory of Spinning Detonation // J. of Chem. Phys., 1952, 20, p. 942.
15. Nigmatullin R.I. Dynamics of multi phase media, Part 1 - Moscow: Nauka, 1987. (in Russian).
16. Pokhil P.F., Belyaev A.F., Frolov Yu.V., Logachev V.S., Korotkov A.I. Burning of the powder metals in active media - Moscow: Nauka, 1972. (in Russian).
17. Brewer L., Searcy A.W. The Gaseous Species of the $Al-Al_2O$ System // Journal of the American Chemical Society, 1951, 73, pp. 5308-5314.
18. Shlikhting G. Theory of boundary layer.- Moscow: IL, 1956 (in Russian).
19. Thermodynamic properties of individual materials. Reference book. Ed. by V.P.Glushko. - Moscow: Nauka, 1979 (in Russian).
20. Zel'dovich Ya.B., Kompaneyets A.S. Theory of detonations - Moscow: Gostekhizdat, 1955.
21. Borisov A.A., Ermolayev B.S., Khasainov B.A. Non-ideal detonation in the two-fraction dust-suspension // Khimicheskaya Phyzika (Chemical physics), 1983, N 8 (in Russian).

Non-ideal detonation in combustible gaseous mixtures with reactive solid particles.

B. Veyssiere
Laboratoire d'Energetique et de Detonique
U.R.A. C.N.R.S. 193
Ecole nationale Superieure de Mecanique et
d'Aerotechnique. 860034 POITIERS-FRANCE
B. A. Khasainov
Institute of Chemical Physics, USSR Academy
of Sciences, 117977 MOSCOW-USSR

Abstract. In hybrid mixtures made of a combustible gaseous mixture with reactive solid particles in suspension, different propagation regimes of non-ideal detonation may be met: Double Front Detonations (DFD) Single Front Detonations (SFD). Their occurrence is linked to the non-monotonous process of heat release. The problem is studied in the case of Hydrogen Air mixtures containing a suspension of aluminium particles. The domains of existence of those propagation regimes are predicted as function of the mass concentration off aluminium particles.

INTRODUCTION

Non ideal detonations are a class of supersonic propagation regimes of combustion waves, which deviates strongly from the classical theory of so-called Chapman-Jouguet (CJ) detonations. Steady propagation of non-ideal detonations is generally achieved with the help of lateral boundary effects. Due to the process of heat release over a long period of time, in most cases neither mechanical, not thermal, nor chemical equilibrium are achieved at the so-called CJ plane. Hybrid mixtures made

A. A. Borissov (ed.), Dynamic Structure of Detonation in Gaseous and Dispersed Media, 255–266.

of a combustible gaseous mixture with reactive solid particles in suspension, provide a good example of non-ideal detonation propagation: in those mixtures, part of the heat release comes from reactions between gaseous components, the other part comes from reactions between reactive solid particles and gaseous products. The burning time of the solid particles may differ from the reaction time of gaseous components by an order of magnitude or more. Moreover, the process of heat release may be non-monotonous. Thus, different structures of steady non ideal detonation may be exist following the actual sequence of heat release. Experimental evidence has been already obtained in gaseous detonable mixtures containing aluminium particles in suspension /1,2/.

Here, we present a first series of numerical results from calculations performed in view to investigate the different propagation regimes of non-ideal detonation in hydrogen-air mixtures containing a suspension of aluminium particles.

INVESTIGATION OF THE DIFFERENT STRUCTURES

Two main regimes of steady non-ideal detonation propagation can be met. In the first one, heat release from gaseous homogeneous reactions and from heterogeneous reactions between particles and gases occur in two separated zones. This situation gives rise to the so-called "Double Front Detonation" structure (DFD) (see fig.1-a): The first detonation front is supported by the gaseous reactions, whereas the second detonation front is supported by reactions between particles and gases (see also /3/). In the second case, there exists a unique heat release zone, including together homogeneous and heterogeneous reactions, supporting only one detonation front (see fig.1-b): this situation has been called "Single Front Detonation" (SFD).

The problem is modelled using the theory of heterogeneous media /4/. The flow is one-dimensional and the balance equations are written under their differential form separately for gaseous components and solid particles. Interaction terms are introduced in the equations to take account of mass, momentum and energy exchanges between particles and

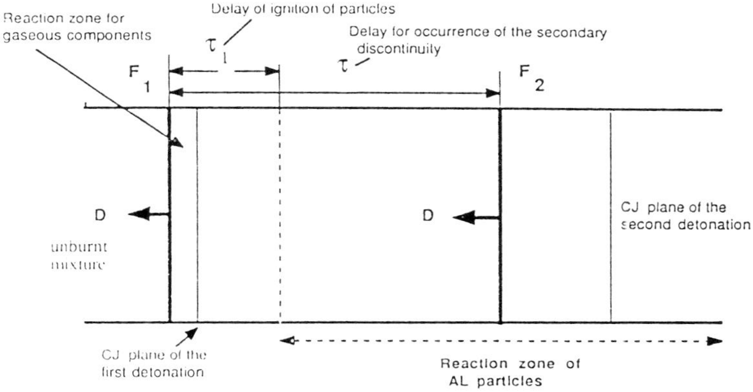

a) Double Front Detonation (DFD)

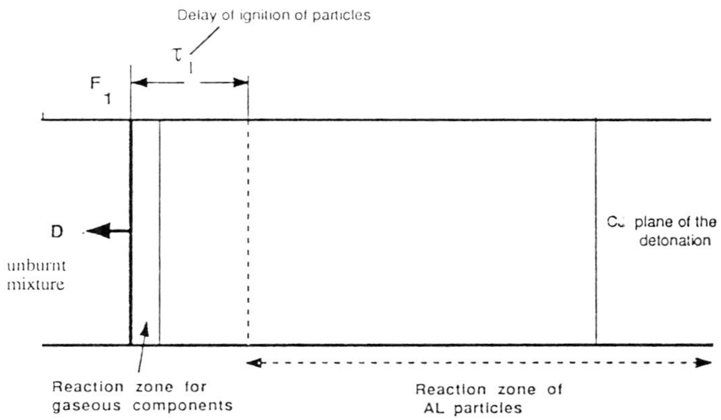

b) Single Front Detonation (SFD)

Fig. 1 Schematic diagrams of DFD and SFD: a) Double Front Detonation (DFD); b) Single Front Detonation (SFD).

gases. Momentum and heat losses to the tube walls are also included. The structure of the detonation wave is described by the ZND model. The gaseous reactions are described by the global empirical model of Korobeinikov et al. /5/ and the heterogeneous reactions between solid particles and gases by an empirical burning law /6/. The equivalent CJ condition is chosen following the classical rule assumed for non-ideal detonations /7/: the conditions $M = 1$ and $\dot{q}^{+} = \dot{q}^{-}$ must be achieved simultaneously (where $\dot{q}^{+}$ and $\dot{q}^{-}$ are respectively the rate of chemical heat release and the effective rate of heat losses, see /6,8/). With those assumptions, the velocity of the steady non- ideal detonation (when it exists) is an eigenvalue of a system of ordinary differential equations.

This system is solved by a numerical code which has been discribed elsewhere (KV code,/9/).The input values used in the present

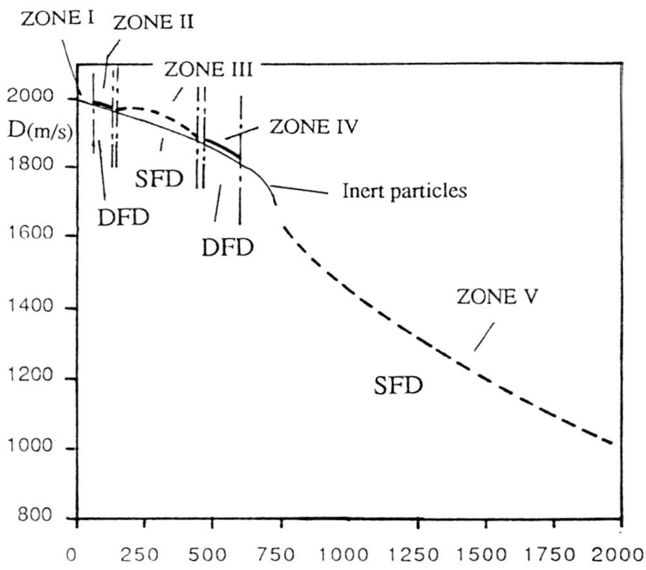

Fig. 2 The different detonation regimes as function of the mass concentration of aluminium particles in the mixture $H_{1.06}$.

work are exactly the same as in previous papers /6,8/.

Computations have been done with several mixtures of hydrogen with air at different equivalent ratios (index after H indicates the equivalent ratio):

H0.6: $H_2 + 0.83O_2 + 3.13N_2$

H0.78: $H_2 + 0.64O_2 + 2.41N_2$

H1.0: $H_2 + 0.50O_2 + 1.88N_2$

H1.06: $H_2 + 0.47O_2 + 1.77N_2$

H1.3: $H_2 + 0.38O_2 + 1.44N_2$

H1.5: $H_2 + 0.33O_2 + 1.25N_2$

H1.8: $H_2 + 0.28O_2 + 1.04N_2$

H3.0: $H_2 + 0.17O_2 + 0.63N_2$

H4.0: $H_2 + 0.13O_2 + 0.47N_2$

The concentration of aluminium particles in the mixtures varied from 0 to 2000 g/m^3.

RESULTS

In Fig. 2 and 3, the evolution of the detonation velocity is plotted as function of the particle mass concentration, in the case of mixture H1.06. For the sake of comparison, the variation of the detonation velocity with the mass concentration of the same particles but considered as inert (no ignition and no heat release), is also plotted. With reactive particles, several domains of steady propagation can be found:

- From σ=0 to σ_1 (zone I of Fig.3), the heat released by the burning of particles is insufficient to generate chocking of the flow downstream of the leading detonation front, and the pressure profile presents a compression wave (see Fig.4a), which correspond to a large-time asymptotic solution. This case has been already discussed in /6/.

- Beyond this concentration σ_1 exits the DFD concentration domain (zone II of Fig.3) investigated in our previous papers /6,8,9/. The characteristic parameter of this structure, the delay τ between the two fronts,decreases with increasing the particle mass concentration σ. An upper limit σ_2 is found beyond which steady DFD cannot exist. Typical pressure profile for this case is shown in Fig.4b.

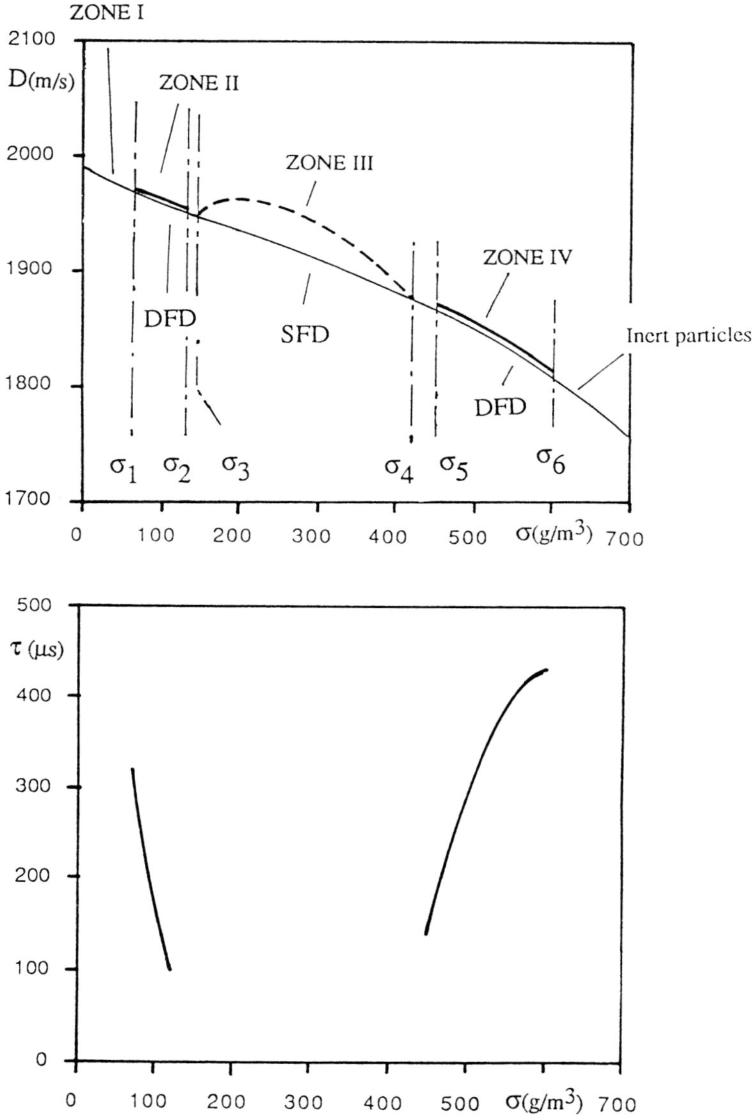

Fig. 3 Variation of the leading front velocity D and of the delay τ between the two detonation fronts (DFD case) as function of the mass concentration of aluminium particles in the mixture H1.06.

- For larger concentrations of aluminium particles (zone III of Fig.3), a steady SFD domain is found. The characteristics of this structure may be seen on the typical pressure profile shown in Fig.4c, where a compression wave is observed to occur before the equivalent CJ condition for non ideal detonations too be reached. In this zone, when increasing the particle mass concentration, the detonation velocity increases, reaches a maximum and then decreases. The range of particle concentration where steady SFD exists, is bounded by a lower limit σ_3 and an upper limit σ_4.

- When increasing further the mass particle concentration σ, a second domain where DFD exists is met. In this zone IV (see Fig.3), the dependence of τ (delay between the two fronts) on σ is the reverse of that off zone II: indeed, when σ increases, τ increases too. Contrarily to the case of zone II, where the dominant effect off increasing σ is to augment the quantity of heat release in the second detonation wave, in zone IV when increasing σ, the heat absorption by particles augments more rapidly than heat addition, so the increase of τ with σ. This second DFD domain is bounded by a lower particle concentration limit σ_5 and an upper one σ_6 (see Fig.3).

- Further increase of σ let pass through a domain without steady propagation regime (several hundred of g/m^3 of particles), followed by a domain of steady SFD propagation again (zone V).As seen in Fig.2, the lower limit σ_7 of this domain seems to correspond to the mass particle concentration for which the gaseous detonation in the mixtures under study is quenched by inert particles. This second SFD domain extends up to large values of σ (1500-2000g/m^3), for which the detonation velocity may become as low as 1000 m/s.

When varying the composition of the gaseous mixture, this situation is changed in the following way:

- For lean compositions (H0.6 and H0.78), the zone IV (second DFD domain) disappears, and the zones III and V are joined to form a unique SFD domain beyond σ_3 (see fig.5).

- For very rich compositions (H3.0 and H4.0), the zone III (first SFD domain) disappears, and the zones II and IV collapse in a unique DFD

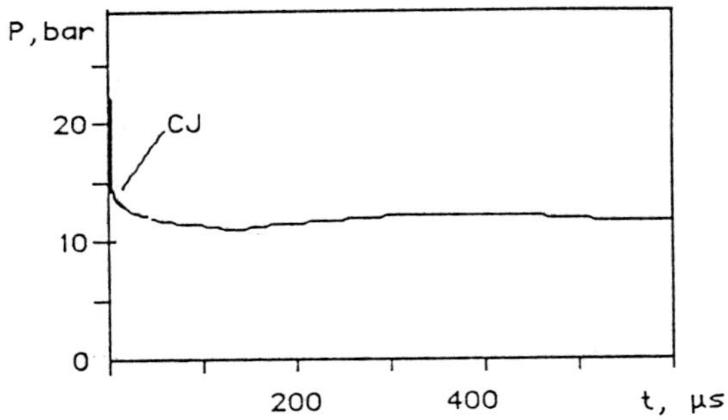

a) in the Zone I

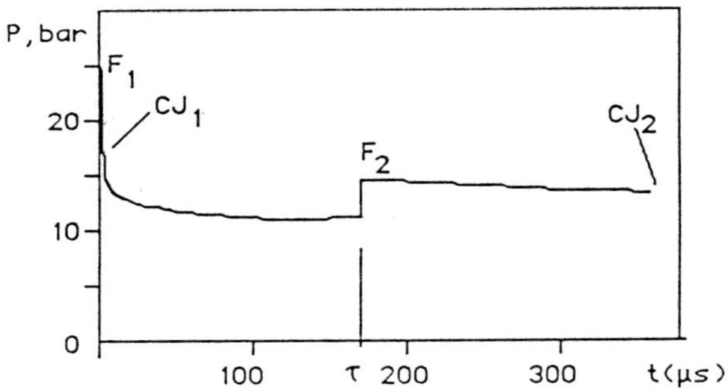

b) Double Front Detonation (**DFD**)

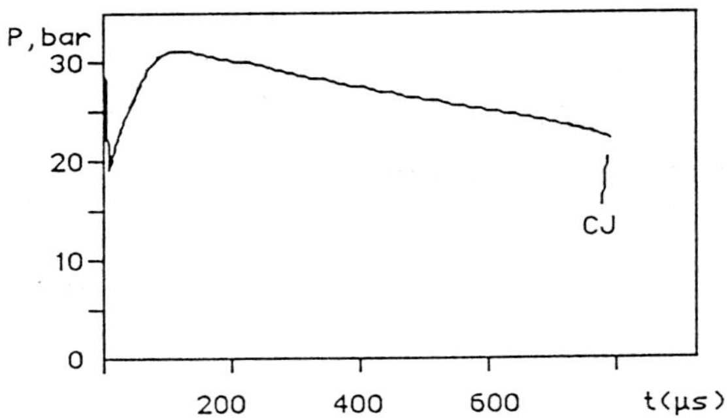

c) Single Front Detonation (**SFD**)

Fig. 4 Typical pressure profiles of the detonation: a) in the Zone I; b) Double Front Detonation (DFD); b) Single Front Detonation (SFD).

domain (see Fig.6): in this case the dependence of τ on σ is no more monotonous: from the lower limit σ_1, the delay τ between the two fronts first decreases with increasing σ, and after

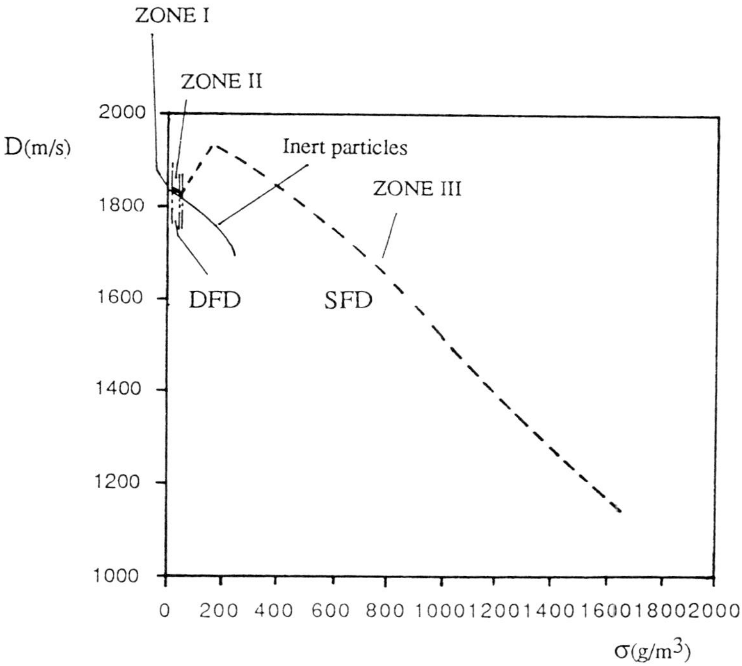

Fig. 5 The different detonation regimes as function of the mass concentration of aluminium particles in the mixture Ho.78.

passing through a minimum increases with σ up to the upper particle mass concentration limit.

DISCUSSION

As shown by those first results, different steady propagation regimes of non ideal detonation can be met in a combustible gaseous mixture containing reactive solid particles in suspension. For a given mixture,

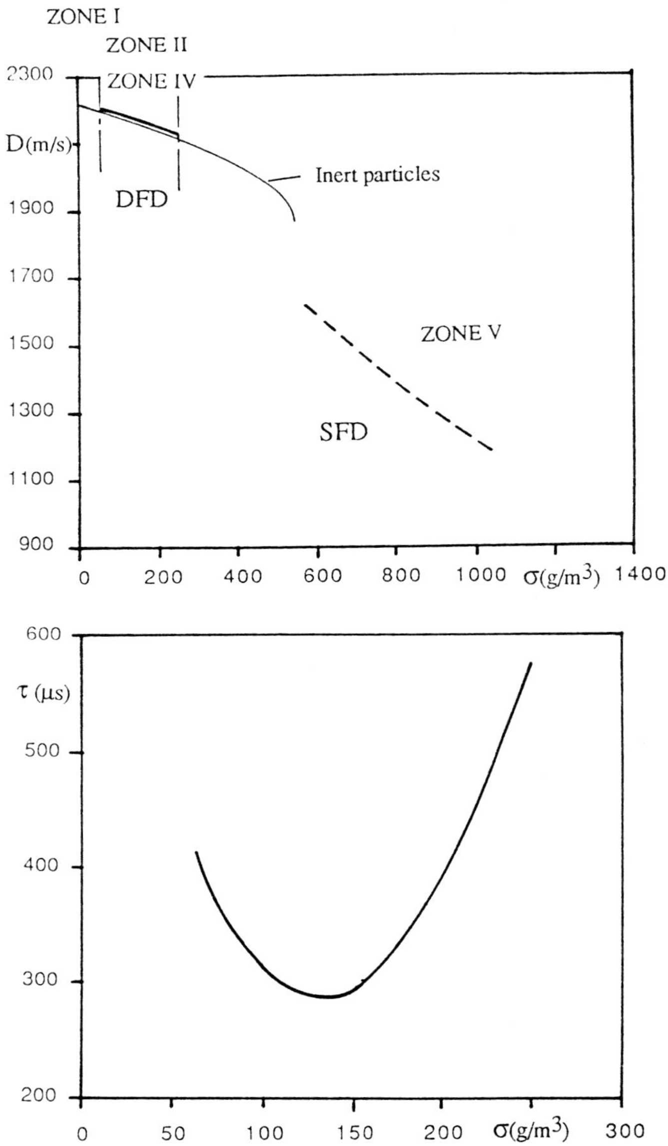

Fig. 6 Variation of the leading front velocity D and of the delay τ between the two detonation fronts (DFD case) as function of the mass concentration of aluminium particles in the mixture H3.0.

those different regimes can be observed successively when increasing the concentration of particles. Indeed, the particles play mainly a double role: they pump heat from the gases to warm up and they release heat when burning; but these effects do not occur in the same time. When increasing the concentration of solid particles, the global heat release time from gaseous reactions is lengthened and the relaxation time and ignition time of solid particles are changed: those modifications are responsible of the shift from one propagation regime to another.

Steady propagation of non ideal detonations is found at high concentration (> 1000 g/m^3) of solid particles (SFD structure), with low velocity of propagation (≅ 1000 m/s). In those conditions, the length of the heat release zone is very large (several tens of centimeters) and present modelling may be thought to be a too rough approximation. However, only suitable experiments can provide definitive arguments to decide of the reality of existence of these propagation regimes at great solid particles concentrations. Further work is needed on this problem, as it appears that the addition of particles could permit to supersonic regimes of combustion to propagate beyond the mass concentration of inert solid particles able to quench the detonation of the same gaseous mixture. Moreover, recent results of calculations indicate that, in the domain of SFD propagation, multiple solutions can be found for the propagation velocity, due to the non monotonous process of heat release.

In many actual situations, instead of pure CJ detonations, one encounters rather non ideal detonation propagation regimes. Thus, from a practical point of view, it is very important to get a better understanding of the fundamental mechanisms of propagation of non ideal detonations.

References

1. Veyssiere B., Manson N.: *Sur l'existence d'un second front de detonation des melanges biphasiques hydrogene-oxygene-azote-particules d'aluminium.* Comptes rendus Acad.Sci., t.295, II, pp. 335-338, 1982.
2. Veyssiere B.: *Structure of the detonations in gaseous mixtures containing aluminium particles in suspensions.* Dynamics Explosions, AIAA Progress in Astronautics and Aeronautics, AIAA, NY, Vol. 106,

pp. 522-544, 1986.

3. Afanasieva E.A., Levin V.A., Tunik Yu.V.: *Multifront combustion two-phase media.* AIAA Progress in Astronautics and Aeronautics, AIAA, NY, Vol 87, pp.394-413, 1983.
4. Nigmatulin R.I.: *Methods of mechanics of a continuous medium for the description of multiphase mixtures.* Prikladnaya matematika i mekhanika, vol.34, n°6, pp. 1097-1112, 1970.
5. Korobeinikov V.P., Levin V.A., Markov V.V., Chernyi G.G.: *Propagation of blast waves in a combustible gas.* Astronautica Acta, Vol.17, pp. 529-537, 1972.
6. Khasainov B.A., Veyssiere B.: *Analysis of the steady double- front detonation structure for a detonable gas laden with aluminium particles.* Archivum Combustionis, vol. 7, n°3-4, pp. 333 - 352, 1987.
7. Zeldovich Ya.B., Kompaneets A.S.: *Theory of detonations.* Academic Press, New York, 1960.
8. Khasainov B.A., Veyssiere B.: *Steady, plane, double-front detonations in gaseous detonable mixtures containing a suspension of aluminium particles.* Dynamics of Explosions, AIAA Progress in Astronautics and Aeronautics, AIAA. NY, Vol 114, pp. 284-299, 1988.
9. Veyssiere B., Khasainov B, A,: *A model for steady, plane, double- front detonations (DFD) in gaseous explosives mixtures with aluminium particles in suspension.* Combustion and Flame (under press), 1990.

DUST RELATED DETONATIONS

C.W. Kauffman and M. Sichel
The University of Michigan, Ann Arbor, U.S.A.

P. Wolański
Warsaw University of Technology, Warsaw, Poland

Abstract. The paper presents a survey of works on the shock wave ignition of dusts, and detonation in dust, hybrid, and gas mixtures containing inert particles. The dust particles ignition in the incident and in the reflected shock waves are presented, and the effect of particles properties on the ignition process is discussed. Self acceleration to detonation, direct initiation of detonation, wave structure, and numerical modelling of detonation in dust mixtures is reviewed. Also the effect of the addition of gaseous fuel to the dust mixtures, and the addition of inert particles to the gaseous mixtures, on the initiation, structure, and limits of detonation in a such mixtures is discussed.

INTRODUCTION

In mining, grain handling and processing facilities, and in many different branches of industry, especially in the rapidly growing powder technologies, serious explosion hazards exist. Many dust air mixtures as well as mixtures of gaseous fuels and inert or combustible particles may cause very dangerous explosions. The majority of such explosions usually involve deflagrations, but some cases clearly involve detonative combus-

A. A. Borissov (ed.), Dynamic Structure of Detonation in Gaseous and Dispersed Media, 267–311.

tion. Detonative explosions create much larger property damage and losses, and they are more difficult to suppress than deflagrative explosions. In some applications detonations of gaseous mixtures with the addition of reacting or inert particles are used in the deposition of protective coating layers on many different metal surfaces, used for high technology products. Thus for all above reasons a good understanding of all aspects of detonation in such mixtures is very essential for the control of initiation, propagation and eventually suppression of detonations in dusty mixtures.

Many studies have been carried out on such dust explosion problems. The earliest investigation came after recognition of the dust explosion hazard in coal mines, the pharmaceutical industry, and in the grain industry. Since that time many papers and books on this subject have appeared, and many references can be found in some basic books concerned with this problem [1]-[4]. But, work on the detonative combustion of dust-air mixtures was not reported until the early seventies [2], [3], [5]. Soon after, much more data was reported on detonation of dust in air and in oxygen [6]-[11]. In this paper we will discuss the problems of shock wave ignition of dust, transition from deflagration to detonation, and direct initiation of detonation, as well as detonation wave properties, structure and limits in mixtures of combustible dusts and air or oxygen, and hybrid mixtures of combustible and inert dust with gaseous fuel-oxidizer mixtures.

DUST-OXIDIZER MIXTURES

Dust explosions have been observed for many years, especially in mines and in the grain industry, where they have first been recognized as such [12]. Rapidly growing new technologies bring an even bigger dust explosion hazard to the powder industry. Many organic and metallic dusts create very serious explosion hazards, and some accidental industrial explosions show strong evidence of detonation-like combustion, so it is of great interest to consider the detonation processes of mixtures of combustible dust and a gaseous oxidizer in detail. First we will discuss the shock ignition of dust, a problem which is closely related to the detonability and detonation characteristics of dust air or dust oxygen mixtures.

Ignition

Since all detonation processes produce a strong shock wave (at its front), it is essential to understand the problem of the shock ignition of dust, since this process is a key element in the detonation phenomena of dust mixtures. Ignition of dust particles by a shock wave has been studied by many authors [6],[13]-[20], for both incident and reflected shock waves. A few important findings from this research should be mentioned.

Many different particles can be easily ignited by shock waves,and in some cases can be more readily ignited than some gaseous mixtures even as sensitive as a stoichiometric hydrogen-oxygen mixture [18]-[21]. This strange result is a consequence of the high temperature region which exist between the particle and the bow shock which forms ahead of the particle in the supersonic flow behind the incident shock wave as shown in Fig. 1. The high temperature gases behind the bow shock cause rapid heating and an increase in the particle surface temperature. Simultaneously the aerodynamic drag causes the particles to accelerate, thus reducing the velocity of the gas with respect to the particle with an accompanying decrease in the stagnation temperature and the rate of convective heating. When the particle surface temperature increases to a critical value, the rapid chemical reaction will lead to ignition. The rate of particle heating depends on particle size and other properties. Very small particles can be heated more rapidly than larger ones, but the smaller particles more rapidly accelerate to the convective flow velocity, so that the high temperature stagnation region will only exist for a short time. Larger particles accelerate more slowly, but also absorb more heat prior to ignition. So one is led to the conclusion that there is an optimum particle diameter providing the most suitable conditions for ignition. Experimental data confirm these expectations.

Measured ignition delay times for particles of different kinds and sizes are shown in Fig. 2. In Fig. 2(a) it can be seen that the ignition

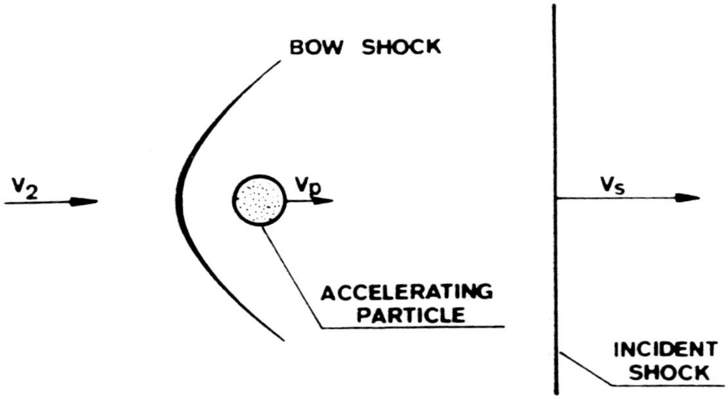

a

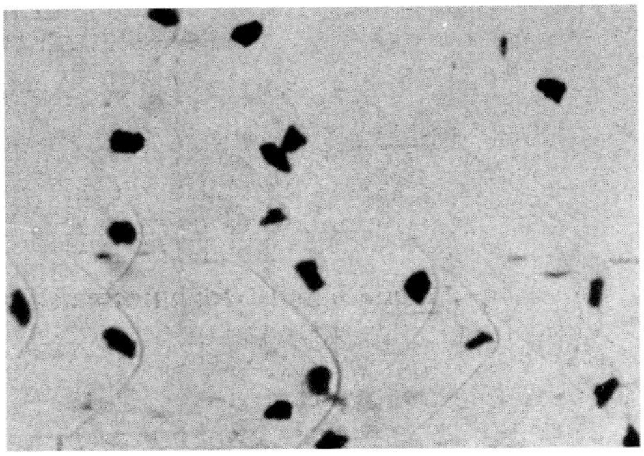

Figure 1. Particle-gas interaction in supersonic flow. (a) schematic diagram, (b) Schlieren picture of particles in the supersonic flow induced by a shock wave.

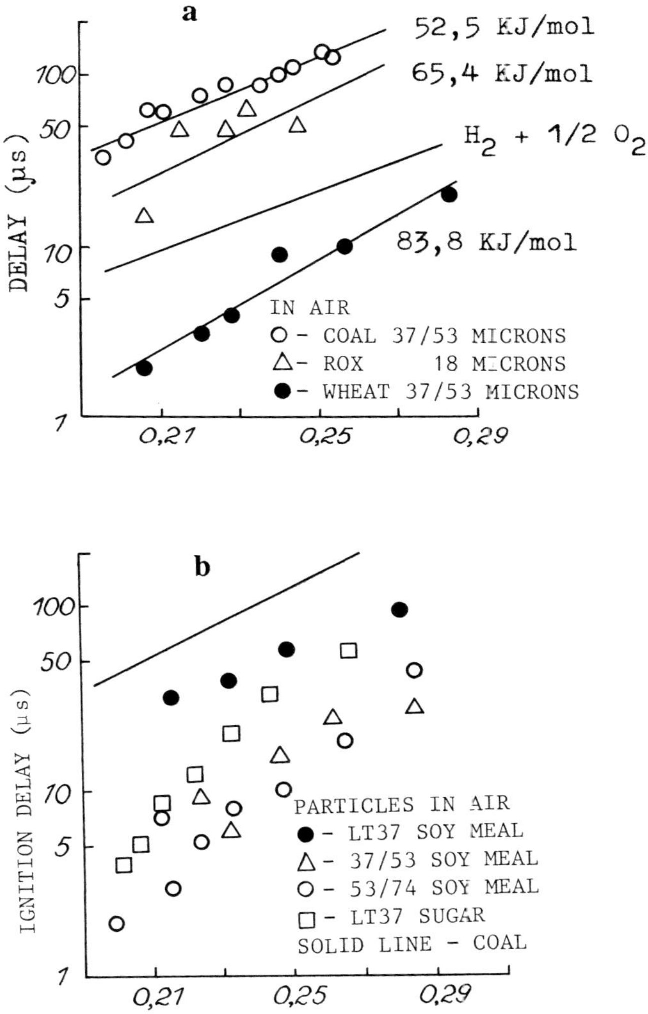

Figure 2. Ignition delay data for particles behind an incident shock wave: (a) for coal, RDX, and wheat dust, and for a stoichiometric hydrogen-oxygen mixture; (b) for soy meal and sugar (the solid line represent coal data).

delay time for wheat dust is actually less than that of a stoichiometric H_2-O_2 mixture, while Fig. 2(b) shows that the 37/53 μm soy meal particles have a somewhat longer ignition delay than the larger 53/74 μm particles. Fig. 3. shows the computed variation of ignition delay time of RDX particles as a function of particle diameter for different particle loading ratios, and clearly shows the existence of an optimum particle diameter for which the ignition delay time has a minimum value.

It was also found that differences in the techniques used for particle dispersal had a relatively small effect on the measured particle ignition delay [18],[21]. However, there is a significant difference in the ignition delay times observed in shock tube experiments behind an incident shock and behind the shock after it has reflected from the end of the shock tube. Ignition delay times measured behind reflected shock waves are an order of magnitude longer than those observed behind incident shock waves as is evident from the data shown in Fig. 4. This difference is due to the absence of the stagnation temperature associated with the bow shock, since there is no convective flow around the particles behind the reflected shock. Behind the reflected shock wave the temperature around particle is actually the same as the static temperature, while in the incident shock the static temperature is much lower than the temperature in the stagnation region between the particle and the bow shock.

The process of dust particle ignition in shock waves was analyzed by many authors [17],[21]-[25]. Very detailed calculations of particle ignition were first made by Ural [22],and Medviediev et al. [23]. Ignition of particles was also the subject of analysis in modeling dust detonation structure [24],[25]. Taking into account the variation of the stagnation temperature at the particle surface during acceleration, the time history of the temperature variation within particle can be computed using the heat conduction equation with a reactive source term, and gives a typical set of temperature contours is shown in Fig. 5. It can be seen, that at ignition there is a very sudden temperature increase within a relatively thin particle surface layer. Below this layer the particle temperature remains constant during ignition. From the

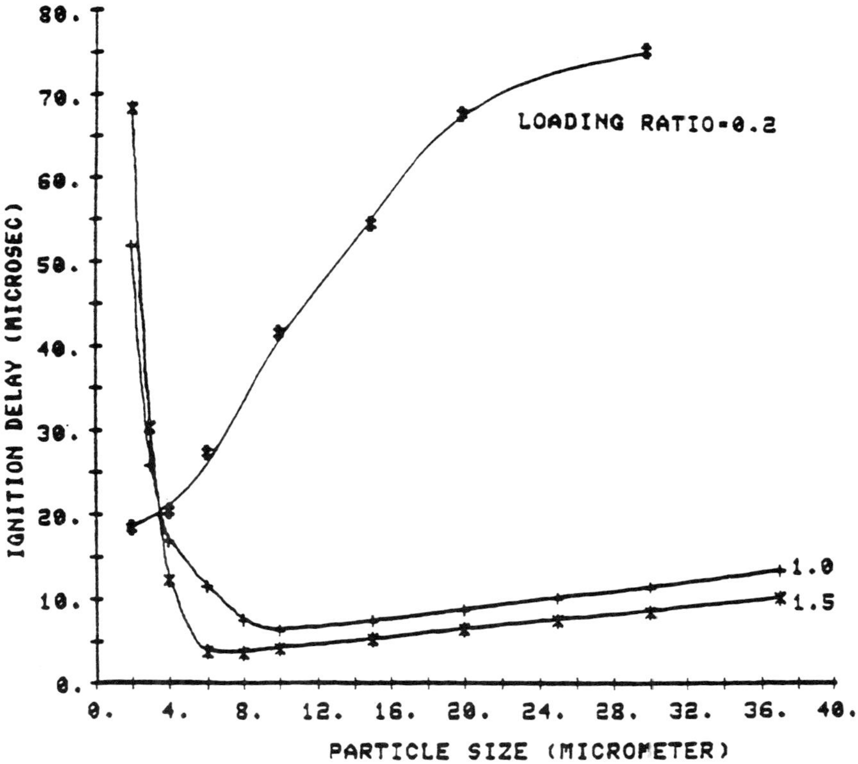

Figure 3. Ignition delay versus particle size for RDX in an air/ O_2 (88/12) mixture behind a Mach 5.1 shock wave[21].

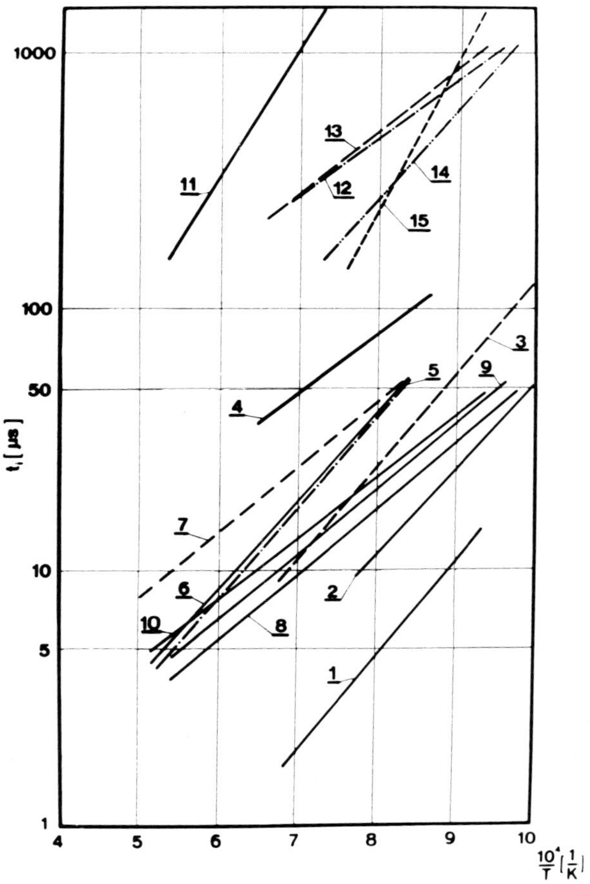

Figure 4. The ignition delay of dust behind incident and reflective shock waves [18]. Incident shock wave data: 1 - wheat, 2 - milo, 3 - wheat flour, 4 - coal, 5 - lycopodium (prism), 6 - wheat (prism), 7 - charcoal (prism), 8 - lycopodium in oxygen, 9 - wheat in oxygen, 10 - charcoal in oxygen; Reflected shock wave data: 11 - coal (layer), 12 - wheat, 13 - corn, 14 - milo, 15 - soybean.

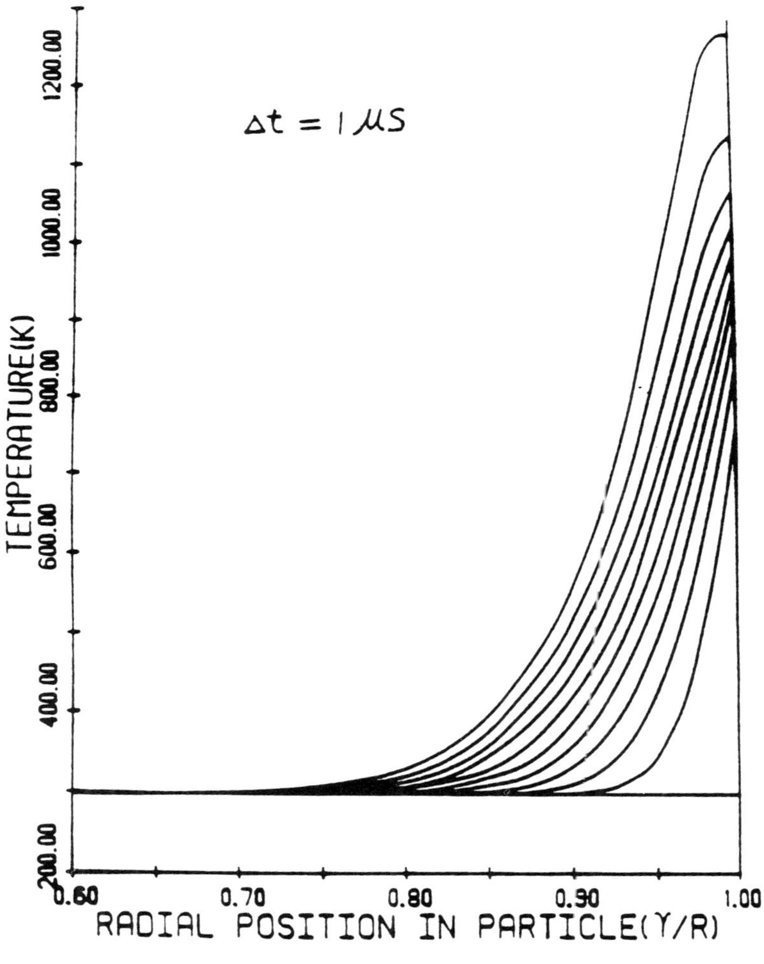

Figure 5. Computed temperature profiles near the surface of 37 μm particles during the induction period [25].

analysis it can be inferred that particles of irregular shape (or porous), will be more suitable for ignition and rapid combustion than spherical particles.

Lack of kinetic data for most dusts is a major problem in computing ignition delay times, but such calculations do provide a reasonable framework for fitting computed and measured results by a suitable choice of kinetic constants. It has been shown [17], that the activation energies for ignition are not very different for a variety of organic dusts (coal,wheat, oats. etc.), suggesting that ignition of dust particles is governed by the physical process of particles heating rather than by chemical reactions at the particle surface.

Ignition data obtained from shock tube experiments can be very useful in evaluating the detonability of dust oxidizer mixtures. Dusts which are more readily ignited by incident shock waves are more likely to support detonative combustion. From experiments, which have been conducted, it is evident that particles having high ballistic coefficient and high porosity are most suitable for ignition. Among the more detonable dusts are some grain dusts,relatively young coal dusts, and some metallic particles (eg. aluminium, magnesium, and others).

Initiation of Detonation

Detonation can be initiated in dust mixtures as a result of continuous flame acceleration, that is by transition from deflagration to detonation (DDT), or it can be directly initiated by a strong shock wave. Both cases are important from the practical point of view. Self-acceleration to detonation is important in the cases in which the flame can propagate in long ducts or tunnels (as in pneumatic or conveyor belt transportation systems, galleries in mines, etc.). Direct initiation may happen when a strong shock wave is generated as a result of a strong primary explosion (eg. explosion of condensed material during mining, explosion of a high pressure gas cylinder, etc.).

Deflagration-Detonation Transition in Dust Mixtures

Flame acceleration in dust mixtures and eventual transition to detonation has been studied by many authors [5], [7-9], [26]. Research studies of flame acceleration in oxygen, and in oxygen enriched and plain air, have mainly been conducted in relatively small laboratory scale apparatus. Only limited data for DDT in dust air mixtures are available at the large scales often encountered in practice.

Laboratory scale flame acceleration tests have been carried out in relatively small vertical or horizontal tubes. The tube diameters range from a few centimeters to a dozen centimeters. In vertical tubes the dust mixture can be created by feeders (located at the top end of the tube), or by special system of dispersing tubes or nozzles (located along the whole tube). Also, dust can be deposited on the tube walls to study layered dust explosions, that is DDT in dust layers. Usually the mixture is ignited by a relatively weak ignitor (electrical spark, chemical match, small gas explosion, etc.). Flame acceleration and eventual transition to detonation has been monitored using direct streak photography (if one the tube walls are provided with windows), or with photodiodes which monitor the radiation emitted from the tube interior. Pressure variations are recorded by pressure transducers, while pressure switches can be used to measure shock or detonation wave velocity.

It is relatively easy to obtain detonation transition for many dusts when the gaseous oxidizer is oxygen. Typical streak photographs of such transitions are shown in Fig. 6. Unlike gaseous mixtures, in which strong transition to detonation is usually observed (the so called "explosion within an explosion"), in dust mixtures a nearly continuous transition to detonation is observed (usually with very weak evidence of the presence of a retonation wave). Dusts composed of less reactive or coarser particles (eg. wood dust) have a relatively large transition distance and may never reach detonative velocities at the small scale of most laboratory experiments [9].

DDT in dust air mixtures can be obtained in the laboratory only in relatively large tubes. Klemens et al. [27] have shown that transition to detonation is possible in a 70 m long, 0.305 m diameter tube, even when the dust is initially layered on the bottom of the tube. In this case an initial dispersion and initiation was generated using a relati-

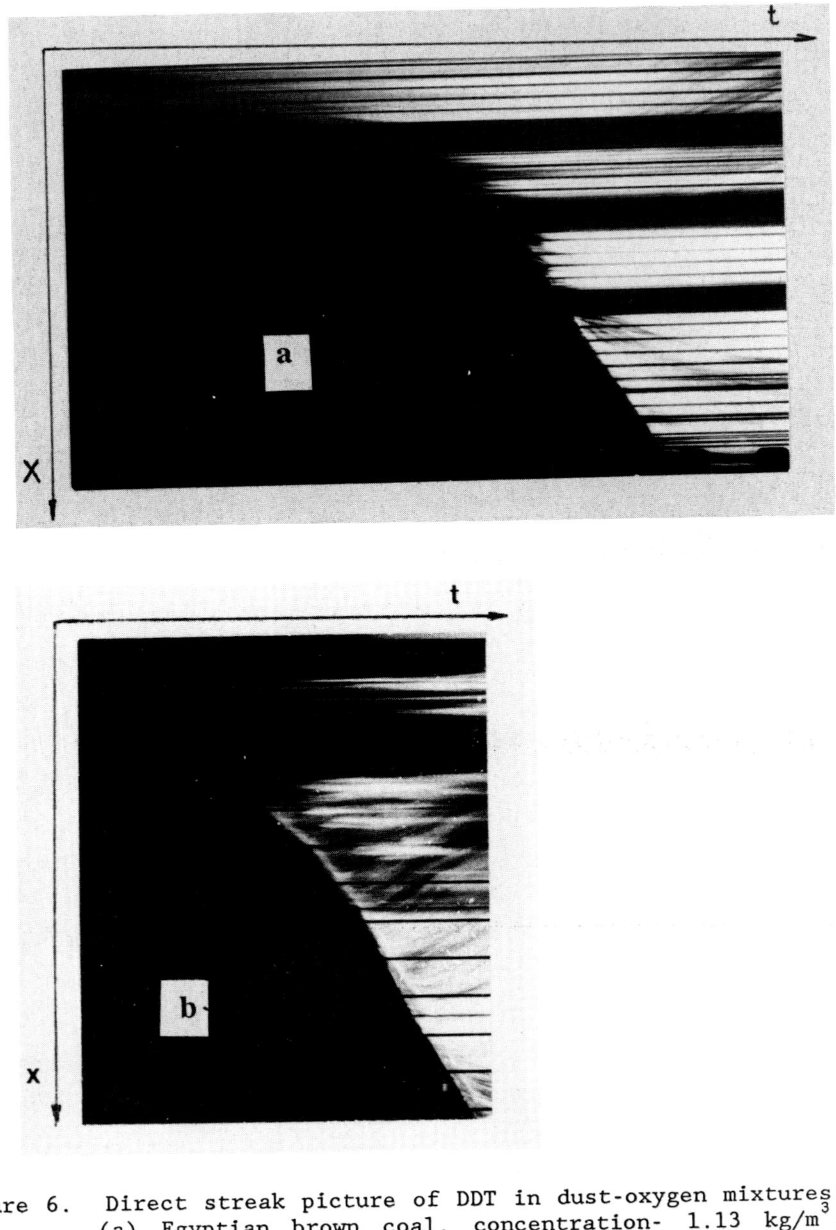

Figure 6. Direct streak picture of DDT in dust-oxygen mixtures: (a) Egyptian brown coal, concentration- 1.13 kg/m^3, (detonation), (b) Egyptian brown coal dust (layer), concentration-0.35kg/m^3, (quasi-detonation).

vely small primary explosion. Typical pressure traces at different stations along the tube with curves indicating the positions of the initial and reflected shock, the flame and the detonation wave are shown on Fig. 7. It should be mentioned that even under such conditions a rather smooth transition is observed, and only a very weak retonation wave is observed. Gardener et al. [8] obtained transition to detonation for suspended coal dust air mixture in a 42m long, 0.6 m diameter tube. In that case maximum velocity observed was 2850 m/s, and the associated pressure reached 81.5 bars. Such conditions are typical for DDT.

In the large scale experiments in the Barbara experimental mine Cybulski [5], showed that the standard"Barbara" coal dust can facilitate transition to detonation. Recently, in the same Experimental Mine, a velocity of 2 km/s was observed during transition to detonation in grain dust at the relatively low concentration of 0.2 kg/m^3. These experiments, which were conducted on a large scale, prove that in real conditions detonative combustion in dust mixtures is likely to occur.

Direct Initiation of Dust Detonation

Studies of the direct initiation of detonations in dust mixtures were carried out in vertical tubes [6], [28], [29-33],for many different dusts and different gaseous oxidizers, ranging from air to the pure oxygen. The initiation is provided by a strong shock wave, generated by a small auxiliary detonation tube, attached to one end of the tube with the dust oxidizer mixture (or close to it). Description of these experimental facilities can be found else where [6],[28]. Depending on the initiation energy, which determines the strength of the initiating shock, and on the mixture properties, three basically different combustion modes were observed. For very high initiation energies direct initiation of detonation is obtained. In this case the velocity of the initiating shock monotonically decreases from a very high value to the steady detonation velocity. This mode of direct initiation is called supercritical. For very low initiation energies, the shock velocity rapidly decreased to the sonic velocity, as the distance between the shock front and flame continuously increases. Detonation is not initiat-

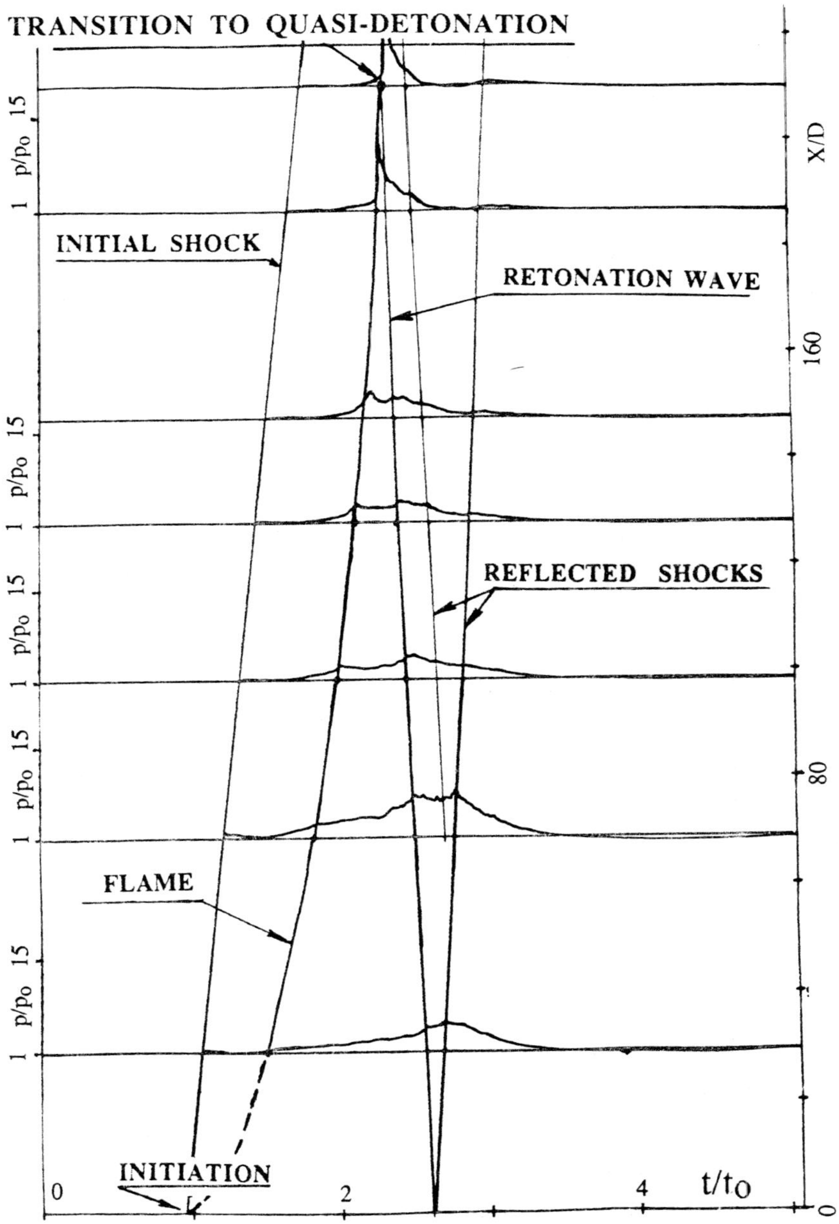

Figure 7. Pressure variation at different locations in the Flame Acceleration Tube for layered corn dust with a concentration of 0.5 kg/m^3. The positions of the initial shock, flame, and transition to quasi-detonation are indicated.

ed, and this mode is therefore called subcritical. For initiation energies in the critical range, the velocity of the initiating shock first drops, below the steady detonation velocity, but then slowly increases to the steady state value. A sketch showing the variation of the shock velocity for these three conditions is shown in Fig. 8. In some cases, when the initiation energy is in the critical range, instability in the transition region has been observed.

For direct initiation of detonation, different energies are required for different dusts and different mixture compositions. A typical example of the variation of the initiation energy with dust concentration, for different oxygen concentrations, is shown in fig. 9. This figure shows the variation of the pressure required in the H_2/O_2 initiator to induce a detonation in the dust-oxidizer mixture with the dust equivalence ratio for different oxygen volume fractions. It is evident from this graph that the detonation limits increase with increasing oxygen concentration, while the critical initiation energy simultaneously decreases. It is obvious from the above and other results that many dusts can easily detonate in oxygen, but that only a limited number will detonate in air. However, when the tube (or duct)diameter increases detonation can be obtained for a wider range of dust-air mixtures [5], [27], [29], [30].

Only limited data is available for the direct initiation of spherical detonations in dust mixtures. Tulis [10] studied the direct initiation of spherical detonations in an aluminum-air mixture, and Veyssiere et al. [11] demonstrated the possibility of the initiation of spherical detonations in mixtures of Egyptian brown coal dust with oxygen. Fig. 10 shows the variation of the pressure recorded at two different distances from the initiation point in such brown coal oxygen detonations. It should be noted that the pressure intensity is the sane at the two locations. Also the propagation velocity observed in the spherical case is very close to that measured in a tube for the same mixture. While these experiments strongly support the possibility of the existence of spherical detonations in dust mixtures, further experiments are necessary to support these findings.

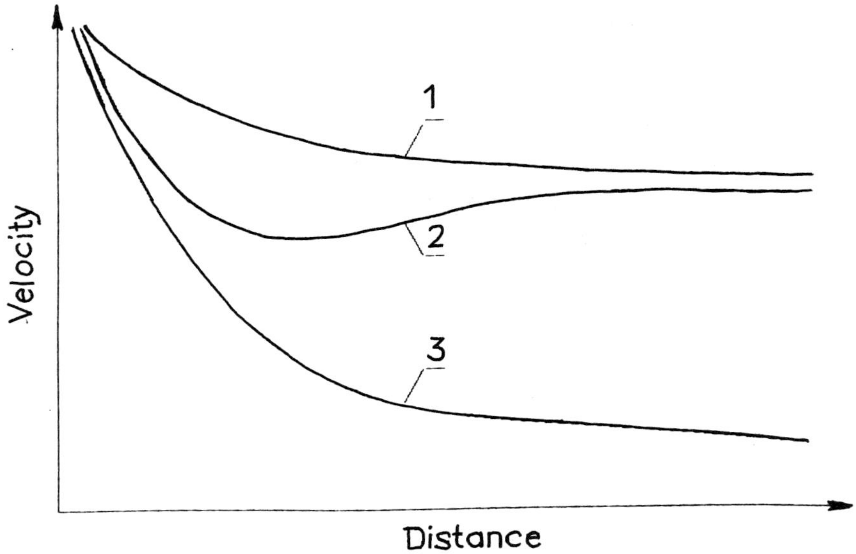

Figure 8. Variation of the shock wave velocity for different initiation energies: 1- supercritical, 2- critical, 3- subcritical.

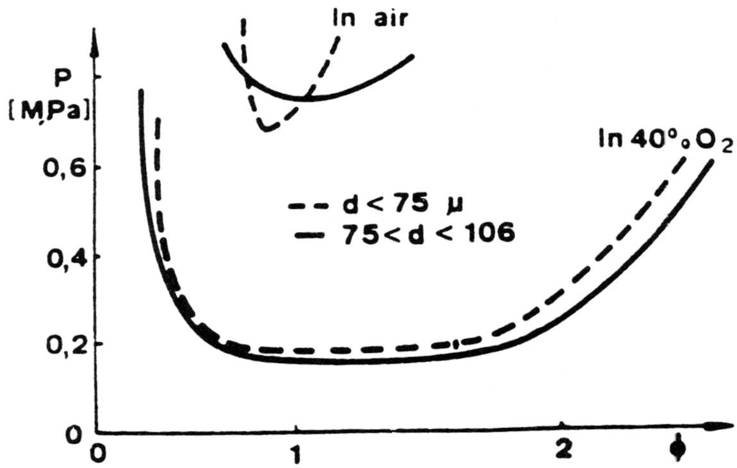

Figure 9. Detonation limits for oats dust mixtures as a function of the relative initiation energy (i.e. the initial pressure of H_2+F(1,2)O_2 mixture in the initiator).

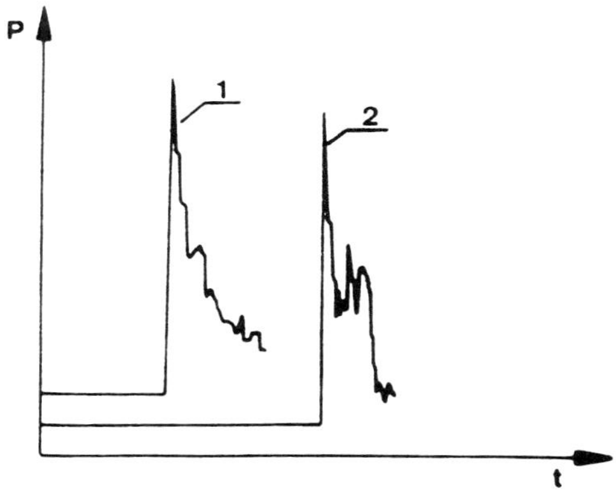

Figure 10. Pressure records for spherical detonation in Egyptian brown coal dust-oxygen mixture. Dust concentration, c= 1.17kg/m^3. Detonation velocity, v= 1670 m/s. 1 -pressure at the distance of 0.355 m from the center, 2 - pressure at the distance of 0.455 m from the center.

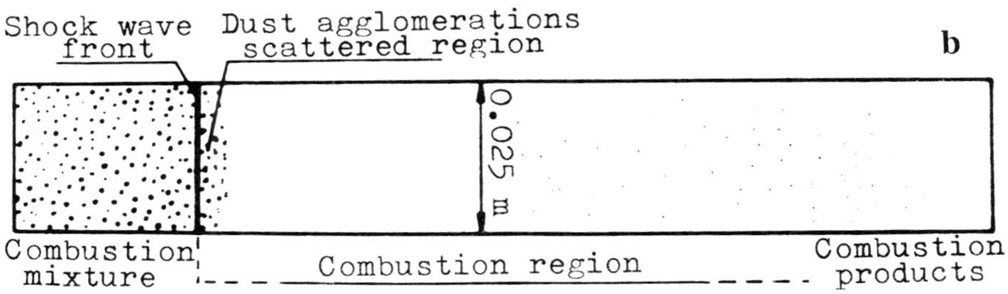

Figure 11. Schlieren pictures of detonation waves in dust mixtures: a - streak picture of detonation of soybean-oxygen mixture (one can notice the variation of ignition delay due to spin mode of detonation), b - structure of oats dust - air detonation drawn schematically on the basis of laser schlieren photographs.

Detonation Structure and Limits

The structure of dust detonations has been studied both experimentally and theoretically. The first information about structure was obtained from pressure measurements within the detonation front, from Schlieren visualization, from direct streak photographs, and from the temperature measurements within the detonation front [6], [7], [9],[28]. The first instantaneous and streak Schlieren pictures revealed several important features of dust detonations [6]. It was found that the leading shock breaks up dust agglomerations causing the mixture to become so opaque that not even intense laser radiation can pass through it as illustrated by Fig. 11. Only when the dust concentration decreases due to intensive burning, does the mixture become transparent again. Pressure and temperature measurements as well as direct streak pictures made it possible to establish the length of the reaction zone, and provided some early evidence of spin structure [6], [7], [9], [28-31]. The recent work of Zhang and Groenig [32], [33], and Dahab et al. [31], fully established the existence of spin structure within detonation fronts in organic dust - oxygen mixtures. A schematic picture of the spin structure in a dust detonation, based on pressure traces at several locations, is shown in Fig. 12. All these measurements show that the reaction zone length is between 0.25-0.5 m, and that the maximum particle temperature in the detonation front does not exceed 3000 K for dust - air mixtures, and 3550 K for dust - oxygen mixtures. Up to now only single head spin has been observed in dust detonation, but further research may reveal more details and possibly multi-headed spin, especially for very reactive mixtures.

Numerical modeling provides another way to examine the structure of dust detonations. The structure of steady state detonation waves in dust mixtures was first calculated by Wolanski et al. [34], and later extended by Lee and Sichel [24], and Fan and Sichel [25]. These numerical codes make it possible to calculate the influence of different parameters on the structure, and to examine the processes of heating, ignition, and particle burning within the detonation wave. The calculations show

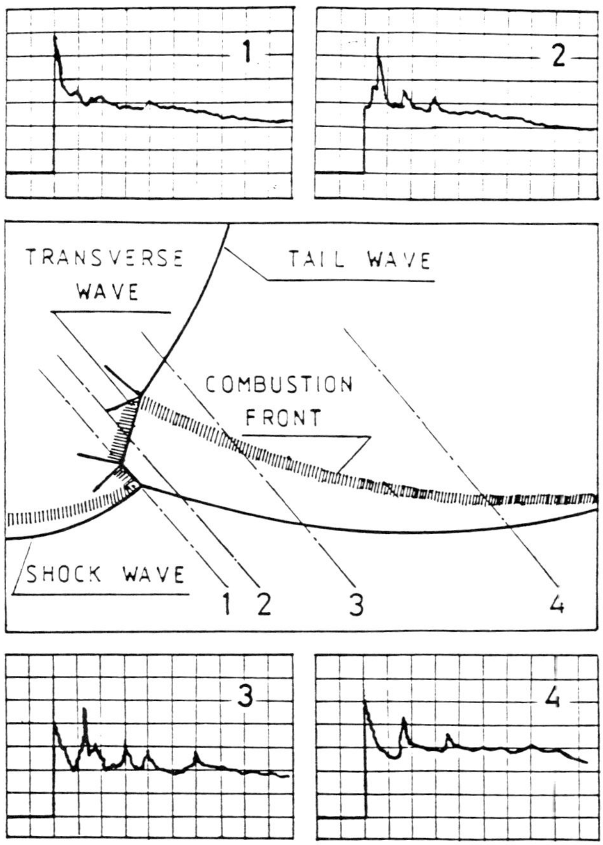

Figure 12. Schematic drawing of the spin structure of a dust detonation as determined from pressure records from the detonation egyptian brown coal dust in oxygen [31].

that local oxygen enrichment occurs just behind the leading shock front due to the rapid compression of the gas phase and the relatively slow acceleration of the particles as indicated in Fig. 13a,b which shows the computed dust and oxygen concentrations and particle and gas velocities. This effect enhances ignition and the initial stages of combustion. Due to highly nonequilibrium conditions within the detonation front, particle parameters differ significantly from the parameters of gas phase as indicated in Fig. 14, which shows the variation of the particle surface and gas phase temperatures.

The best way to confirm the validity of numerical calculations is by the comparison of computed and measured parameters. The most readily available experimental parameters are pressure profiles and the detonation velocity. There was relatively good agreement between the computed and measured variation of these parameters so that other quantities, difficult or impossible to measure in experiments, can be obtained with good confidence from numerical calculations. Numerical calculations can be used to estimate the losses due to the finite detonation tube diameter. It was found that even in small tubes the total heat losses are not higher than 5 % of the heat released between the leading shock and the C-J plane [34]. For much larger tube diameters, or in real conditions, heat losses to the walls should be insignificant.

One dimensional numerical models similar to that used by Fan and Sichel [25]provide a reasonable description of the major feature of detonation wave structure. However, as is well known from the theory of gaseous detonations, transverse cellular structure plays a critical role in self-sustaining detonation, so additional two, and even three dimensional calculations of the detonation processes in dust mixtures may be necessary for a complete explanation of the nature of such processes.

Only very limited data are available for detonation limits in dust mixtures[19],[21],[28],[29],[31], in part because these limits are very narrow, especially in dust air mixtures and in small tubes. The addition of oxygen widens such limits significantly as is evident from Fig. 9. Very reactive dusts, such as the Egyptian brown coal dust, can detonate in oxygen for a very wide range of concentrations even for a relatively

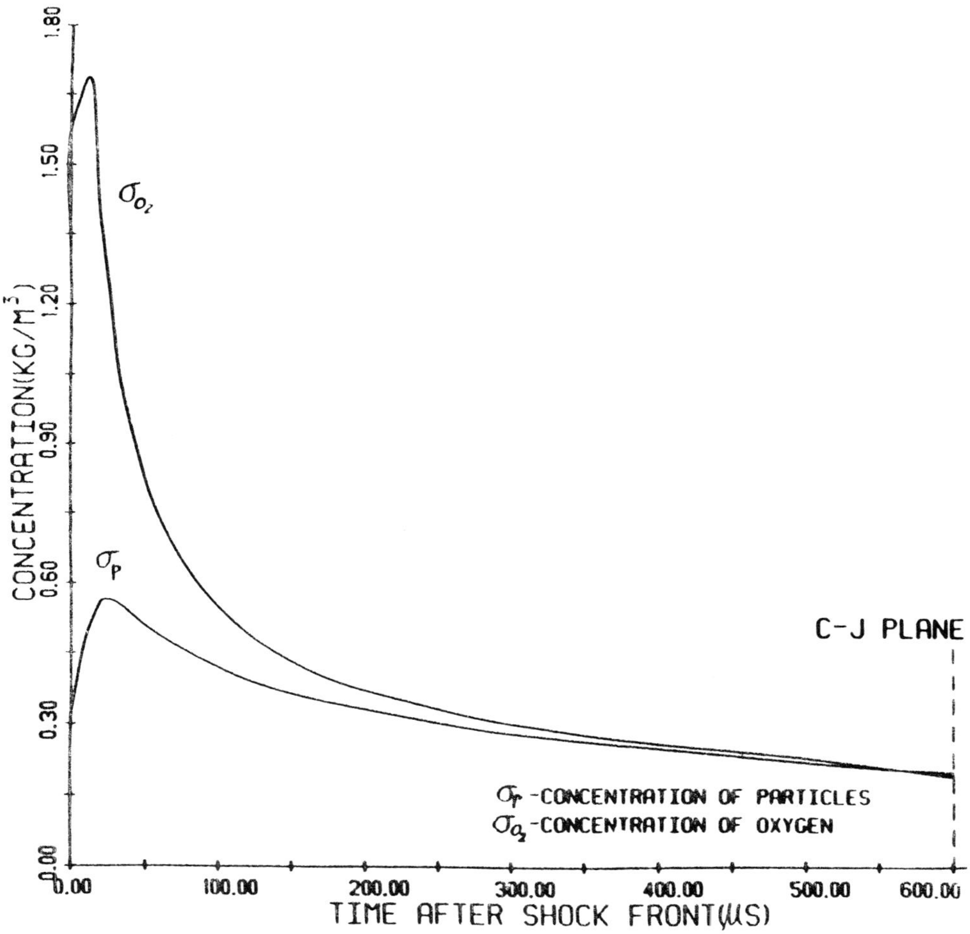

Figure 13. Variations of parameters between the leading shock and C-J plane in a dust detonation: a -concentration of particles and oxygen.

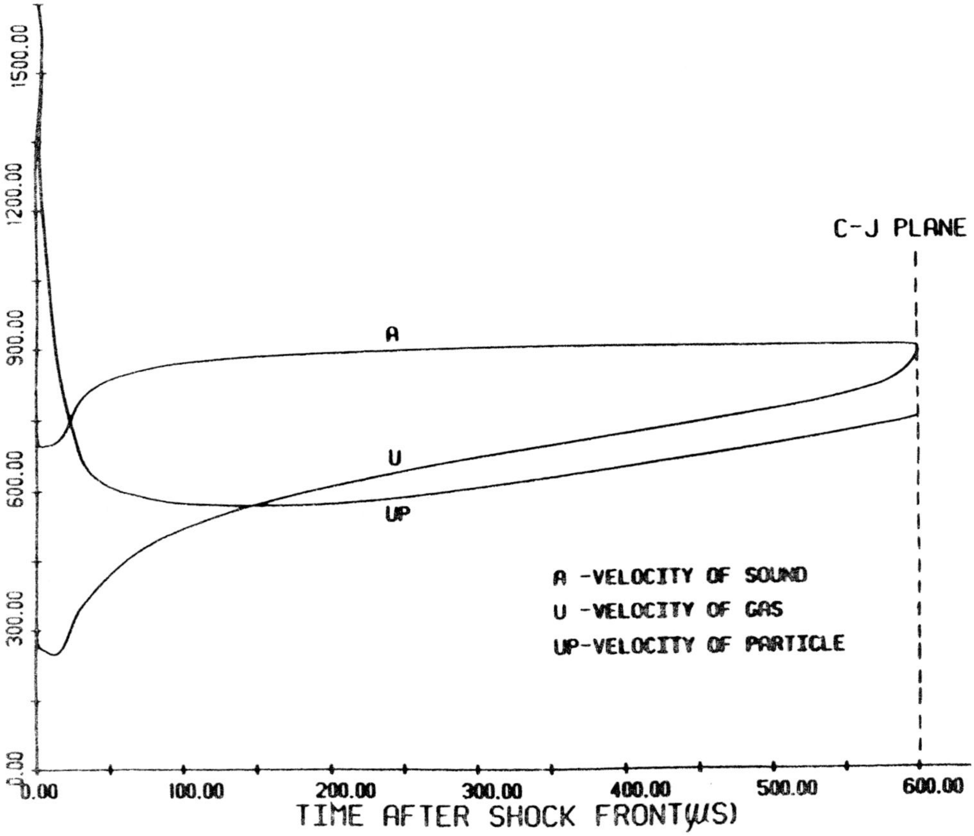

Figure 13. Variations of parameters between the leading shock and C-J plane in a dust detonation: b - gas and particle velocities [25].

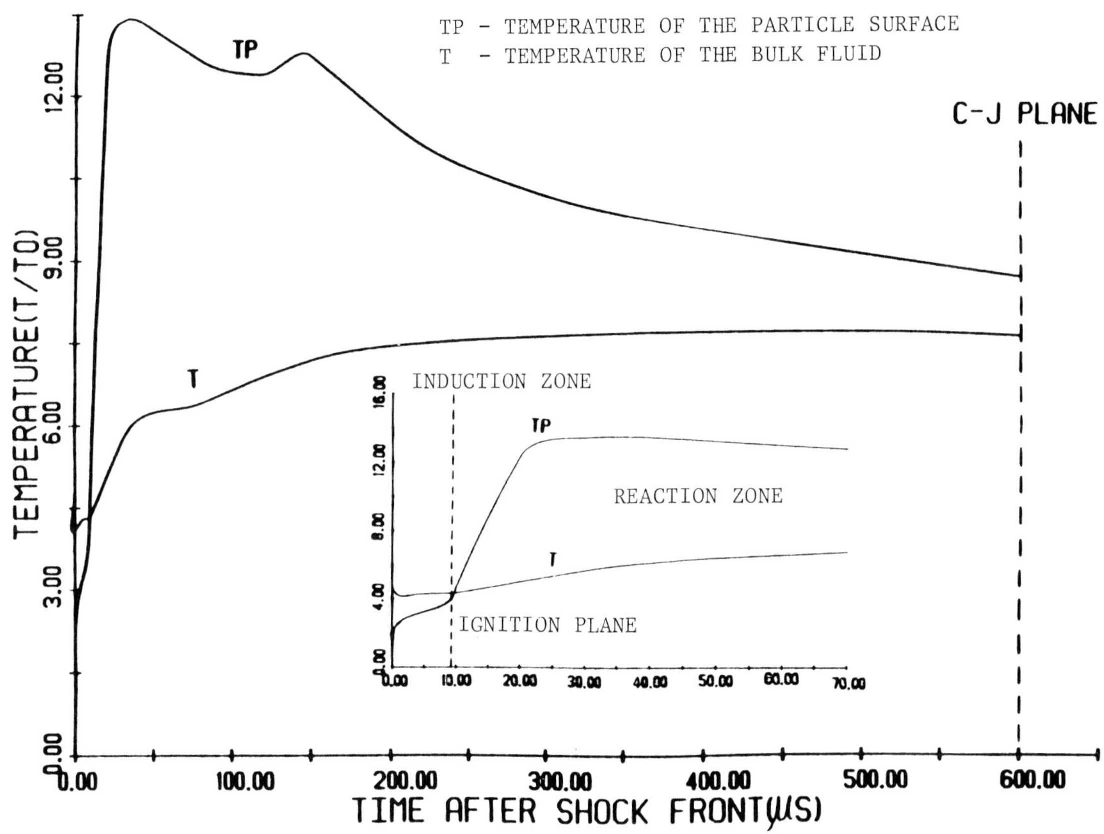

Figure 14. Variation of the particle surface and the gas temperatures within the detonation front [25].

large particle size as is evident from Fig. 15, which shows the variation of detonation velocity with dust concentration for particles in different size ranges.

The majority of the experimental data concerning the initiation, and propagation of detonations in dust mixture a reobtained for the dust concentrations close to the "optimum" one,and many additional experiments (usually difficult and expensive)are necessary to collect more data, since such data is essential for better protection against uncontrolled detonative explosions.

HYBRID MIXTURES

Mixtures containing both combustible dust and combustible gas are relatively common in practice. Typical examples are coal dust mixed with methane in mines, or metallic dusts and gaseous combustible mixtures used in some metal coating processes. Smoldering combustion of organic dust can also create such mixtures,when smoldering combustion products, containing CO, H_2 , and some hydrocarbons, mix with air and dust. The addition of gaseous fuel to combustible dust mixtures, or of combustible dust to gaseous fuel mixtures widens the flammability limit and frequently makes such mixtures more explosive, thus making the detonative combustion of such mixtures more likely than for the dust or gaseous fuel alone.

Detonation of Hybrid Mixtures

Some experimental and theoretical studies of hybrid detonations have been carried out . Kauffman et al. [28], Veyssiere [35-38], Veyssiere et al. [39], and Kauffman and Nicholls [40], studied such processes experimentally, while Afanasieva et al. [41], and Khasainov and Veyssiere [42],conducted theoretical studies of such phenomenon.

Detonations in hybrid mixtures depend, mainly, on the reactivity of the gaseous mixture and the solid particles, the mixture composition (stoichiometric factor), and the particle size. Generally two extreme cases can be observed, the first when the dust is much more reactive

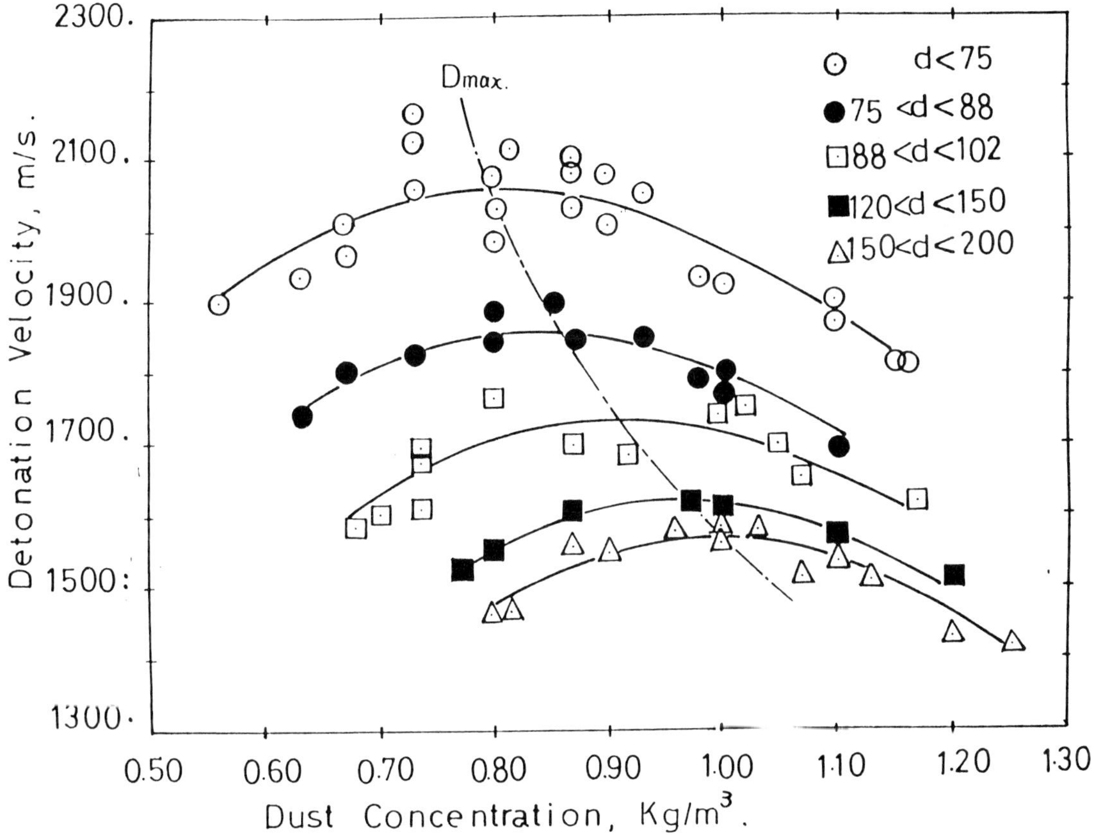

Figure 15. Variation of the detonation velocity of Egyptian brown coal dust with concentration for different particle diameters [31].

than the gaseous mixture, and reverse when the gas mixture is highly reactive relative to the dust. In the first case the behavior can be very similar to that of dust detonations, while the inverse case behaves more like a gaseous detonation in the presence of inert particles (such detonations will be discussed below). However, there are many intermediate cases, and in some situations, while the detonation of a mixture consisting of only one component is impossible, the combination of both solid and gaseous combustible components result in a detonable mixture.

The behavior of the detonation of a typical example of a hybrid mixture of coal dust and methane is illustrated by the streak Schlieren photograph shown in Fig. 16. The differences between a spinning detonation in pure gas, and in this hybrid mixture can be clearly recognized. The major difference is in the ignition delay behind the leading front. Dust particles are ignited relatively easily even by the weak segment of the leading spin shock wave. Successive ignition of the gas mixture (as may be recalled from the discussion above,certain dust particles are more readily ignited than some gaseous mixtures), and the combined combustion of gaseous and solid fuel then drive the detonation wave. In more reactive gaseous mixtures with combustible particles, it is very easy to distinguish the difference between the leading gaseous detonation front and particle ignition behind this front as illustrated by the streak Schlieren photograph of a detonation propagating through a mixture of ethylene and aluminum dust shown in Fig. 17.

For very reactive gas mixtures, the addition of reactive particles does not change the propagation velocity significantly[36]. For rich gas mixtures with an equivalence ratio greater than unity, the addition of particles usually results in a decrease in the detonation velocity. For lean gaseous the addition of reactive particles can result in an increase in the detonation velocity. In very reactive mixtures with extremely reactive particles the pressure profiles within the detonation differ significantly from those recorded for purely gaseous mixtures. Double-shock fronts are usually observed in such mixtures as illustrated by Fig. 18, which shows the pressure trace behind a detonation propagating through a hydrogen-air-aluminum dust mixtures. The second shock may

Figure 16. Streak schlieren picture of a detonation wave in the hybrid mixture: coal dust (d<74μm, c= 0.1kg/m^3), dispersed in a 7% methane - air mixture.

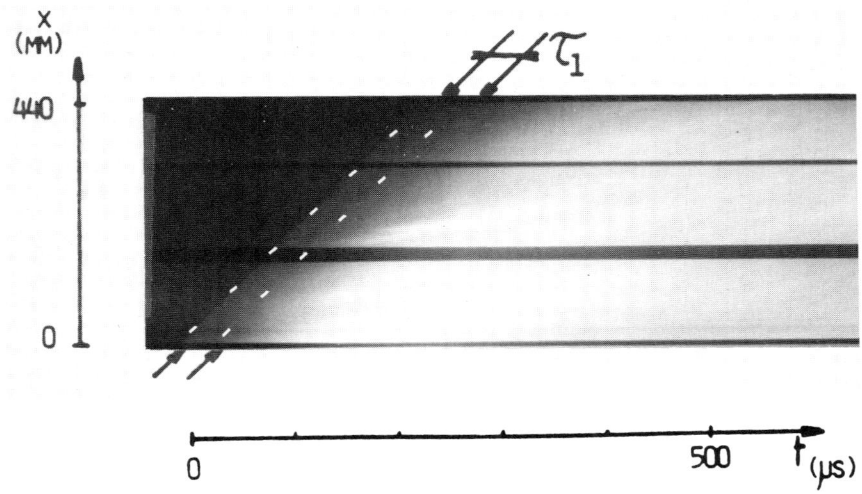

Figure 17. Streak picture of a detonation in an ethylene/N_2/O_2 mixture with aluminum particles(c=0.03 kg/m^3), [38].

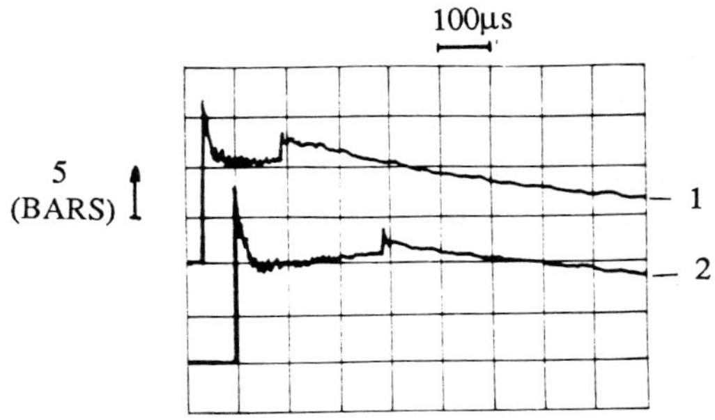

Figure 18. Pressure profiles in a detonation of a hydrogen -air mixture containing aluminium particles at different tube locations: 1 - 1.945 m from ignition point, 2 - 4.175 m from ignition point [36].

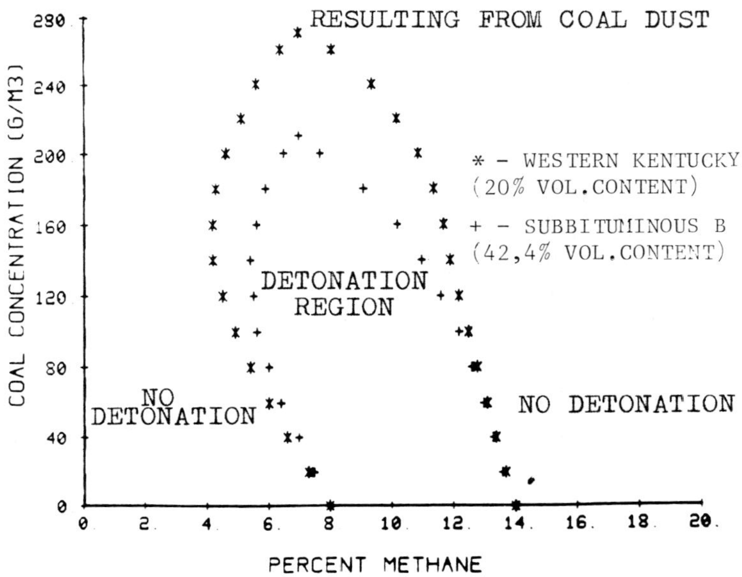

Figure 19. Detonation limits for coal dust/methane/air mixtures.

accelerate, can be nearly stationary, or may decay depending on the mixture properties. Up to now there is no conclusive explanation for this behavior, although some theoretical analyses do predict the existence of a stationary double shock structure in hybrid mixtures [41], [42]. It is interesting to note that such double shock behavior also has been observed in the detonation of mixtures of gaseous and liquid spray fuels [43]. Double-shock stationary systems tend to be unstable and thus will be very sensitive to a small disturbance, so that it will be very difficult to maintain stationary double structure in very long tubes. This problem is, however, mainly of academic interest, and will have a marginal effect on the practical applications of these results.

The addition of less reactive particles to a highly reactive gaseous mixture does not appreciably effects the propagation of a detonation front, but a second shock is still observed [42]. The addition of reactive particles to less reactive gaseous mixtures can effect not only the propagation velocity but also the detonation limits. For example, the addition of coal particles to methane - air mixtures can significantly change the detonation limit. Thus, as shown in Fig. 19 the lower detonation limit can be extended from 8% CH_4 by volume to about 4% CH_4. On the other hand, the upper detonation limit of methane - air mixtures decreases with the addition of coal particles.

In some cases combustible particles can behave like inert particles within a detonation wave. This can happen when relatively large and not very reactive particles are placed in very reactive gaseous mixture or when detonation products of the gaseous mixture cannot react with particles. In such cases the detonation will propagate in the same way, as it would propagate in gaseous mixture in the presence of inert particles.

DETONATION OF GASEOUS MIXTURES WITH INERT PARTICLES

The addition of fine dust (powder) to combustible mixtures is a technique commonly used to suppress deflagrative or detonative combustion, and there are many publications dealing with this subject. The suppression of gaseous detonation by a fine inert dust was first reported by Laffitte and Bouchet [44], but the influence of relatively large inert particles on the initiation and propagation of gaseous detonation was studied only recently[28], [45-48]. Wolinski et al. [45] showed that the addition of relatively large inert particles speeds up transition to detonation. In other experiments, the influence of inert particles on the detonation properties, structure, and limits were investigated.

Initiation of Detonation

The enhancement of flame acceleration in duct by the addition of inert particles was first reported by Jarosinski et al. [49]. The influence of very fine and coarse dust on the propagation of a methane air flame near the lower flammability limit was investigated. Surprisingly it was found, that even the addition of a very fine dust can result in a considerable increase in the flame propagation velocity. Further research carried out by Wolinski and Wolanski [44], [46], and by Wolinski et al. [45],demonstrated the considerable influence of relatively large inert particles on DDT in gaseous mixtures. The experiments were carried out in vertical detonation tubes of circular, square,and square with rounded corners cross sections. Tube diameters ranged from 0.05 m to 0.08m, with lengths varying from 3.2m, to approximately 6m. Hydrogen-air mixtures of different compositions and mixtures of methane or LPG with oxygen enriched air were used in the experiments. Combustion was always initiated using relatively weak ignition sources. Detailed descriptions of the experimental arrangement can be found in Refs. [45-47].

Typical streak photographs showing the flame acceleration, and pressure variation at the center of the detonation tube for different inert particle concentrations are shown in Fig. 20. The detonation tube, in this case, was circular with a diameter 0.08 m, and a length of 4.5 m. A stoichiometric hydrogen-air mixture was used. For the case of flame acceleration without the addition of inert particles, only a relatively small velocity increase can be observed in Fig. 20a. The

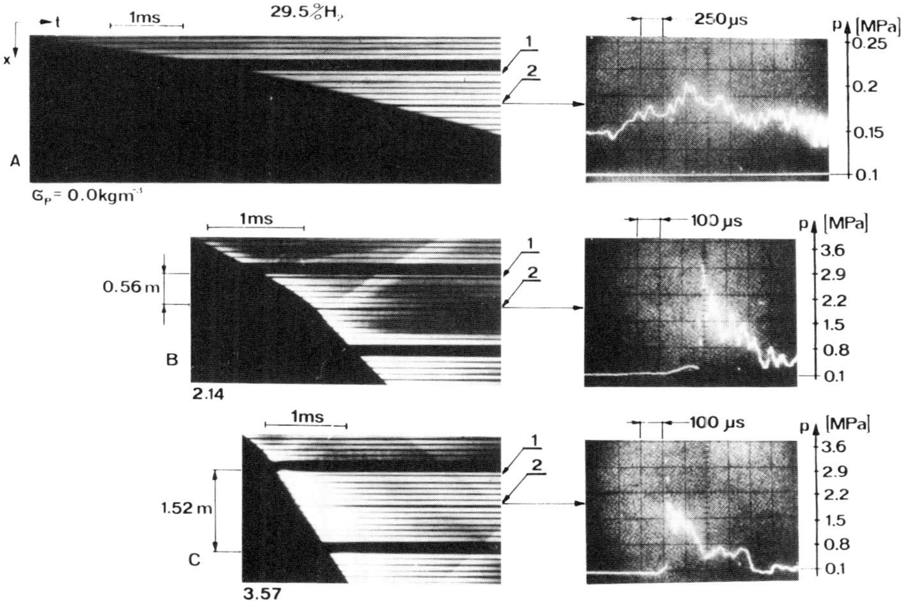

Figure 20. Direct streak pictures of flame acceleration in a stoi chiometric hydrogen - air mixture, and oscilloscope traces of the relative pressure variation in the middle of the tube, for different particle loadings of 0.5 mm sand : A-without particles, B - particle concentration = 2.14kg/m^3, C - particles concentration = 3.57 kg/m^3, 1- location of the oscilloscope triggering device, 2 - location of the pressure transducer, [47].

addition of sand particles to this mixture significantly increased the flame acceleration resulting in DDT (Fig. 20B,C). Similar effects have also been observed in other mixtures. Figure 21 illustrate the influence of the concentration of sand particles on the distance and time required for transition to detonation for different mixtures. Evidently, for the studied concentration range of 25-37% of H_2 in air, the addition of 0.5 mm sand particles always resulted in a decrease in the transition distance and time, even for sand particles concentration up to nearly 7 kg/m^3. But for other gaseous mixtures, such as methane or LPG mixed with oxygen enriched air, after an initial positive influence, a detrimental effect on transition was always observed. This means that for higher concentrations of inert particles the flame is slowed down, and with a sufficiently high particle concentration the flame will eventually be quenched.

The mechanisms by which the presence of large inert particles influence flame acceleration are relatively well understood, and involve many different elements. At the beginning flame acceleration is basically due to the generation of small scale turbulence by the falling particles, and subsequently it is promoted further by the interaction of the flame induced flow and the particles. The strength of the flame-particle interaction increases with an increase in the velocity of the flame and the associated flow. Due to their relatively large mass the particles accelerates lowly, so that when the induced flow velocity becomes supersonic small bow shocks are formed around particles. Under such conditions the elements of the gaseous mixture interact continuously with the particles, particularly for large particle loading ratios. Such interactions are highly nonisentropic processes (the gas is repeatedly compressed by bow shocks and expanded in the particle wake), so that a rapid temperature increase can be expected [50]. In addition, when the induced flow reaches a high enough Mach number, forced ignition of the gaseous mixture in the region between the particle and the bow shock can be expected. Interferometric pictures showing the interaction of supersonic flow with a single particle are shown in Fig. 22. Ignition of the gas mixture in the vicinity of the particles will significantly speed up

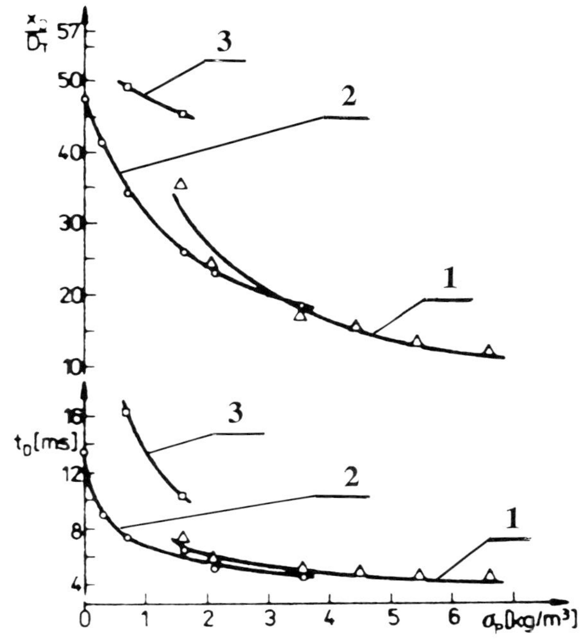

Figure 21. Influence of the concentration of sand particles on dimensionless DDT transition distance, and transition time, for different mixtures: 1- stoichiometric hydrogen / air mixture, 2 - methane/40% oxygen enriched air, 3 - LPG /30% oxygen enriched air, [47].

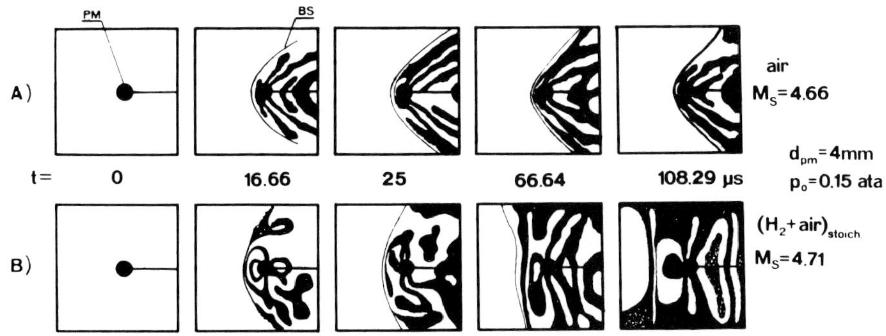

Figure 22. Interferometric pictures of the interaction of a super sonic flow with a single 4 mm diameter spherical particle: A - in air, B - in a stoichiometric hydrogen/air mixture, PM - spherical particle, BS - bow shock.

transition to detonation. The interactions are stronger, and DDT occur more rapidly for relatively high particle concentrations. The transition to detonation becomes smoother with increasing particle concentration. In such cases many particles interacting with the flow significantly increase the flame propagation velocity, ease multipoint ignition, and decrease the strength of "the explosion within an explosion" characteristic of DDT, thus also decreasing the intensity of the retonation wave. Smoothing of the transition to detonation in gaseous mixtures with increasing inert particle concentration is in many ways similar to the relatively smooth character of DDT in mixtures of combustible dust and gaseous oxidizer.

Detonation Properties, Structure, and Limits

The addition of inert particles to gaseous fuel oxidizer mixtures influence not only the initiation of the detonation, but also the detonation properties, the structure, and even the detonation limits. Unlike their usual positive influence transition to detonation, inert particles always result in a decrease in the detonation velocity. Typical examples of the dependence of the detonation velocity on the inert particles concentration for different mixtures, are shown in Fig. 23. The detonation velocity decreases almost linearly with increasing particle concentration for the cases shown. The rate of decrease depends on the mixture properties and the particle size, and becomes larger as the mixture reactivity and particle size decrease. It should also be noted that negative effect on detonation velocity is related to the mixture cell size. Thus hydrogen - air mixtures, with a relatively small cell size, have significantly smaller detonation velocity losses than methane -air mixtures (having the marginal spinning propagation mode) for the same inert particle loading.

Numerical calculations of the inert particle-gaseous detonation interaction by Wolinski and Wolanski [47] showed that particle parameters change very little within the reaction zone of the detonation. The particles, essentially, remain stationary, relative to the supersonic gas

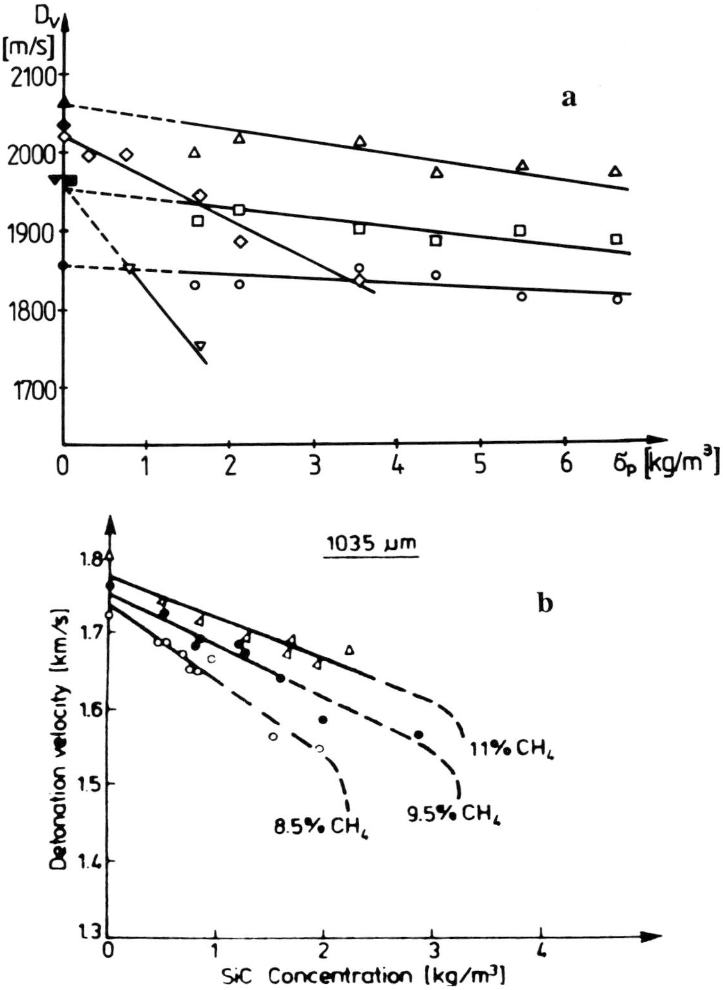

Figure 23. Influence of inert particle concentration on detonation velocities in gaseous mixtures: a - sand particles of 0.5 mm diameter in: o-25% H_2/air, □-29.5% H_2/air,Δ- 37% H_2/air, ◊-stoichiometric CH_4/air enriched to 40% oxygen, ∇- stoichiometric LPG/air enriched to 30% oxygen; b - silicon carbide particles in the different CH_4-air mixtures.

stream behind the leading shock resulting in the formation of bow shocks around the particles, which ease ignition and speed up combustion of the gaseous mixture. The decrease in the detonation velocity is due, mainly, to momentum losses of the gaseous flow to particles. A schematic diagram of a spinning detonation including the trajectories of the inert particles is shown on Fig. 24.

Inert particles can also influence the detonability limits. Extensive research on this subject was carried out at the University of Michigan [47]. The influence of inert particles on the detonation limits was evaluated for methane - air mixtures in a vertical detonation tube. The detonation was initiated in the same way as in the mixtures without particles using an oblique detonation tube at the top of the main tube filled with an H_2-O_2 mixture [49]. Four different particles sizes were used: 0.014; 0.45; 1.035; and 3.46 mm. It was found, that the smaller particles effectively quench the detonation, as shown in [43]. The influence of the larger particles depends on the particle concentration and the composition of the gaseous mixture. There is always some optimum particle concentration for which the lean limit is extended slightly. However, the addition of particles always decreases the upper detonability limit (as in hybrid mixtures). For higher particle concentrations the detonation limit always decreases, and beyond a critical concentration detonation will be no longer possible. However this latter limit is not a sharp one, and there is an area where there is a degree of uncertainty in the results. This behavior is illustrated in Fig. 25 which shows the influence of inert particle size on the detonation limits of a stoichiometric methane-air mixture.

For relatively large inert particles, the most important parameter affecting the detonation velocity and limits is the total particle surface area per unit volume, i.e. the specific volumetric particle surface area in m^2/m^3, which is directly related to the concentration/particle diameter ratio c/d in kg/m^4. As shown in Fig. 26, when the detonation velocity for many different inert particle sizes and concentrations is plotted versus the c/d ratio all the results fall on a single curve. The solid line on this figure represents the region where stable detonation

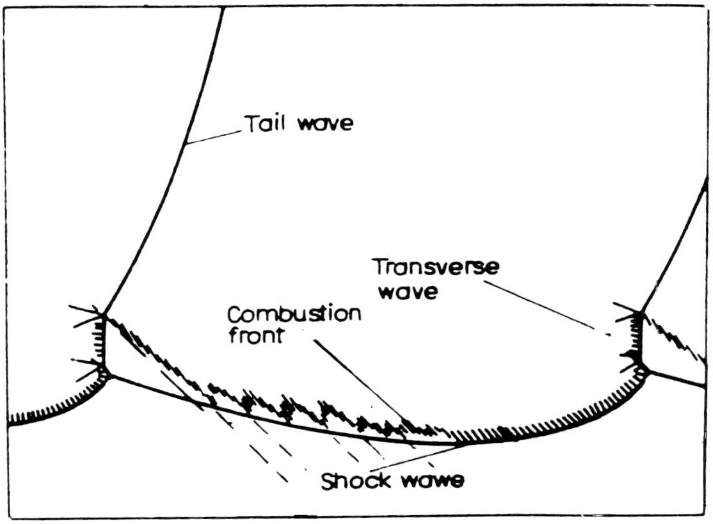

Figure 24. Schematic diagram of a spinning detonation in a methane/air mixture with relatively large inert particles (based on the schlieren streak pictures) [48]. Dashed lines indicate the trajectories of the particles involved in the enhanced ignition of the gaseous mixture.

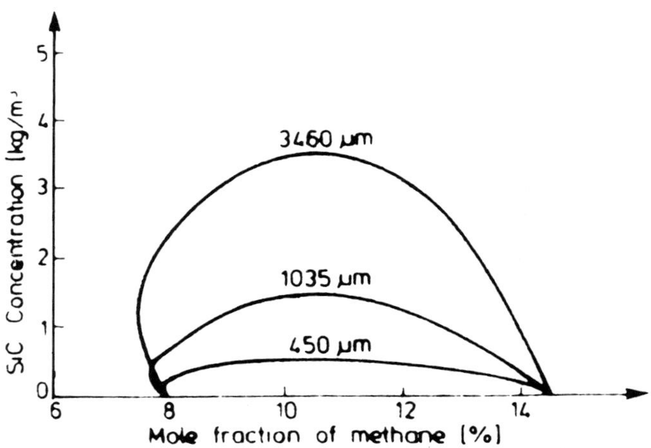

Figure 25. Influence of the particle size on the detonability limits of methane/air mixtures.

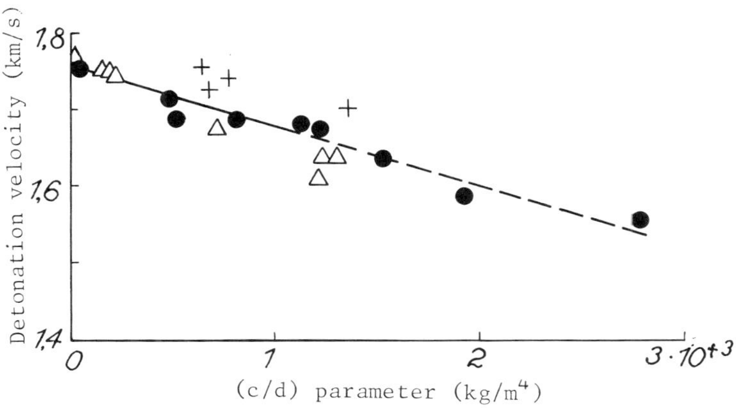

Figure 26. Detonation velocity in a stoichiometric methane/ air mixture as a function of the parameter c/d(specific volumetric surface area). Data obtained for different particle diameters: + - 0.45mm; o - 1.035mm; Δ-3.46 mm.

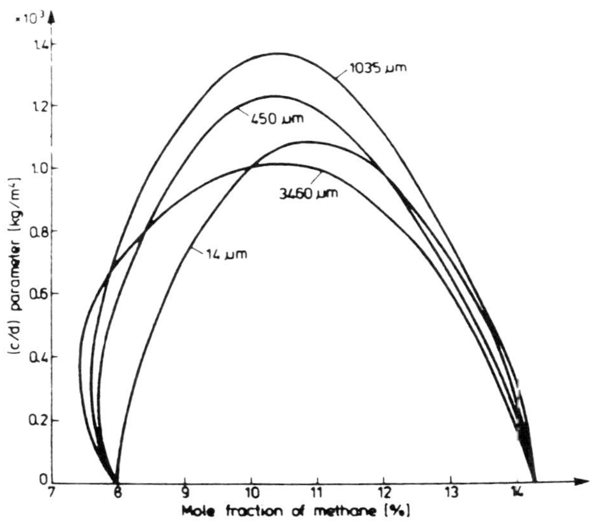

Figure 27. Detonability limits in methane/air mixture as a function of the parameter c/d (specific volumetric surface area), for different particle diameters [48].

was observed, while the dashed line represent an uncertain region, where galloping detonation is most likely to occur. Figure 27 shows the variation of the detonation limit of methane - air mixtures as a function of the c/d parameter for different inert particle diameters, and it can be seen that the results correlate fairly well with c/d. The correlation with the c/d parameter supports the explanation of the role of the inert particles and justifies for some extrapolation of the results to other particle sizes.

CONCLUSIONS

Many different dusts, both reactive and inert can be involved in detonation, and there are many practical reasons why the knowledge of such detonative processes is very important.

Many organic and metallic dusts can be readily ignited by shock waves, even at relatively low Mach number, in some cases more easily than purely gaseous mixtures. The relative ease with which particles are ignited behind incident shock waves facilitates the detonative combustion of such mixtures. It has been shown experimentally that many dust mixtures can detonate when mixed with oxygen or with air. It was shown that many different dust oxygen mixtures will detonate, and there is increasing evidence supporting the existence of detonative combustion of dusts in air. Larger and longer tubes or ducts will widen the detonability range of dusts in air. From an industrial point of view the connection between dust and detonations can not be overemphasized. It is not uncommon to find layered dust in the work place and most installed industrial safety systems are totally ineffectual in suppressing detonations.

Detonations in dust mixtures can propagate with velocities ranging from about 1200 m/s, to more than 2200 m/s. The pressure behind the leading shock can be as high as 60 bars,and the temperature can exceed 3500 K. The length of the reaction zone is usually between 0.25 - 0.5 m. Some mixtures have a spinning structure, and in some other cases there is strong evidence of the existence of spin.

Numerical calculations can provide estimates of particle ignition delay times in good agreement with experimental data, and are very useful in understanding the details of one-dimensional detonation structure. More complex two and three dimensional calculations will be required to simulate the actual processes which occur within the structure of dust detonations, particularly in the case of spin.

In hybrid detonations the influence of reacting particles depends very much on the reactivity of the gas phase. Reactive particles enhance gas detonation parameters when the gaseous mixture is lean, but for stoichiometric and rich mixtures the addition of combustible particles usually results in an attenuation of the detonation parameters. The addition of combustible particles to less reactive gaseous mixtures will result in an increase in detonations parameters and an extension or modification of limits. Very reactive particles usually result in a "double - shock" structure, and the second shock may be accelerating, or decelerating. Some observation of a stable "double - shock" structure require more data for definite confirmation, since such processes within detonations are usually unstable.

The addition of inert, and relatively large particles to detonable gaseous mixtures can speedup DDT and extend lean detonation limit, but it has a detrimental effect on the detonation velocity and the rich detonability limit. The increase in flame acceleration during DDT is caused by particle generated micro turbulence, and flame induced supersonic flow-particle interaction (the formation of bow shocks and intensive energy dissipation near the particles). The DDT is usually smoother, the "explosion within an explosion" less intense, and as a result the retonation wave is much weaker, than in a purely gaseous systems. The influence of relatively large inert particles on detonation parameters has been shown to depend only on the concentration particle diameter ratio so that the available data can be used to estimate the effects of particles other than those for which specific data is available. Very small particles will result in detonation quenching.

During the last two decades significant progress has been made in the understanding dust and detonations, but there are still many question which must be answered before this very complicated phenomenon is fully understood.

REFERENCES

1. Price, D.J., and Brown, H.H. "Dust Explosions," National Fire Protection Association", 1922.
2. Cybulski, W., "Coal Dust Explosions and Their Suppression", Warsaw, 1975.
3. Bartknecht,W., "Explosions, Course, Prevention, Protection", Springer-Verlag, Berlin, Heidelberg, New York, 1981.
4. Nagy, J., and Verakis, H.C., "Development and Control of Dust Explosion", Marcel Dekker Inc., New York and Basel, 1983.
5. Cybulski, W., "Detonation of Coal Dust", Bulletin de L Academie Polonaise des Sciences, 19 (1971), pp.37-41.
6. Kauffman, C.W.,Wolanski, P., Ural, E., Nicholls, J.A., and Van Dyk, R., "Shock Wave Initiated Combustion of Grain Dust", Proceedings of the International Symposium on Grain Dust, Manhattan, Kansas, 1979, pp. 164-190.
7. Wojcicki, S., and Zalesinski, M., "The Mechanism of Transition from Combustion to Detonation in a Mixture of Coal Dust and Gaseous Oxidizer", Proceedings of the Ninth International Shock Tube Symposium, Stanford University Press, 1979.
8. Gardener, B.R., Winter, R.I., and Moore, M.J., "Tests of Explosion Development in Coal Dust/Air Suspension", Twenty-First Symposium International on Combustion, Munich, 1986.
9. Fangrat,J., Glinka, W., Wolanski, P., and Wolinski, M., "Detonation Structure in Organic Dust-Oxygen Mixtures", Archivum Combustionis, 7, (1987), pp. 321-332.
10. Tulis, A.I., "On the Unconfined Detonation of Aluminum Powder-Air Cloud", First International Colloquium on Explosibility of Industrial Dusts, Baranow,1984, Book of Papers, Part-1, pp. 178-186.
11. Veyssiere,B., Desbordes, D., and Wolanski, P., "Preliminary Investigation of Direct Initiation of Spherical Detonation in the Brown Coal-Oxygen Mixture", Private Communication.
12. Verakis, H.C.,and Nagy, J., "A Brief History of Dust Explosions", Industrial Dust Explosions, ASTM STP 958. K.L. Cashdollar and M. Hertzberg, Eds., American Society for Testing and Materials, Philadelphia,1987, pp. 342-350.
13. Nettleton, M.A., and Stirling, R., "The Ignition of Clouds of Particles in Shock Heated Oxygen", Proc.Roy. Soc. London, V. 300 (1967), pp. 62-67.
14. Nettleton, M.A.,and Stirling, R., "The Influence of Additives on Burning of Cloud of Coal Particles in Shocked Gases", Combustion and Flame, 22, (1974), pp. 543-549.
15. Fox, T.W., Rackett, C.W., and Nicholls, J.A., "Shock Wave Ignition of Magnesium Powders", Proceedings of the 11th International Symposium on Shock Tubes and Waves, pp.262-268.
16. Boiko, V.M., Papyrin, A.N., and Poplavskij, S.V., "Ignition of Metallic Suspension Behind the Shock Waves", (in Russian), First International Colloquium on Explosibility of Industrial Dusts,

Baranow,1984, Book of Papers, Part-2, pp.188-195.
17. Sichel, M., Baek, S.W., Kauffman, C.W.,Marker, B., Nicholls, J.A., and Wolanski, P., "The Shock Wave Ignition of Dusts", AIAA Journal, 23, (1985), pp. 1374-1380.
18. Wolanski, P., "Ignition and Combustion of Organic Dusts in Shock Waves", (in Polish), Biuletyn Informacyjny ITC PW, Nr. 65, (1984), pp. 15-32.
19. Wolanski,P., "Problems of Dust Explosions", First Specialists Meeting (International) of the Combustion Institute,Bordeaux, 1981, pp. 497-502.
20. Borisov, A.A., Gelfand, B.E.,Timofeev, E.I., Tsyganov, S.A., and Khomic, S.V., "Ignition of dust Suspensions Behind Shock Waves", Dynamics of Shock Waves, Explosions, and Detonations, Progress in Aeronautics and Astronautics, Vol.94, 1984, pp. 332-339.
21. Lee, F.P., Kauffman,C.W., Sichel, M.,and Nicholls, J.A., "Detonability of RDX Dust in Air/Oxygen Mixtures", AIAA Journal, 24, (1986), pp. 1811-1816.
22. Ural, E., "Shock Wave Ignition of Pulverized Coal", Ph.D. Thesis, The University of Michigan, 1981.
23. Medviediev, A.E.,Fedorov, A.V.,Fomin, V.M. Teodorczyk, A., and Wolanski,P., "Mathematical Simulation of the Ignition of Aerodispersed Systems", (in Russian), III-th International School on Explosibility of Industrial Dusts, Turawa, 1982,pp.65-74.
24. Lee, D., and Sichel, M., "The Chapman-Jouguet Conditions and Structure of Detonation in Dust-Oxidizer Mixtures", Dynamics of Explosions, Progress of Astronautics and Aeronautics, Vol. 106, 1986, pp. 505-52.
25. Fan, B.C.,and Sichel, M., "A Comprehensive Model for the Structure of Dust Detonation", Twenty-Second Symposium (International) on Combustion, The Combustion Institute, 1988, pp.1741-1750.
26. Srinath, S.R., Kauffman, C.W., Nicholls, J.A., and Sichel, M., "Flame Propagation Due to Layered Combustion Dusts", Dynamics of Reactive Systems, Progress in Astronautics and Aeronautics, Vol.105, Part II, 1986, pp. 196-216.
27. Klemens, R., Johnston, V.,Alexander, C., Li, Y., Kauffman, C.W., and Sichel, M., "Flame Acceleration in a Grain Dust - Air Mixtures in a Long Horizontal Tube", Fourth International Colloquium on Dust Explosions, Porabka - Kozubnik, 1990.
28. Kauffman, C.W., Wolanski, P.,Arisoy, A., Adams, P.R., Maker, B.N., and Nicholls, J.A., "Dust,Hybrid, and Dusty Detonations", Dynamics of Shock Waves, Explosions, and Detonations, Progress in Astronautics and Aeronautics, Vol. 94, 1984, pp. 221-224.
29. Wolanski, P.,Sacha, W.,and Zalesinski, M., "Effect of Dust Concentration on Detonation Parameters in Grain Dust-Air Mixtures", Fourth International Colloquium on Dust Explosions, Porabka-Kozubnik,1990.
30. Wolanski, P., "Deflagration and Detonation Combustion of Dust Mixtures", Dynamics of Explosions, Progress in Astronautics and Aeronautics, Vol.133, 1990, (in press).
31. Dahab, O., Kapuscinski, M., and Wolanski, P., "Influence of Dust Parameters on Detonation Velocity, Structure and Limits", Dynamics of Explosions, Progress in Astronautics and Aeronautics,Vol. 134,

1990, (in press).

32. Zhang, F., and Groenig, H., "Investigation of Detonation of Corn-Starch-Oxygen Mixtures", Archivum Combustionis, Vol.9, 1989 (in press).
33. Zhang, F.,and Groenig, H., "Detonation Structure of Corn Starch Particles-Oxygen Mixture", Dynamics of Explosions, Progress in Astronautics and Aeronautics, Vol. 134, 1990, (in press).
34. Wolanski, P.,Lee, D., Sichel, M., Kauffman, C.W., and Nicholls,J.A., "The Structure of Dust Detonations", Dynamics of Shock Waves, Explosions, and Detonations, Progress in Astronautics and Aeronautics, Vol.94, 1984, pp. 241-263.
35. Veyssiere,B., "Ignition of Aluminium Particles in a Gaseous Detonation", Shock Waves, Explosions, and Detonations, Progress in Astronautics and Aeronautics, Vol. 87, 1983, pp. 363-375.
36. Veyssiere, B., "Double Front Detonation in Gas-Solid Particles Mixtures", Dynamics of Shock Waves, Explosions, and Detonations, Progress in Astronautics and Aeronautics, Vol. 94, 1984, pp. 264-276.
37. Veyssiere, B., "On the Detonation of Gaseous Mixtures Containing Aluminium Particles in Suspension", First International Colloquium on Explosibility of Industrial Dusts, Baranow,1984, Book of Papers, Part -1, pp. 160-166.
38. Veyssiere B., "Structure of the Detonation in Gaseous Mixtures Containing Aluminium Particles in Suspension", Dynamics of Explosions. Progress in Astronautics and Aeronautics, Vol. 106, 1986, pp. 522-544.
39. Veyssiere, B., Bouriannes, R., and Manson, M., "Detonation Characteristics of Two Ethylene-Oxygen-Nitrogen Mixtures Containing Aluminium Particles in Suspensions", Gasdynamics of Detonations and Explosions, Progress in Astronautics and Aeronautics, Vol. 74, 1981, pp. 423-438.
40. Kauffman, C.W., and Nicholls, J. A., "The Dust Explosion Research at the University of Michigan", Fuel - Air Explosions, University of Waterloo Press, 1982, pp. 623-643.
41. Afanasieva, L.A., Levin, V.A., and Tunik, Y.V., "Multifront Combustion in Two Phase Media, Shock Waves, Explosions, and Detonations", Progress in Astronautics and Aeronautics, Vol.87, 1983, pp. 394-413.
42. Khasainov B.A., and Veyssiere,B., "Steady, Plane, Double- Front Detonations in Gaseous Detonable Mixtures Containing a Suspension of Aluminium Particles", Dynamics of Explosions, Progress in Astronautics and Aeronautics, Vol. 114, 1988, pp. 284-299.
43. Laffitte, P., and Bouchet,R., "Suppression of Explosion Wave in Gaseous Mixtures by Means of Fine Powders", 7th Symposium (International) on Combustion, The combustion Institute, 1958, pp.504-508.
44. Pierce, T.H., "Experimental and Theoretical Study of the Structure of Two-Phase Detonations in Sprays", PhD thesis, The University of Michigan,Ann Arbor, MI, 1972.
45. Wolinski, M., and Wolanski,P., "Transition into Detonation in the Gaseous Mixtures in the Presence of Inert Particles", First International Colloquium on Explosibility of Industrial Dusts, Baranow, 1984, Book of papers, Part - 2, pp. 196-207.

46. Wolinski, M., Wolanski, P.,Myers, R.M., Kauffman, C.W., Sichel, M., and Nicholls,J.A., "Influence of Inert Particles on Deflagration - Detonation Transition in Gas Mixtures", Ninth International Pyrotechnics Seminar, Colorado Springs, 1984, pp. 861-872.
47. Wolinski, M.,and Wolanski, P., "Gaseous Detonation Processes in Presence of Inert Particles", Archivum Combustionis, Vol.7, 1987, pp. 353-370.
48. Wolanski, P., Liu, J.C., Kauffman. C.W., Nicholls,J.A., and Sichel, M., "The Effects of Inert Particles on Methane- Air Detonation", Archivum Combustionis, Vol.8, 1988, pp. 15-32.
49. Jarosinski, J., Klemens, R., and Wolanski, P., "Investigation of the Inert Particles Influence on Gaseous Flame Structure Near the Lower Flammability Limit", First International Colloquium on Explosibility of Industrial Dusts, Baranow, 1984, Book of papers, Part - 1, pp. 30-38.
50. Kuhl, A.L., Private Communication, 1986.
51. Wolanski, P., Kauffman, C.W., Sichel,M., and Nicholls, J.A., "Detonation of Methane - Air Mixtures", 18th Symposium (International) on Combustion, The Combustion Institute, 1981, pp.1651-1670.

Mechanics

FLUID MECHANICS AND ITS APPLICATIONS

Series Editor: R. Moreau

Aims and Scope of the Series

The purpose of this series is to focus on subjects in which fluid mechanics plays a fundamental role. As well as the more traditional applications of aeronautics, hydraulics, heat and mass transfer etc., books will be published dealing with topics which are currently in a state of rapid development, such as turbulence, suspensions and multiphase fluids, super and hypersonic flows and numerical modelling techniques. It is a widely held view that it is the interdisciplinary subjects that will receive intense scientific attention, bringing them to the forefront of technological advancement. Fluids have the ability to transport matter and its properties as well as transmit force, therefore fluid mechanics is a subject that is particularly open to cross fertilisation with other sciences and disciplines of engineering. The subject of fluid mechanics will be highly relevant in domains such as chemical, metallurgical, biological and ecological engineering. This series is particularly open to such new multidisciplinary domains.

1. M. Lesieur: *Turbulence in Fluids*. 2nd rev. ed., 1990 ISBN 0-7923-0645-7
2. O. Métais and M. Lesieur (eds.): *Turbulence and Coherent Structures*. 1991 ISBN 0-7923-0646-5
3. R. Moreau: *Magnetohydrodynamics*. 1990 ISBN 0-7923-0937-5
4. E. Coustols (ed.): *Turbulence Control by Passive Means*. 1990 ISBN 0-7923-1020-9
5. A. A. Borissov (ed.): *Dynamic Structure of Detonation in Gaseous and Dispersed Media*. 1991 ISBN 0-7923-1340-2

Kluwer Academic Publishers – Dordrecht / Boston / London

Mechanics

SOLID MECHANICS AND ITS APPLICATIONS

Series Editor: G.M.L. Gladwell

Aims and Scope of the Series

The fundamental questions arising in mechanics are: *Why?*, *How?*, and *How much?* The aim of this series is to provide lucid accounts written by authoritative researchers giving vision and insight in answering these questions on the subject of mechanics as it relates to solids. The scope of the series covers the entire spectrum of solid mechanics. Thus it includes the foundation of mechanics; variational formulations; computational mechanics; statics, kinematics and dynamics of rigid and elastic bodies; vibrations of solids and structures; dynamical systems and chaos; the theories of elasticity, plasticity and viscoelasticity; composite materials; rods, beams, shells and membranes; structural control and stability; soils, rocks and geomechanics; fracture; tribology; experimental mechanics; biomechanics and machine design.

1. R.T. Haftka, Z. Gürdal and M.P. Kamat: *Elements of Structural Optimization*. 2nd rev.ed., 1990 ISBN 0-7923-0608-2
2. J.J. Kalker: *Three-Dimensional Elastic Bodies in Rolling Contact*. 1990 ISBN 0-7923-0712-7
3. P. Karasudhi: *Foundations of Solid Mechanics*. 1991 ISBN 0-7923-0772-0
4. N. Kikuchi: *Computational Methods in Contact Mechanics*. (forthcoming) ISBN 0-7923-0773-9
5. Y.K. Cheung and A.Y.T. Leung: *Finite Element Methods in Dynamics*. (forthcoming) ISBN 0-7923-1313-5
6. J.F. Doyle: *Static and Dynamic Analysis of Structures*. With an Emphasis on Mechanics and Computer Matrix Methods. 1991 ISBN 0-7923-1124-8; Pb 0-7923-1208-2
7. O.O. Ochoa and J.N. Reddy: *Finite Element Modelling of Composite Structures*. (forthcoming) ISBN 0-7923-1125-6
8. M.H. Aliabadi and D.P. Rooke: *Numerical Fracture Mechanics*. ISBN 0-7923-1175-2
9. J. Angeles and C.S. López-Cajún: *Optimization of Cam Mechanisms*. 1991 ISBN 0-7923-1355-0
10. D.E. Grierson, A. Franchi and P. Riva: *Progress in Structural Engineering*. 1991 ISBN 0-7923-1396-8

Kluwer Academic Publishers – Dordrecht / Boston / London

Mechanics

From 1990, books on the subject of *mechanics* will be published under two series:
***FLUID* MECHANICS AND ITS APPLICATIONS**
Series Editor: R.J. Moreau
***SOLID* MECHANICS AND ITS APPLICATIONS**
Series Editor: G.M.L. Gladwell

Prior to 1990, the books listed below were published in the respective series indicated below.

MECHANICS: DYNAMICAL SYSTEMS
Editors: L. Meirovitch and G.Æ. Oravas

1. E.H. Dowell: *Aeroelasticity of Plates and Shells.* 1975 ISBN 90-286-0404-9
2. D.G.B. Edelen: *Lagrangian Mechanics of Nonconservative Nonholonomic Systems.* 1977 ISBN 90-286-0077-9
3. J.L. Junkins: *An Introduction to Optimal Estimation of Dynamical Systems.* 1978 ISBN 90-286-0067-1
4. E.H. Dowell (ed.), H.C. Curtiss Jr., R.H. Scanlan and F. Sisto: *A Modern Course in Aeroelasticity.* *Revised and enlarged edition see under Volume 11*
5. L. Meirovitch: *Computational Methods in Structural Dynamics.* 1980 ISBN 90-286-0580-0
6. B. Skalmierski and A. Tylikowski: *Stochastic Processes in Dynamics.* Revised and enlarged translation. 1982 ISBN 90-247-2686-7
7. P.C. Müller and W.O. Schiehlen: *Linear Vibrations.* A Theoretical Treatment of Multi-degree-of-freedom Vibrating Systems. 1985 ISBN 90-247-2983-1
8. Gh. Buzdugan, E. Mihăilescu and M. Radeş: *Vibration Measurement.* 1986 ISBN 90-247-3111-9
9. G.M.L. Gladwell: *Inverse Problems in Vibration.* 1987 ISBN 90-247-3408-8
10. G.I. Schuëller and M. Shinozuka: *Stochastic Methods in Structural Dynamics.* 1987 ISBN 90-247-3611-0
11. E.H. Dowell (ed.), H.C. Curtiss Jr., R.H. Scanlan and F. Sisto: *A Modern Course in Aeroelasticity.* Second revised and enlarged edition (of Volume 4). 1989 ISBN Hb 0-7923-0062-9; Pb 0-7923-0185-4
12. W. Szemplińska-Stupnicka: *The Behavior of Nonlinear Vibrating Systems.* Volume I: Fundamental Concepts and Methods: Applications to Single-Degree-of-Freedom Systems. 1990 ISBN 0-7923-0368-7
13. W. Szemplińska-Stupnicka: *The Behavior of Nonlinear Vibrating Systems.* Volume II: Advanced Concepts and Applications to Multi-Degree-of-Freedom Systems. 1990 ISBN 0-7923-0369-5
Set ISBN (Vols. 12–13) 0-7923-0370-9

MECHANICS OF STRUCTURAL SYSTEMS
Editors: J.S. Przemieniecki and G.Æ. Oravas

1. L. Frýba: *Vibration of Solids and Structures under Moving Loads.* 1970 ISBN 90-01-32420-2
2. K. Marguerre and K. Wölfel: *Mechanics of Vibration.* 1979 ISBN 90-286-0086-8

Mechanics

3. E.B. Magrab: *Vibrations of Elastic Structural Members.* 1979 ISBN 90-286-0207-0
4. R.T. Haftka and M.P. Kamat: *Elements of Structural Optimization.* 1985
   Revised and enlarged edition see under Solid Mechanics and Its Applications, Volume 1
5. J.R. Vinson and R.L. Sierakowski: *The Behavior of Structures Composed of Composite Materials.* 1986 ISBN Hb 90-247-3125-9; Pb 90-247-3578-5
6. B.E. Gatewood: *Virtual Principles in Aircraft Structures.* Volume 1: Analysis. 1989
   ISBN 90-247-3754-0
7. B.E. Gatewood: *Virtual Principles in Aircraft Structures.* Volume 2: Design, Plates, Finite Elements. 1989 ISBN 90-247-3755-9
   Set (Gatewood 1 + 2) ISBN 90-247-3753-2

MECHANICS OF ELASTIC AND INELASTIC SOLIDS

Editors: S. Nemat-Nasser and G.Æ. Oravas

---

1. G.M.L. Gladwell: *Contact Problems in the Classical Theory of Elasticity.* 1980
   ISBN Hb 90-286-0440-5; Pb 90-286-0760-9
2. G. Wempner: *Mechanics of Solids with Applications to Thin Bodies.* 1981
   ISBN 90-286-0880-X
3. T. Mura: *Micromechanics of Defects in Solids.* 2nd revised edition, 1987
   ISBN 90-247-3343-X
4. R.G. Payton: *Elastic Wave Propagation in Transversely Isotropic Media.* 1983
   ISBN 90-247-2843-6
5. S. Nemat-Nasser, H. Abé and S. Hirakawa (eds.): *Hydraulic Fracturing and Geothermal Energy.* 1983 ISBN 90-247-2855-X
6. S. Nemat-Nasser, R.J. Asaro and G.A. Hegemier (eds.): *Theoretical Foundation for Large-scale Computations of Nonlinear Material Behavior.* 1984 ISBN 90-247-3092-9
7. N. Cristescu: *Rock Rheology.* 1988 ISBN 90-247-3660-9
8. G.I.N. Rozvany: *Structural Design via Optimality Criteria.* The Prager Approach to Structural Optimization. 1989 ISBN 90-247-3613-7

MECHANICS OF SURFACE STRUCTURES

Editors: W.A. Nash and G.Æ. Oravas

---

1. P. Seide: *Small Elastic Deformations of Thin Shells.* 1975 ISBN 90-286-0064-7
2. V. Panc: *Theories of Elastic Plates.* 1975 ISBN 90-286-0104-X
3. J.L. Nowinski: *Theory of Thermoelasticity with Applications.* 1978
   ISBN 90-286-0457-X
4. S. Łukasiewicz: *Local Loads in Plates and Shells.* 1979 ISBN 90-286-0047-7
5. C. Fiřt: *Statics, Formfinding and Dynamics of Air-supported Membrane Structures.* 1983 ISBN 90-247-2672-7
6. Y. Kai-yuan (ed.): *Progress in Applied Mechanics.* The Chien Wei-zang Anniversary Volume. 1987 ISBN 90-247-3249-2
7. R. Negruţiu: *Elastic Analysis of Slab Structures.* 1987 ISBN 90-247-3367-7
8. J.R. Vinson: *The Behavior of Thin Walled Structures.* Beams, Plates, and Shells. 1988
   ISBN Hb 90-247-3663-3; Pb 90-247-3664-1

Mechanics

MECHANICS OF FLUIDS AND TRANSPORT PROCESSES
Editors: R.J. Moreau and G.Æ. Oravas

1. J. Happel and H. Brenner: *Low Reynolds Number Hydrodynamics*. With Special Applications to Particular Media. 1983 ISBN Hb 90-01-37115-9; Pb 90-247-2877-0
2. S. Zahorski: *Mechanics of Viscoelastic Fluids*. 1982 ISBN 90-247-2687-5
3. J.A. Sparenberg: *Elements of Hydrodynamics Propulsion*. 1984 ISBN 90-247-2871-1
4. B.K. Shivamoggi: *Theoretical Fluid Dynamics*. 1984 ISBN 90-247-2999-8
5. R. Timman, A.J. Hermans and G.C. Hsiao: *Water Waves and Ship Hydrodynamics*. An Introduction. 1985 ISBN 90-247-3218-2
6. M. Lesieur: *Turbulence in Fluids*. Stochastic and Numerical Modelling. 1987 ISBN 90-247-3470-3
7. L.A. Lliboutry: *Very Slow Flows of Solids*. Basics of Modeling in Geodynamics and Glaciology. 1987 ISBN 90-247-3482-7
8. B.K. Shivamoggi: *Introduction to Nonlinear Fluid-Plasma Waves*. 1988 ISBN 90-247-3662-5
9. V. Bojarevičs, Ya. Freibergs, E.I. Shilova and E.V. Shcherbinin: *Electrically Induced Vortical Flows*. 1989 ISBN 90-247-3712-5
10. J. Lielpeteris and R. Moreau (eds.): *Liquid Metal Magnetohydrodynamics*. 1989 ISBN 0-7923-0344-X

MECHANICS OF ELASTIC STABILITY
Editors: H. Leipholz and G.Æ. Oravas

1. H. Leipholz: *Theory of Elasticity*. 1974 ISBN 90-286-0193-7
2. L. Librescu: *Elastostatics and Kinetics of Aniosotropic and Heterogeneous Shell-type Structures*. 1975 ISBN 90-286-0035-3
3. C.L. Dym: *Stability Theory and Its Applications to Structural Mechanics*. 1974 ISBN 90-286-0094-9
4. K. Huseyin: *Nonlinear Theory of Elastic Stability*. 1975 ISBN 90-286-0344-1
5. H. Leipholz: *Direct Variational Methods and Eigenvalue Problems in Engineering*. 1977 ISBN 90-286-0106-6
6. K. Huseyin: *Vibrations and Stability of Multiple Parameter Systems*. 1978 ISBN 90-286-0136-8
7. H. Leipholz: *Stability of Elastic Systems*. 1980 ISBN 90-286-0050-7
8. V.V. Bolotin: *Random Vibrations of Elastic Systems*. 1984 ISBN 90-247-2981-5
9. D. Bushnell: *Computerized Buckling Analysis of Shells*. 1985 ISBN 90-247-3099-6
10. L.M. Kachanov: *Introduction to Continuum Damage Mechanics*. 1986 ISBN 90-247-3319-7
11. H.H.E. Leipholz and M. Abdel-Rohman: *Control of Structures*. 1986 ISBN 90-247-3321-9
12. H.E. Lindberg and A.L. Florence: *Dynamic Pulse Buckling*. Theory and Experiment. 1987 ISBN 90-247-3566-1
13. A. Gajewski and M. Zyczkowski: *Optimal Structural Design under Stability Constraints*. 1988 ISBN 90-247-3612-9

Mechanics

MECHANICS OF FRACTURE
Editors: G.C. Sih

1. G.C. Sih (ed.): *Methods of Analysis and Solutions of Crack Problems.* 1973 ISBN 90-01-79860-8
2. M.K. Kassir and G.C. Sih (eds.): *Three-dimensional Crack Problems.* A New Solution of Crack Solutions in Three-dimensional Elasticity. 1975 ISBN 90-286-0414-6
3. G.C. Sih (ed.): *Plates and Shells with Cracks.* 1977 ISBN 90-286-0146-5
4. G.C. Sih (ed.): *Elastodynamic Crack Problems.* 1977 ISBN 90-286-0156-2
5. G.C. Sih (ed.): *Stress Analysis of Notch Problems.* Stress Solutions to a Variety of Notch Geometries used in Engineering Design. 1978 ISBN 90-286-0166-X
6. G.C. Sih and E.P. Chen (eds.): *Cracks in Composite Materials.* A Compilation of Stress Solutions for Composite System with Cracks. 1981 ISBN 90-247-2559-3
7. G.C. Sih (ed.): *Experimental Evaluation of Stress Concentration and Intensity Factors.* Useful Methods and Solutions to Experimentalists in Fracture Mechanics. 1981 ISBN 90-247-2558-5

MECHANICS OF PLASTIC SOLIDS
Editors: J. Schroeder and G.Æ. Oravas

1. A. Sawczuk (ed.): *Foundations of Plasticity.* 1973 ISBN 90-01-77570-5
2. A. Sawczuk (ed.): *Problems of Plasticity.* 1974 ISBN 90-286-0233-X
3. W. Szczepiński: *Introduction to the Mechanics of Plastic Forming of Metals.* 1979 ISBN 90-286-0126-0
4. D.A. Gokhfeld and O.F. Cherniavsky: *Limit Analysis of Structures at Thermal Cycling.* 1980 ISBN 90-286-0455-3
5. N. Cristescu and I. Suliciu: *Viscoplasticity.* 1982 ISBN 90-247-2777-4

Kluwer Academic Publishers - Dordrecht / Boston / London

Mechanics

MECHANICS: ANALYSIS

Editors: V.J. Mizel and G.Æ. Oravas

1. M.A. Krasnoselskii, P.P. Zabreiko, E.I. Pustylnik and P.E. Sbolevskii: *Integral Operators in Spaces of Summable Functions.* 1976 ISBN 90-286-0294-1
2. V.V. Ivanov: *The Theory of Approximate Methods and Their Application to the Numerical Solution of Singular Integral Equations.* 1976 ISBN 90-286-0036-1
3. A. Kufner, O. John and S. Pučík: *Function Spaces.* 1977 ISBN 90-286-0015-9
4. S.G. Mikhlin: *Approximation on a Rectangular Grid.* With Application to Finite Element Methods and Other Problems. 1979 ISBN 90-286-0008-6
5. D.G.B. Edelen: *Isovector Methods for Equations of Balance.* With Programs for Computer Assistance in Operator Calculations and an Exposition of Practical Topics of the Exterior Calculus. 1980 ISBN 90-286-0420-0
6. R.S. Anderssen, F.R. de Hoog and M.A. Lukas (eds.): *The Application and Numerical Solution of Integral Equations.* 1980 ISBN 90-286-0450-2
7. R.Z. Has'minskiĭ: *Stochastic Stability of Differential Equations.* 1980 ISBN 90-286-0100-7
8. A.I. Vol'pert and S.I. Hudjaev: *Analysis in Classes of Discontinuous Functions and Equations of Mathematical Physics.* 1985 ISBN 90-247-3109-7
9. A. Georgescu: *Hydrodynamic Stability Theory.* 1985 ISBN 90-247-3120-8
10. W. Noll: *Finite-dimensional Spaces.* Algebra, Geometry and Analysis. Volume I. 1987 ISBN Hb 90-247-3581-5; Pb 90-247-3582-3

MECHANICS: COMPUTATIONAL MECHANICS

Editors: M. Stern and G.Æ. Oravas

1. T.A. Cruse: *Boundary Element Analysis in Computational Fracture Mechanics.* 1988 ISBN 90-247-3614-5

MECHANICS: GENESIS AND METHOD

Editor: G.Æ. Oravas

1. P.-M.-M. Duhem: *The Evolution of Mechanics.* 1980 ISBN 90-286-0688-2

MECHANICS OF CONTINUA

Editors: W.O. Williams and G.Æ. Oravas

1. C.-C. Wang and C. Truesdell: *Introduction to Rational Elasticity.* 1973 ISBN 90-01-93710-1
2. P.J. Chen: *Selected Topics in Wave Propagation.* 1976 ISBN 90-286-0515-0
3. P. Villaggio: *Qualitative Methods in Elasticity.* 1977 ISBN 90-286-0007-8